Mixture Modelling for Medical and Health Sciences

Chapman & Hall/CRC Biostatistics Series

Recently Published Titles

Bayesian Methods for Repeated Measures
Lyle D. Broemeling

Modern Adaptive Randomized Clinical Trials: Statistical and Practical Aspects
Oleksandr Sverdlov

Medical Product Safety Evaluation: Biological Models and Statistical Methods
Jie Chen, Joseph Heyse, Tze Leung Lai

Statistical Methods for Survival Trial Design
With Applications to Cancer Clinical Trials Using R
Jianrong Wu

Bayesian Applications in Pharmaceutical Development
Satrajit Roychoudhury, Soumi Lahiri

Platform Trials in Drug Development: Umbrella Trials and Basket Trials
Zoran Antonjevic and Robert Beckman

Innovative Strategies, Statistical Solutions and Simulations for Modern Clinical Trials
Mark Chang, John Balser, Robin Bliss and Jim Roach

Cost-effectiveness Analysis of Medical Treatments: A Statistical Decision Theory Approach
Elias Moreno, Francisco Jose Vazquez-Polo and Miguel Angel Negrin-Hernandez

Analysis of Incidence Rates
Peter Cummings

Mixture Modelling for Medical and Health Sciences
Shu Kay Ng, Liming Xiang, Kelvin Kai Wing Yau

For more information about this series, please visit: https://www.crcpress.com/go/biostats

Mixture Modelling for Medical and Health Sciences

Shu Kay Ng
Liming Xiang
Kelvin Kai Wing Yau

CRC Press is an imprint of the
Taylor & Francis Group, an **informa** business

CRC Press
Taylor & Francis Group
6000 Broken Sound Parkway NW, Suite 300
Boca Raton, FL 33487-2742

First issued in paperback 2020

ISBN-13: 978-1-4822-3675-0 (hbk)
ISBN-13: 978-0-367-72933-2 (pbk)

Library of Congress Cataloging-in-Publication Data

Names: Ng, Shu-Kay (Shu-Kay Angus), author. | Xiang, Liming, author. | Yau, Kelvin Kai Wing, author.
Title: Mixture modelling for medical and health sciences / Shu-Kay Ng, Liming Xiang, and Kelvin Kai Wing Yau.
Description: Boca Raton : CRC Press, Taylor & Francis Group, 2019.
Identifiers: LCCN 2018061440 | ISBN 9781482236750 (hardback : alk. paper) | ISBN 9781482236774 (ebook)
Subjects: LCSH: Medical sciences--Mathematics. | Biometry.
Classification: LCC R853.M3 N4 2019 | DDC 610.1/5195--dc23
LC record available at https://lccn.loc.gov/2018061440

Visit the Taylor & Francis Web site at
http://www.taylorandfrancis.com

and the CRC Press Web site at
http://www.crcpress.com

To Regin
and Ludwig.

Contents

Preface

Finite mixture models are powerful statistical tools with a sound theoretical basis for modelling the distribution of a heterogeneous population as well as providing a clustering of sample data to extract useful information on the group structure of the underlying population. Over the past few decades, there has been an ever-increasing use of finite mixture models to provide a model-based approach to the analysis of data arising in a wide variety of scientific fields, including bioinformatics, biometrics/biostatistics, data science, economics, health, image analysis, medicine, and psycho-oncology, among many others.

Mixture models with normal component density functions are the most widely used mixture models, given its computational tractability. There are, however, many significant advancements in extending (normal) mixture models to address real-world problems that demand highly novel methods due to new data sources, applications, and study designs for their analysis and understanding. This book presents some of these advanced mixture models developed for tackling problems that present significant methodological challenges to standard mixture modelling approaches, especially in the fields of medical and health sciences. A key aspect of these advanced approaches is the use of random-effects models to account for complex correlation structures among observations, providing a flexible modelling approach for data that are not independent. Indeed, many real-life data have been collected in settings where observations are highly correlated; these include multilevel data, longitudinal data, multivariate dependent data, multiple-source data, and "Big Data".

Chapter 1 of the book presents fundamental concepts in mixture modelling, parameter estimation using the maximum likelihood method via the expectation-maximisation (EM) algorithm, and the overview of the book. We begin with normal mixture models in Chapter 2 to highlight the applications of mixture models with symmetric component distributions and how useful information can be extracted and interpreted through mixture modelling. Step-by-step we then work on mixture models for analyzing non-normal data, count or categorical data, and censored survival data in Chapters 3 to 5, respectively. Chapter 6 introduces the fundamental concepts of random-effects models and the following Chapters 7 and 8 describe advanced mixture models for analyzing multilevel and repeated measured data, as well as correlated multivariate continuous data. We conclude the book with a brief account of some topical areas including the handling of missing data in the applications of mixture models, followed by cluster analysis of "Big Data" using mixture models in Chapters 9 and 10, respectively.

We hope that this book will be useful to applied and theoretical statisticians, as well as to investigators in substantive and diverse fields who will make relevant use of mixture models for the statistical analyses of their own data. While it is assumed that the reader has a fair grounding in statistics, we intend to introduce important concepts in mixture modelling by exploring challenges arising from real problems. Hence the emphasis on practice is a key feature of this book, where the primary focus is on comprehensive analyses of practical examples taken from real-life research problems in the fields of medical and health sciences. Most of these existing data sets are publicly available, as are the computing programs in R and Fortran provided in an accompanying website (`https://www.crcpress.com/9781482236750`). We hope this will stimulate readers to reproduce the examples given in the book in an endeavour to begin exploring the use of mixture models to analyze their own data.

We wish to thank Geoffrey McLachlan, who introduced us to the field of mixture models and with whom we share a long-lasting collaboration and many helpful and insightful discussions. One of the authors (SK Ng) would like to gratefully acknowledge financial support received from the Australian Research Council. Thanks must also be extended to the authors and owners for their permission to use copyrighted material to reproduce tables and figures. We would like to acknowledge Kui Wang for his help to convert some computing programs from Fortran to R. We also wish to thank Erin Pitt, Richard Tawiah, and Jane Ou for their constructive comments on drafts of the manuscript and assistance in preparing the manuscript.

Brisbane, Australia	Shu Kay Ng
Singapore	Liming Xiang
Hong Kong	Kelvin Kai Wing Yau

1

Introduction

This chapter provides a general introduction to mixture modelling including its fundamental theoretical concepts, relevant computational procedures and available software packages.

1.1 Why Mixture Modelling is Needed

Mixture modelling provides a flexible and powerful probabilistic approach in major areas of statistics such as cluster analysis, where the primary goal is to identify the group structure of the population under study. Fields in which mixture models have been successfully applied include chemistry (Yu and Qin, 2009), engineering (Hafemeister et al., 2011), genetics (McLachlan, Bean, and Peel, 2002), medicine (Windle et al., 2004), physics (Hao et al., 2009), criminology (Neema and Böhning, 2010), psychiatry (Wessman et al., 2009), psycho-oncology (Chambers et al., 2017), education (Palardy and Vermunt, 2010), and economics (Stull, Wyrwich, and Frueh, 2009), among many other fields in the medical and health sciences (Tentoni et al., 2004; Fahey et al., 2012; Ji et al., 2012). In applications within the context of cluster analysis, the grouping of entities into a number of clusters is obtained via mixture models so that entities in the same cluster are more alike to one another than entities from different clusters. Such model-based clustering approaches are undertaken on the basis where observed feature data are measured on variables associated with each entity, without any prior knowledge on the group structure of the underlying population (this is referred to as "unsupervised classification" in the related fields of pattern recognition). Mixture models not only provide a sound mathematical framework for clustering, but also allow formal statistical inferences to be made about the determination of the number of components and assessment of classification uncertainty, as illustrated by McLachlan and Peel (2000, Chapters 1 and 6) and Zhu and Melnykov (2015).

In wider contexts of applied statistics, the contemporary approach to data analysis has relied on the assumption that the observed data are collected from a single population. This assumption of homogeneity is often unrealistic or inadequate to accurately represent the population under study. Mixture models play an important role in these applications as they provide a model-based approach to account for the existence of heterogeneous sub-populations with different parameters (even though identification of the group structure may not be the primary interest of the studies). We illustrate this with a widely used example in regression analysis available at the UCLA Statistical Consulting Group website at `https://stats.idre.ucla.edu/other/annotatedoutput/`. Data formats are available in Stata, SAS, and SPSS.

1.1.1 Example 1.1: UCLA Example Data Set

The UCLA example data set "hsb2" contains scores on various school tests including science, mathematics, reading, writing, and social studies from 200 high school students, as well as an indicator of the gender of the students. Table 1.1 presents the results of the regression analysis on the science scores. It can be seen that there are significant positive associations between the science score with reading, writing, and mathematics scores (all with $p < 0.001$), but not the social studies score. Moreover, female students generally have a significantly lower science score compared to male students (mean difference 3.576; 95% confidence interval (CI): 1.48 to 5.67; $p = 0.001$). These findings are based on the assumption of homogeneity; that there is a single population of high school students.

An obvious follow-up question on the results displayed in Table 1.1 would be: "Are the relationships between the science score and the independent variables (other school test scores and gender) the same for students with low science scores compared to those with high scores?". To answer this question, we fitted a two-component mixture of normal distributions with unequal component-variances (mixtures of normal distributions will be described in details in Chapter 2) to the science scores of 200 students. We obtained two clusters of students: the low-score cluster consists of 54 (27%) students who have a lower estimated mean science score of 40.2 with an estimated standard deviation of 5.8; the high-score cluster contains 146 (73%) students who have a higher estimated mean science score of 56.0 with a larger estimated standard deviation of 7.4. Table 1.2 presents the regression results of sub-group analyses separately for the two clusters. It can be seen that the relationships between the science score and the independent variables are different for students with low science scores. Specifically, none of the independent variables are significantly associated with the science score for students belonging to the low-score cluster (this is not due to a smaller sample size, as the estimated coefficients (effect sizes) and the adjusted R^2 are actually small). In addition, it was found that female students and social studies scores were positively associated with the science scores. This is in contrast to the results obtained for the high-score cluster or for all students (see Tables 1.1 and 1.2). Although the coefficients corresponding to "social studies score" and "female" for the low-score cluster were not statistically significant, this interesting finding regarding reverse signs of the coefficients, compared to the high-score cluster, warrants further study.

The distinct results between the low-score and high-score clusters presented in Table 1.2 demonstrate why it is necessary in common statistical data analyses (such as regression analysis) to explore heterogeneity in the population under study. Even though the primary aim of the study is not the identification of the group structure, the

TABLE 1.1
Results of the regression analysis on the science scores for Example 1.1 ($n = 200$).

Variable	Coefficient	95% CI for Coefficient	*p*-value
Constant	9.472	(3.24, 15.70)	0.003
Reading score	0.271	(0.129, 0.413)	<0.001
Writing score	0.313	(0.158, 0.468)	<0.001
Mathematics score	0.285	(0.135, 0.435)	<0.001
Social studies score	-0.026	(-0.150, 0.098)	0.684
Female	-3.576	(-5.67, -1.48)	0.001

R^2 = 52.8% and adjusted R^2 = 51.6%.

TABLE 1.2
Sub-group analyses: Results of the regression analysis on the science scores for Example 1.1 (low-score and high-score clusters).

Variable	Low-score Cluster ($n = 54$) Coefficient (95% CI)	*p*-value	High-score Cluster ($n = 146$) Coefficient (95% CI)	*p*-value
Constant	29.20 (18.19, 40.21)	<0.001	31.60 (24.70, 38.49)	<0.001
Reading score	0.092 (-0.119, 0.302)	0.386	0.159 (0.030, 0.288)	0.016
Writing score	-0.062 (-0.265, 0.142)	0.545	0.144 (-0.004, 0.293)	0.057
Mathematics score	0.048 (-0.208, 0.305)	0.707	0.212 (0.081, 0.343)	0.002
Social studies score	0.122 (-0.044, 0.288)	0.145	-0.035 (-0.148, 0.078)	0.541
Female	1.380 (-1.376, 4.137)	0.319	-3.165 (-5.097, -1.233)	0.002
R^2; Adjusted R^2	13.1%; 4.0%		34.6%; 32.3%	
Range of science scores	26-44		45-74	

findings displayed in Tables 1.1 and 1.2 illustrate how mixture models can be utilized to provide additional information that is useful in identifying subtle results and new interpretations for the study (in Example 1.1, the association between science scores and explanatory variables is different for students with low or high scores and some effects may even be in opposite directions). Although the number of clusters can be determined using a penalized form of the log-likelihood or a formal hypothesis test (this will be described in Section 1.2.3), the emphasis of the cluster analysis in Example 1.1 is on the grouping of the students based on their science scores so as to explore the differences (if any) in characteristics between the low-score and high-score students. There is no implication that the resulting clusters are in any sense a natural division of the data, as described in McLachlan, Bean, and Ng (2008).

In Table 1.2, the sub-group regression analyses were based on the "hard" classification of students into either the low and high science score sub-groups. This outright classification of students provides an alternative way other than arbitrary approaches to divide students into groups (such as, using the median as a cut-off) for exploring potential group differences in students' characteristics via sub-group analyses. In other applications, a g-component mixture regression model may be used to identify heterogeneous sub-populations of students in terms of associations between an outcome variable (y) and covariates (x). That is,

$$Y = W\beta_h + \varepsilon_h \qquad \varepsilon_h \sim N(0, \sigma_h^2) \tag{1.1}$$

for Y belonging to the hth component ($h = 1, \ldots, g$), where β_h is a vector of regression coefficients and W is a design matrix containing covariates x with the first column of W assumed to be 1 to account for (constant) intercept terms. This approach thus attempts to answer a research question different from that answered through the results in Table 1.2. The clustering result of fitting a two-component mixture regression model to the hsb2 data set is given in Table 1.3. It can be seen that Cluster 1 corresponds to the majority of students (93.5%) having a pattern indicating significant associations of covariates with the science scores similar to that displayed in Table 1.1 for all students. Not surprisingly, results in Table 1.3 are different from those obtained in sub-group analyses in Table 1.2 because the former attempts to address heterogeneity in the regression coefficients but not in the science scores; see (1.1). To illustrate this, the clustering of students into two groups using the two-component mixture regression model is displayed in Figure 1.1. It can be observed that the two clusters of students obtained were not based on the science scores themselves (indeed, the estimated conditional mean science score for Cluster 1 was higher than that for Cluster 2 in 137 out of 200 students). Rather, the clustering in Figure 1.1 was based on the heterogeneous association between science scores and covariates x. For example, Cluster 2 in Figure 1.1 shows a negative association (slope of -1.183) between science and social studies scores, which is distinguishable from Cluster 1 as shown in Table 1.3. The diversity in the clustering results presented in Tables 1.2 and 1.3 also highlights the importance of understanding how different formulations of a mixture model for cluster analysis affect the interpretation of clustering results.

It is also worth mentioning that the sub-group regression analysis concerning the disparity in relationships between the science score and other scores and gender

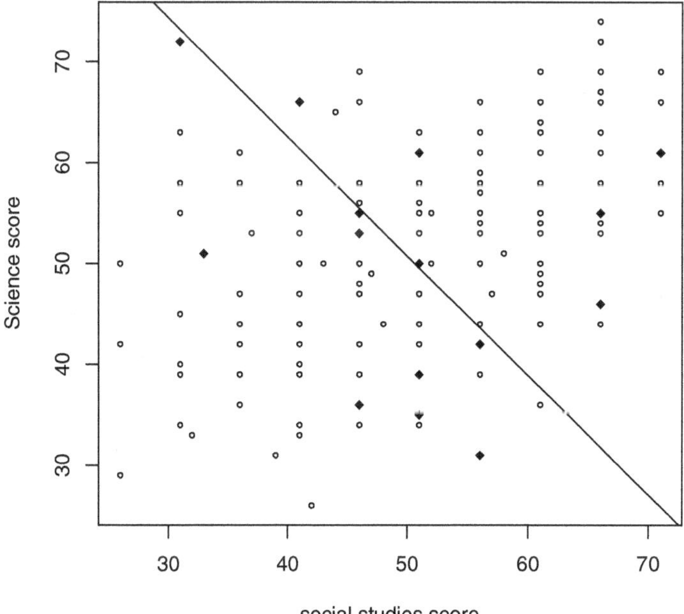

FIGURE 1.1
A two-component mixture regression model for the UCLA example – two clusters represented by ○ (93.5%) and ♦ (6.5%) (intercept of the fitted regression line is computed at the means of other covariates; note the negative association between science and social studies scores for the smaller cluster).

TABLE 1.3
Estimates and standard errors (in parentheses) for the two-component mixture regression model on the science scores (Example 1.1).

Variable	Cluster 1	Cluster 2
Constant	6.99* (2.91)	42.55* (0.48)
Reading score	0.237* (0.068)	1.255* (0.017)
Writing score	0.356* (0.078)	-0.642* (0.015)
Mathematics score	0.259* (0.070)	0.695* (0.015)
Social studies score	0.041 (0.064)	-1.183* (0.013)
Female	-3.660* (0.864)	-1.704* (0.218)
Mixing proportion	0.935	0.065
Variance	39.10	0.006

* Significance at the level of 0.05.

for students with different abilities in science (Table 1.2) is a different approach to the quantile regression analysis widely used in statistics and econometrics literature (Koenker and Hallock, 2001). The quantile regression method attempts to make inferences about conditional quantile functions of the outcome variable (see, for example, Amugsi et al. (2016) and Olsen et al. (2012)), by minimizing the sum of (asymmetrically) weighted absolute values of the residuals. To illustrate this difference between the two methods, we present the results of a quantile regression and sub-group regression analyses with five independent variables on the same data set in Figures 1.2(a) and (b), respectively. It can be seen that the association between the science and mathematics scores at different quantiles of the science score are generally similar except for the 10th quantile, which shows a larger slope. In contrast, the sub-group analysis shows that the adjusted effect of mathematics scores on science scores is relatively smaller for students with low science scores, as shown in Table 1.2. Another major difference comes from the fact that quantile regression analysis indeed considers all the observed data regardless of the student's science score. As cautioned by Firpo, Fortin, and Lemieux (2009) and Borah and Basu (2013), the conditional quantile depends on the characteristics of individual subjects and thus certain quantile regression estimates may be a "mixed" impact for subjects well above or below a specific quantile threshold of the target sub-population (for example in Figure 1.2, the 10th quantile of science scores for students with high mathematics scores is well above the low science score cut-off).

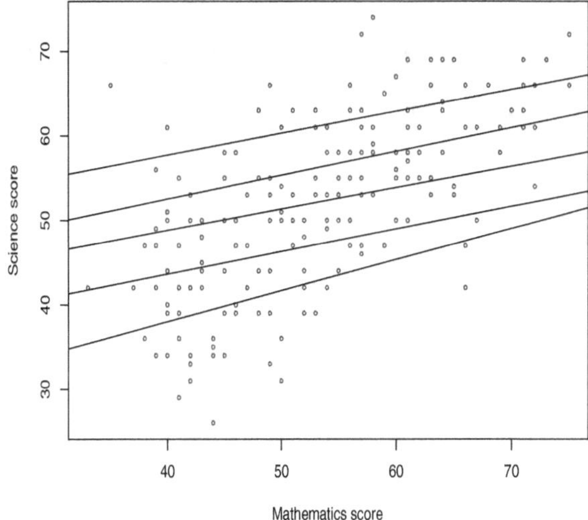

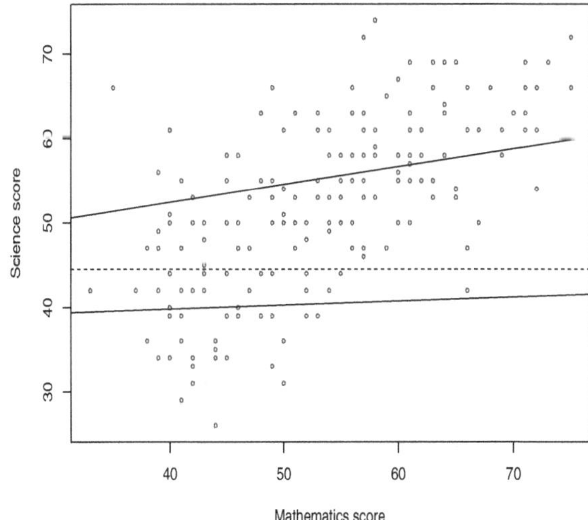

FIGURE 1.2
Science score versus mathematics score (UCLA example): the top graph shows the results of quantile regression at the 10th, 25th, 50th, 75th, and 90th quantiles (five estimated regression lines from the bottom to the top, respectively); the bottom graph presents the sub-group regression lines for students with low science scores (bottom part) and high science scores (top part) where the dashed line indicates the cut-off science score between the two sub-groups. In both graphs, intercepts are computed at the means of other covariates.

1.2 Fundamental Concepts of Finite Mixture Models

We let the p-dimensional vector $y_j = (y_{1j}, \ldots, y_{pj})^T$ contain the observed values of p variables measured on the jth entity $(j = 1, \ldots, n)$, where n is the total number of (independent) entities and the superscript T denotes vector transpose. Within the mixture modelling framework, $y_1, \ldots, y_n$ are assumed to be an observed random sample from a mixture of a finite number, say g, of components in some unknown proportions $\pi_1, \ldots, \pi_g$ that sum to one. The mixture density of y_j is written as

$$f(y_j; \Psi) = \sum_{h=1}^{g} \pi_h f_h(y_j; \psi_h) \qquad (j = 1, \ldots, n), \tag{1.2}$$

where the component-density $f_h(y_j; \psi_h)$ is specified up to a vector ψ_h of unknown parameters $(h = 1, \ldots, g)$ and Ψ is the vector of unknown parameters containing the mixing proportions $\pi_1, \ldots, \pi_{g-1}$, and the vectors of component parameters $\psi_1, \ldots, \psi_g$. In (1.2), we consider the observed vector y_j to be a continuous random sample, but we can still view $f(y_j; \Psi)$ as a density in the case where y_j is discrete by the adoption of a counting measure. As we will be focusing on finite mixture models in this book, we shall refer to finite mixture models as just mixture models in the sequel.

The parameter vector Ψ can be estimated by maximum likelihood. We let $\widehat{\Psi}$ denote a maximum likelihood estimate of Ψ. The mixture modelling approach provides a probabilistic clustering of the data into g clusters in terms of estimated posterior probabilities of component membership,

$$\tau_h(y_j; \widehat{\Psi}) = \frac{\hat{\pi}_h f_h(y_j; \hat{\psi}_h)}{f(y_j; \widehat{\Psi})} = \frac{\hat{\pi}_h f_h(y_j; \hat{\psi}_h)}{\sum_l^g \hat{\pi}_l f_l(y_j; \hat{\psi}_l)}, \tag{1.3}$$

where $\tau_h(y_j; \Psi)$ is the posterior probability that the jth entity y_j belongs to the hth component of the mixture $(h = 1, \ldots, g;\ j = 1, \ldots, n)$. An outright or hard clustering of the data can be effected by assigning each y_j to the component to which it has the highest estimated posterior probability of component membership.

The following sub-sections describe the maximum likelihood estimation of the parameter vector Ψ in more detail and the approaches used to determine the number of components g of the mixture. The estimation of Ψ on the basis of the observed data y_j $(j = 1, \ldots, n)$ is only meaningful if Ψ in the mixture model (1.2) is identifiable. The concept of identifiability in mixture distributions is presented in Section 1.2.4.

1.2.1 Maximum Likelihood Estimation

With likelihood-based inference, the aim is to find an estimate $\widehat{\Psi}$ that maximizes the likelihood function, or equivalently, the log likelihood function for Ψ which is given by:

$$\log L(\Psi) = \log \prod_{j=1}^{n} f(y_j;\Psi) = \sum_{j=1}^{n} \log f(y_j;\Psi) = \sum_{j=1}^{n} \log \left\{ \sum_{h=1}^{g} \pi_h f_h(y_j;\psi_h) \right\}. \quad (1.4)$$

The likelihood function $L(\Psi)$ can be expressed as multiplication of individual density functions $f(y_j;\Psi)$ with an assumption commonly used in mixture modelling that individual observations y_j $(j = 1,\ldots,n)$ are independent of one another. With the maximum likelihood estimation, the estimate $\widehat{\Psi}$ defines a sequence of roots of the log likelihood equation

$$\frac{\partial \log L(\Psi)}{\partial \Psi} = 0 \quad (1.5)$$

over the parameter space Θ that is consistent and asymptotically efficient under suitable regularity conditions, as described in Ng (2013). These roots correspond to local maxima of the log likelihood function in the interior of Θ with the probability tending to one. In applications where the log likelihood has a global maximum in the interior of Θ, the maximum likelihood estimate is obtained by taking it to be the root $\widehat{\Psi}$ that globally maximizes $\log L(\Psi)$ with the desired asymptotic properties. In situations where the log likelihood is unbounded (such as in the case of mixtures of multivariate normal distributions with unequal covariance matrices) or $\widehat{\Psi}$ may not globally maximize the log likelihood, there may still exist, under the usual regularity conditions, local maxima with the properties of consistency, efficiency, and asymptotic normality; see, for example McLachlan and Peel (2000, Chapter 2). In these latter situations, we shall refer to the largest local maximizer $\widehat{\Psi}$ as the maximum likelihood estimate.

The expectation-maximization (EM) algorithm (Dempster, Laird, and Rubin, 1977) is widely applied to iteratively compute the roots of Equation (1.5) corresponding to local maximizers of $\log L(\Psi)$. Section 1.3 describes the EM algorithm in detail.

1.2.2 Spurious Clusters

With many applications of mixture models, the log likelihood function has multiple local maxima. The EM algorithm should be implemented from a wide choice of initial parameter values in an attempt to search for all local maxima. As described in Section 1.2.1, the largest local maximizer with the desired asymptotic properties defines the maximum likelihood estimate $\widehat{\Psi}$. However, in practice, consideration must be given to the problem of relatively large local maxima of the log likelihood function. This occurs when a fitted component converges to a cluster containing only a few entities relatively close together or entities lying in almost a lower-dimensional sub-space in the case of multivariate data, referred to as the "empty cluster" and the "collapse cluster" problems, respectively; see, for example, Wang, Ng, and McLachlan (2009). In cases of continuous component densities such as normal and t distributions, the issue of singularity in component-covariance matrices may be encountered. Figures 1.3 and 1.4 display simulated two-dimensional data (both with $n = 1{,}000$

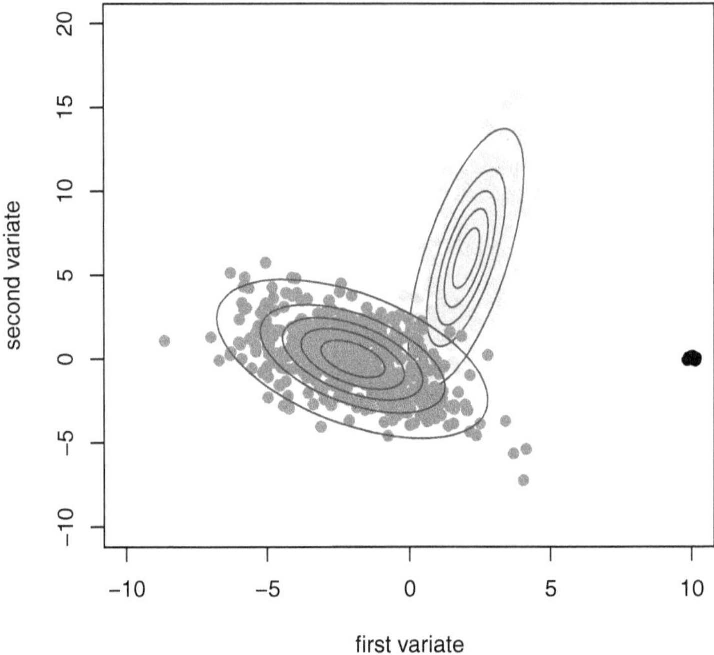

FIGURE 1.3
Simulated data set 1: A three-component mixture of normal distributions with a few data points (in black colour) close together for the third cluster.

data points) corresponding to these two situations. For simulation data set 1, the third cluster contains only $n_3 = 10$ data points. For simulation data set 2, the third cluster contains $n_3 = 50$ data points lying on a line (a lower-dimensional sub-space of the original bivariate data).

To define the maximum likelihood estimate, it is thus necessary to monitor the variance of each component, such as the eigenvalue of the component-covariance matrix for normal mixtures, and the relative size of the fitted mixing proportions $\pi_1, \ldots, \pi_g$. A lower bound of π_h (such as 1%) may be set up to avoid clustering solutions that occur as a consequence of the spurious local maximizers as displayed in Figures 1.3 and 1.4.

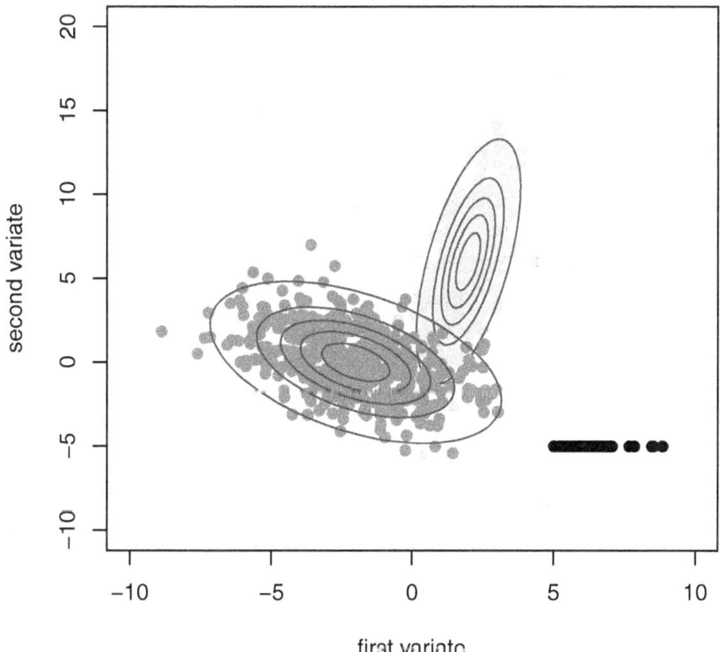

FIGURE 1.4
Simulated data set 2: A three-component mixture of normal distributions with a collapse cluster (with data points in black colour).

1.2.3 Determination of the Number of Components

In the absence of any prior information as to the number of clusters present in the data, the number of components in the mixture model to be adopted has to be inferred from the data. We can monitor the increase in log likelihood function as the value of g increases in order to determine an appropriate value of g in a mixture model. At any stage, the choice of $g = g_0$ versus $g = g_0 + 1$ can be determined by using either a penalized form of the log likelihood or a hypothesis test via bootstrapping.

With the first approach, the log likelihood is penalized by the subtraction of a term that "penalizes" the complex model for the number of parameters in it. Many information-based criteria for the selection of g use this approach, such as the Akaike Information Criterion (AIC), the Bayesian Information Criterion (BIC), the consistent AIC (CAIC), the Laplace-Empirical Criterion (LEC), and the Integrated Classification Likelihood (ICL) criterion; see Schwarz (1978), Biernacki, Celeux, and Govaert (1998), McLachlan and Peel (2000, Chapter 6), Fraley and Raftery (2002), and Ng and McLachlan (2014b). These criteria adopt different degrees of penalty for model complexity. For instances, AIC selects the model that minimizes

$$\text{AIC}(g) = -2\log L(\widehat{\Psi}) + 2d(g), \tag{1.6}$$

where $d(g)$ is the number of free parameters in the mixture model with g components. The first term in (1.6) is the maximum likelihood of the model representing the lack of fit, while the second term is the penalty term that measures the compensation for the bias in the first term when maximum likelihood estimators are used. The other commonly used criterion, BIC, is expressed as

$$\text{BIC}(g) = -2\log L(\widehat{\Psi}) + d(g)\log(n), \tag{1.7}$$

where n is the total number of observations. With the ICL criterion (Biernacki, Celeux, and Govaert, 1998), we have

$$\text{ICL}(g) = -2\log L(\widehat{\Psi}) + 2EN(\hat{\tau}) + d(g)\log(n), \tag{1.8}$$

where

$$EN(\tau) = -\sum_{h=1}^{g}\sum_{j=1}^{n} \tau_{hj}\log\tau_{hj} \tag{1.9}$$

is the entropy of the fuzzy classification matrix defined by $((\tau_{hj}))$ and where $\tau^T = (\tau_1^T, \ldots, \tau_n^T)$ and $\tau_j = (\tau_1(y_j;\widehat{\Psi}), \ldots, \tau_g(y_j;\widehat{\Psi}))^T$; see, for example, McLachlan and Peel (2000, Section 6.10). The minimum value of the entropy is zero when $g = 1$ corresponding to a single population. The value of the entropy is close to zero when the components of the mixture model are well separated; its value increases when the mixture components are poorly separated. Comparing (1.6) with (1.7) and (1.8), the criteria differ only in the penalty terms on model complexity as well as how well separated are the fitted mixture components. Comparison of these penalized likelihood criteria in the general context of mixture models has been reported in McLachlan and Peel (2000, Section 6.11); see also Keribin (2000) and Leroux (1992).

TABLE 1.4
Number of components selected using various methods for Example 1.1.

No. of components (g)	Log likelihood	AIC	BIC	ICL	p-value[a]
1	-741.81	1487.6	1494.2*	1508.8*	0.07*
2	-738.22	1486.4*	1502.9	1539.4	0.55
3	-737.27	1490.5	1516.9	1575.3	–

* Number of components selected based on the criterion.
[a] p-value using bootstrap likelihood ratio statistic for g versus $g+1$.

The other approach is to perform a hypothesis test, using the likelihood ratio as the test statistic. However, with mixture models, regularity conditions do not hold for $-2\log\lambda$ to have its usual asymptotic null distribution of chi-squared with degrees of freedom equal to the difference between the number of parameters under the null (H_0: $g = g_0$) and alternative (H_a: $g = g_0 + 1$) hypotheses. A bootstrap resampling approach can be adopted to assess the null distribution (and hence the p-value) of the likelihood ratio test statistic, as discussed in McLachlan and Peel (2000, Sections 6.4-6.7). With this approach, bootstrap samples are generated from the mixture model fitted under H_0 with g_0 components, where the vector of unknown parameters Ψ is replaced by its maximum likelihood estimate obtained from the original data under H_0. The process is repeated independently B times and the replicated values of $-2\log\lambda$, obtained by fitting mixture models for $g = g_0$ and $g = g_0 + 1$ to each bootstrap sample in turn, provide an assessment of the null distribution of $-2\log\lambda$. It thus enables an approximation to the p-value of significance of the value of $-2\log\lambda$ computed from the original data. In general, the empirical approximation using the maximum likelihood estimate for Ψ tends to overestimate the p-value of tests based on this statistic; see McLachlan and Peel (2000, Sections 6.6-6.7) for details.

Table 1.4 presents the comparison of some penalized criteria and the bootstrap method for selecting the number of components for Example 1.1 corresponding to Table 1.2. It can be seen that the AIC indicates there are two components, while the BIC and ICL penalize complex models more heavily than the AIC and result in the selection of a single component. For the bootstrap method, it is found that the p-value for assessing $g = 1$ versus $g = 2$ is 0.07, supporting a single component as selected by BIC and ICL.

For the simulated data sets given in Figures 1.3 and 1.4, the number of components selected by some penalized criteria and the bootstrap method are provided in Table 1.5. For simulation data set 1, it can be seen that the AIC and the bootstrap method indicate there are four components, while the BIC and ICL penalize complex models more heavily and result in the selection of a three-component solution. For simulation data set 2, all four methods support a three-component solution. As described in Section 1.2.2, consideration regarding the number of components should be given when a fitted component converges to a cluster containing only a few data points relatively close together or data points lying in almost a lower-dimensional sub-space.

TABLE 1.5
Number of components selected using various methods for simulated data sets.

No. of components (g)	Log likelihood	AIC	BIC	ICL	p-value[a]
(a) Simulation data set 1					
1	-5040.3	10091	10115	10115	0.00
2	-4550.3	9123	9177	9257	0.00
3	-4390.5	8815	8899*	8971*	0.01
4	-4377.8	8802*	8915	9304	–
(b) Simulation data set 2					
1	-5291.7	10593	10618	10618	0.00
2	-4611.4	9245	9299	9352	0.00
3	-4257.4	8549*	8632*	8694*	0.89*
4	-4255.5	8557	8670	8887	–

* Number of components selected based on the criterion.
[a] p-value using bootstrap likelihood ratio statistic for g versus $g+1$.

When determining the number of components in a mixture model, it is sensible, in practice, to assess the smallest number of components that is compatible with the data (especially when various criteria lead to different values of the number of components) and in which each component contains no fewer than 1% of the overall data (see Section 1.2.2). With applications to large data sets, penalized likelihood criteria will be less computationally demanding than the bootstrap resampling approach.

1.2.4 Identifiability of Mixture Distributions

Suppose that the component densities come from the same parametric family indexed by a parameter vector ψ and that the vector of unknown parameters Ψ containing the mixing proportions and the component parameters belong to a parameter space Θ. This class of mixture models is identifiable for $\Psi \in \Theta$ if the equality of any two representatives in this class

$$f(y_j;\Psi) = \sum_{h=1}^{g} \pi_h f_h(y_j;\psi_h) \quad \text{and} \quad f(y_j;\Psi') = \sum_{h=1}^{g'} \pi'_h f_h(y_j;\psi'_h) \tag{1.10}$$

implies that $g = g'$ and the component labels can be permuted such that

$$\pi_h = \pi'_h \quad \text{and} \quad \psi_h = \psi'_h \qquad (h = 1,\ldots,g). \tag{1.11}$$

In (1.11), it is necessary to eliminate the problem due to permutation of component labels because $f(y_j;\Psi)$ is invariant under the g permutations of the component labels in Ψ; see McLachlan and Peel (2000, Section 1.14) and Boldea and Magnus (2009). Taking Example 1.1 concerning a mixture of univariate normal distributions as an example, the parameter vector Ψ is given by

$$\Psi = (\pi_{ls}, \mu_{ls}, \sigma_{ls}, \mu_{hs}, \sigma_{hs})^T,$$

where μ denotes the mean, σ denotes the standard deviation, and the subscripts ls and hs represent the low-score cluster and the high-score cluster, respectively. When these two cluster labels are interchanged in Ψ, the equality of the mixture densities in (1.10) holds but the parameter vector is not identifiable as $\Psi \neq \Psi'$, where

$$\Psi' = (\pi_{hs}, \mu_{hs}, \sigma_{hs}, \mu_{ls}, \sigma_{ls})^T.$$

To allow for permutation of component labels in (1.11), we need to ensure that only one labelling is possible and hence the parametric model (1.2) defines a unique model. In this book, we report the results for only one of the possible permutations of the elements of the parameter vector Ψ, which makes it easier for readers to interpret the findings.

Another identifiability issue concerns potential overfitting, as shown by Crawford (1994). Considering a g-component mixture distribution defined as in the left-hand side of (1.10), Crawford (1994) stated that any mixture distribution with $(g-1)$ components of densities from the same parametric family defines a nonidentifiability subset in the larger parameter space Θ_g corresponding to g-component mixture distributions, where either one component is empty or two components are found equal; see also Frühwirth-Schnatter (2006, Section 1.3.2). Again this identifiability issue due to potential overfitting could be handled by imposing formal identifiability constraints, such as with $\pi_g > \ldots > \pi_1 > 0$ applied in a Bayesian inference of mixture models in Lenk and DeSarbo (2000), or thorough consideration in determining the number of components in the mixture model as discussed in Sections 1.2.2 and 1.2.3.

From (1.10) and (1.11), it can be seen that the estimation of the parameter vector Ψ on the basis of the observed data y_j $(j = 1, \ldots, n)$ is meaningful only if Ψ is identifiable, representing a distinct member in the class of mixture models. Indeed, identifiability of a mixture model is fundamental for its parameter estimation and is a necessary condition for the asymptotic theory to hold. Information on addressing the more formal issue of identifiability of mixture distributions can be obtained in the monographs by Titterington, Smith, and Makov (1985, Section 3.1), McLachlan and Basford (1988, Section 1.5), McLachlan and Peel (2000, Section 1.14), Frühwirth-Schnatter (2006, Section 1.3) and the references therein.

In many cases of practical interest, identifiability of mixture models is obtained, including mixtures of most continuous densities (including exponential and Gamma distributions), and Poisson and negative binomial distributions (for count data). Examples of non-identifiable mixture models are mixtures of uniform distributions, mixtures of g binomial distributions with $n < (2g-1)$ and multivariate generalized Bernoulli distributions with $p < 2\log_d g + 1$, where d is the number of categories (see Allman, Matias, and Rhodes (2009) and Ng (2015)).

1.3 EM Algorithm

The EM algorithm is widely applied to the iterative computation of maximum likelihood estimates in a wide variety of incomplete data problems; see the monograph by McLachlan and Krishnan (2008) and the book chapters by Ng, Krishnan, and McLachlan (2012) and Ng (2013). In some situations, the incompleteness of data may not be so obvious, such as statistical models involving mixtures, random effects, and latent class structures (McLachlan and Peel, 2000). With the EM algorithm, it is possible for these statistical models to formulate an associated statistical problem that is obtained by augmenting the original observed variables (the incomplete data) with additional variables that are unobservable (the missing data), and from which maximum likelihood estimation is simpler analytically and/or computationally. The EM algorithm takes an iterative approach and is closely related to the *ad hoc* approach to estimation with missing data. The parameters are estimated after filling in initial values for the missing data. The latter are then updated by their predicted values using these initial parameter estimates; re-estimation of the parameters proceeds iteratively until convergence (Ng, Krishnan, and McLachlan, 2012).

The EM algorithm possesses a number of desirable properties, such as its numerical stability, reliable global convergence, and simplicity of implementation. The following subsections describe the basic principles of the EM algorithm, how mixture models can be formulated as an incomplete-data problem, and advanced developments of the EM algorithm in the context of mixture models and in its applications to the related fields of medical and health sciences.

1.3.1 Basic Principles of the EM Algorithm

The EM algorithm is formulated under an incomplete-data framework, where there are unobservable (or missing) data z and the observed data y can be considered as incomplete data. The corresponding log likelihood based on y given in (1.4) is referred to as incomplete-data log likelihood. As described in Section 1.2.1, the EM algorithm is applied to iteratively compute roots of the incomplete-data log likelihood equation (1.5) corresponding to the local maximizer of (1.4). The mechanism of missing data is conceptualized in the EM framework to simplify maximum likelihood estimation from data that can be viewed as being incomplete. It is not necessary that data are missing in a practical sense, thus the phrase "incomplete data" is used quite broadly to represent a variety of data models including mixtures, censoring, clustered, and missing observations; see, for example, McLachlan and Krishnan (2008, Chapter 3) and Ng (2013).

The EM algorithm works on the complete data expressed as

$$(y^T, z^T). \tag{1.12}$$

Let $f((y^T, z^T)^T; \Psi)$ denotes the probability density function (p.d.f.) of the random vector corresponding to the complete data. The complete-data log likelihood for Ψ

can then be formed if the complete data (1.12) were fully observed. That is,

$$\log L_c(\Psi) = \log f((y^T, z^T)^T; \Psi). \tag{1.13}$$

For many statistical problems, the complete-data log likelihood has a nice form and its equation is easier to solve. The EM algorithm obtains roots of (1.5) indirectly by proceeding iteratively in terms of the complete-data log likelihood $\log L_c(\Psi)$. In each iteration, there are two steps, the expectation step (E-step) and the maximization step (M-step); see McLachlan and Krishnan (2008, Section 1.8) for a brief history of the EM algorithm. On the $(k+1)$th iteration of the EM algorithm, the E-step computes the so-called Q-function

$$Q(\Psi; \Psi^{(k)}) = E_{\Psi^{(k)}}\{\log L_c(\Psi)|y\}, \tag{1.14}$$

where $E_{\Psi^{(k)}}$ denotes expectation using the current parameter vector $\Psi^{(k)}$ and the E-step calculates the expectation of $\log L_c(\Psi)$ conditioned on observed data y using the current fit for Ψ. The M-step updates the estimate of Ψ by the new value $\Psi^{(k+1)}$ of Ψ that maximizes the Q-function with respect to Ψ over the parameter space Θ. In many applications of the EM algorithm, the solution of the M-step exists in closed form. For situations where global maximization of the Q-function is not possible, the M-step can be proceeded by updating $\Psi^{(k)}$ to $\Psi^{(k+1)}$ that increases the Q-function over its value at $\Psi = \Psi^{(k)}$. This results in a generalized EM (GEM) algorithm; see Ng, Krishnan, and McLachlan (2012).

The E- and M-steps are alternated repeatedly until the change in the estimates $\Psi^{(k)}$ and $\Psi^{(k+1)}$ or the difference in the log likelihood values is less than some specified threshold. Both the E- and M-steps will have particularly simple forms when the p.d.f. of the complete data, $f((y^T, z^T)^T; \Psi)$, is from an exponential family including most common distributions, such as multivariate normal for continuous data, Poisson for count data, and multinomial for categorical data, as described in McLachlan and Krishnan (2008, Section 1.5).

1.3.2 Formulation of Mixture Modelling as Incomplete-Data Problems

Within the EM framework, mixture models can be conceptually formulated by assuming that each y_j has come from one of the g clusters (1.2). To pose the estimation of mixture models as an incomplete-data problem, we let $z_1, \ldots, z_n$ denote the unobservable component-indicator vectors, where the hth element z_{hj} of z_j is taken to be one or zero accordingly as the jth entity y_j does or does not belong to the hth cluster $(h = 1, \ldots, g)$. Since the component-indicator vectors $z = (z_1^T, \ldots, z_n^T)^T$ are not available, they are treated as missing data and the observed data $y = (y_1^T, \ldots, y_n^T)^T$ as incomplete data. Together, they form the complete data (y^T, z^T) as in (1.12). Assuming that the observations $y_1, \ldots, y_n$ are conditionally independent and identically distributed (i.i.d.) given $z_1, \ldots, z_n$, and $z_1, \ldots, z_n$ themselves are i.i.d. according to a multinomial distribution

$$z_1, \ldots, z_n \overset{i.i.d.}{\sim} \text{Mult}_g(1, \pi_1, \ldots, \pi_g) \tag{1.15}$$

consisting of one draw on g categories, then it follows from (1.2) that the complete-data log likelihood for Ψ is given by

$$\log L_c(\Psi) = \log f((y^T, z^T)^T; \Psi) = \sum_{h=1}^{g} \sum_{j=1}^{n} z_{hj} \{\log \pi_h + \log f_h(y_j; \psi_h)\}. \tag{1.16}$$

As the complete-data log likelihood (1.16) is linear in the missing data z_{hj}, the E-step, which involves the computation of the Q-function, is simply equal to $\log L_c(\Psi)$ with each z_{hj} replaced by its conditional expectation given the observed data y using the current fit for Ψ; see (1.14). On the $(k+1)$th iteration of the E-step, that is

$$E_{\Psi^{(k)}}(Z_{hj}|y) = \text{pr}_{\Psi^{(k)}}\{Z_{hj} = 1|y\} = \tau_h(y_j; \Psi^{(k)}) = \frac{\pi_h^{(k)} f_h(y_j; \psi_h^{(k)})}{\sum_{l=1}^{g} \pi_l^{(k)} f_l(y_j; \psi_l^{(k)})}, \tag{1.17}$$

which is the posterior probability that the jth entity y_j belongs to the hth component of the mixture $(h = 1, \ldots, g;\ j = 1, \ldots, n)$. From (1.16) and (1.17), the Q-function is given by

$$\begin{aligned} Q(\Psi; \Psi^{(k)}) &= E_{\Psi^{(k)}}\{\log L_c(\Psi)|y\} \\ &= \sum_{h=1}^{g} \sum_{j=1}^{n} E_{\Psi^{(k)}}(Z_{hj}|y)\{\log \pi_h + \log f_h(y_j; \psi_h)\} \\ &= \sum_{h=1}^{g} \sum_{j=1}^{n} \tau_h(y_j; \Psi^{(k)})\{\log \pi_h + \log f_h(y_j; \psi_h)\}. \end{aligned} \tag{1.18}$$

It can be seen from (1.18) that for i.i.d. data, the Q-function can be decomposed into separate terms corresponding to the mixing proportions $\pi = (\pi_1, \ldots, \pi_{g-1})^T$ and the component parameters $\psi_1, \ldots, \psi_g$. For some typical component densities such as the normal distribution, the component parameters can be further separated such that the M-step, involving the maximization of the Q-function with respect to Ψ, can be performed independently for each component and is in closed form. For dependent data, Equation (1.16) does not hold, however, and different approaches are needed for the estimation of Ψ. This will be discussed further in Chapters 6, 7, and 8.

1.3.3 Convergence and Initialization of the EM Algorithm

An appealing property of the EM algorithm is its numerical stability, where the likelihood function is not decreased after an EM or GEM iteration. That is,

$$L(\Psi^{(k+1)}) \geq L(\Psi^{(k)}) \qquad (k = 0, 1, 2, \ldots), \tag{1.19}$$

which implies the self-consistency of the EM algorithm. For a bounded sequence of likelihood values $\{L(\Psi^{(k)})\}$, $L(\Psi^{(k)})$ converges monotonically to some value L^*. According to Wu (1983), when the complete data $(y^T, z^T)^T$ are from a curved exponential family with compact parameter space and the Q-function satisfies a certain differentiability condition, then any EM sequence $\{\Psi^{(k)}\}$ converges to a stationary point (not necessarily a maximum) of the likelihood function. If $L(\Psi)$ has multiple stationary points, an EM sequence will converge to either a local or global maximizer, or a saddle point, depending on the initial value $\Psi^{(0)}$ for Ψ; see Wu (1983), McLachlan and Krishnan (2008, Sections 3.4-3.5), and Ng (2013). In practice, it is necessary to examine with caution the nature of the stationary point to which an EM sequence converges. Moreover, the EM algorithm should be implemented from a wide choice of initial parameter values $\Psi^{(0)}$ in an attempt to search for all local maxima. Various methods for the specification of initial values have been considered specifically within the mixture models framework, where initialization of the EM algorithm can be achieved by specifying initial estimates of the unknown parameters, or equivalently, by specifying initial grouping of the data y with respect to the components of the mixture model under fit. A commonly used scheme to locate the global maximizer of the log likelihood function is to run the EM algorithm with various initializations and the set of parameters corresponding to the largest likelihood value is then used as initial values for the EM algorithm to obtain the final estimates of parameters; see, for example, McLachlan and Peel (2000, Chapter 2), Biernacki, Celeux, and Govaert (2003), and Maitra and Melnykov (2010). Alternatively, there are variations of the EM algorithm that intend to escape from local maxima by considering a genetic-based EM algorithm (Martinez and Vitria, 2000; Pernkopf and Bouchaffra, 2005) or by performing split and merge operations (Split and Merge EM algorithm); see, for example, Zhang et al. (2003).

One of the main drawbacks of the EM algorithm is that the convergence rate is only linear and is usually slower than the quadratic convergence typically available with Newton-type methods. If $\Psi^{(k)}$ converges to some point Ψ^*, we have in a small neighborhood of Ψ^* that

$$\Psi^{(k+1)} - \Psi^* \approx J(\Psi^*)(\Psi^{(k)} - \Psi^*), \tag{1.20}$$

given that the mapping $M(\Psi)$, defined implicitly by the EM algorithm $\Psi \to M(\Psi)$, is continuous. In (1.20), $J(\Psi^*)$ is the Jacobian matrix for $M(\Psi)$ and can be expressed in terms of the information ratio matrix that measures the proportion of missing information about Ψ; see Dempster, Laird, and Rubin (1977) and McLachlan and Krishnan (2008, Section 3.9). The larger the eigenvalue of $J(\Psi^*)$, the greater the proportion of missing information and hence the slower the rate of convergence. In particular, different components of Ψ may converge to Ψ^* at various rates, which depends on the corresponding eigenvalue of $J(\Psi^*)$. Variants of the EM algorithm have been proposed to speed up the EM algorithm for fitting mixture models to large data sets or for applications where the complete-data maximum likelihood estimation is complicated. These include the incremental EM (IEM)-based algorithms (Ng, 2013), which will be discussed further in Chapter 10, the Expectation-Conditional Maximization (ECM)-based algorithm (see Meng and Rubin (1993), Meng and van Dyk (1997), and He and Liu (2012)), and the P-EM algorithm proposed by Berlinet and Roland (2012).

In some applications, such as fitting mixture models that involve parameter estimation of generalized linear mixed models (GLMMs) (see, for example, the monograph by McCulloch and Searle (2001)), the computation of the expectation of the complete-data log likelihood in the E-step could be too complex. This is because the likelihood function has an intractable integral whose dimension depends on the structure of the random effects (Ng, 2013). In this situation, the Monte Carlo EM (MCEM) algorithm may be used, where the E-step is implemented via a Monte Carlo simulation process, as discussed by Robert and Casella (2004). However, as a Monte Carlo error is introduced in the E-step, the monotonicity property for the EM algorithm is lost. It is thus necessary to specify an appropriate value of the number of independent sets of simulation in approximating the E-step and to monitor convergence; see, for example, Jank (2006). Alternatively, there are modelling approaches that lead to the implementation of the EM algorithm without the need to use Monte Carlo methods in the E-step. These will be described in details in Chapters 6, 7, and 8.

1.3.4 Provision of Standard Errors of Estimates

Another issue associated with the use of the EM algorithm is the provision of standard errors for the maximum likelihood estimates of the parameters Ψ, which are not provided directly using the EM algorithm. With an information-based approach, the asymptotic covariance matrix of the estimates of Ψ can be approximated by the inverse of the observed information matrix $I(\hat{\Psi};y)$, which involves the analytical evaluation of the second-order derivatives of the log likelihood for the normal mixture model; see Roberts et al. (1998), McLachlan and Peel (2000, Sections 2.15-2.16), and Lee et al. (2002). However, as described in Ng, Krishnan, and McLachlan (2012), the standard errors of the ML estimates based on the observed information matrix are valid only when the number of observations n is very large such that the asymptotic theory of ML applies to mixture models. To this end, we adopt a bootstrap resampling approach to obtain the standard errors of the ML estimates of Ψ. The procedure is given as follows:

1. Let $\hat{F}$ denote an estimate of the distribution function of Y formed from the original observed data y. A new set of data, y^*, called the bootstrap sample, is generated according to $\hat{F}$. That is, in the case where y contains the observed values of a random sample of size n, y^* consists of the observed values of the random sample

$$Y_1^*, \ldots, Y_n^* \overset{\text{i.i.d.}}{\sim} \hat{F},$$

where the estimate $\hat{F}$ is held fixed at its observed value.

2. The EM algorithm is applied to the bootstrap sample y^* to compute the ML estimates for this data set, $\hat{\Psi}^*$.

3. Steps 1 and 2 are repeated independently a number of times (say, B) to give B independent realizations of $\hat{\Psi}^*$, denoted by $\hat{\Psi}_1^*, \ldots, \hat{\Psi}_B^*$.

4. The bootstrap covariance matrix of $\hat{\Psi}^*$ is then approximated by the sample covariance matrix of the B bootstrap replications to give

$$\text{Cov}^*(\hat{\Psi}^*) \approx \sum_{b=1}^{B} (\hat{\Psi}_b^* - \overline{\hat{\Psi}}^*)(\hat{\Psi}_b^* - \overline{\hat{\Psi}}^*)^T/(B-1), \tag{1.21}$$

where $\overline{\hat{\Psi}}^* = \sum_{b=1}^{B} \hat{\Psi}_b^*/B$.

The standard error of the ith element of $\hat{\Psi}$ can be estimated by the positive square root of the ith diagonal element of (1.21). We shall adopt $B = 100$ bootstrap replications, which is generally sufficient for standard error estimation, as discussed in Efron and Tibshirani (1993). A nonparametric version of the bootstrap resampling approach is also available for estimating the standard errors of the ML estimates, where in Step 1 above, $\hat{F}$ is replaced by the empirical distribution function formed from the observed data y. With the use of the empirical distribution function, this nonparametric approach allows for the mixing proportions to vary randomly among the B bootstrap replicated samples. This is particularly useful for obtaining the standard errors of the mixing proportion estimates. On the other hand, the parametric approach generates data according to the fitted mixture model and thus allows for extrapolation beyond the range of the original observed data. This offers a better representation of the tails of the component densities; see Zheng and Frey (2004).

1.4 Applications of Mixture Models in Medical and Health Sciences

Mixture models have been of considerable interest in past decades in terms of both methodological development and their applications in a wide range of scientific fields of study. Specifically in medical and health sciences, applications of mixture models have been considered in anatomy (Akram et al., 2014), bioinformatics (Ng et al., 2006; Wang, Ng, and McLachlan, 2012), cell biology (Lo, Brinkman, and Gottardo, 2008; Pyne et al., 2014), chronic diseases (Alava et al., 2013; Rzehak et al., 2013), critical care (Lofgren et al., 2014), genetics (Luan et al., 2009; Lillehammer et al., 2013), geriatrics (Xue et al., 2012), infectious diseases (Pai et al., 2008; Fujii et al., 2014), medical imaging (Ji et al., 2012; Ng and Lam, 2013), nutrition (Fahey et al., 2012; Pryer and Rogers, 2009; Onwezen et al., 2012), obstetrics and gynecology (Ng et al., 2010; Urquia, Moineddin, and Frank, 2012), orthopedics (Riihimaki, Sund, and Vehtari, 2010), pediatrics (Stevens et al., 2012; Ventura, Loken, and Birch, 2006; Haynie et al., 2013), pharmacology (Piper et al., 2008; Wu et al., 2011), psychology and psychiatry (Touchette et al., 2007; DeRoon-Cassini et al., 2010), among others. Mixture models arise in a natural way in that they are modelling unobserved population heterogeneity.

Mixture models also have an important role in the context of discriminant analysis (also referred to as "supervised classification") where the data are classified with respect to a finite number, say g, of known classes. The intent is to construct a nonlinear discriminant rule (or classifier) on the basis of these classified data for assigning an unclassified entity to one of the g classes on the basis of its observed feature variables and for assessing the associated error rate; see, for example, the monograph by McLachlan (1992) and the more recent review in McLachlan (2012).

There are a number of monographs that focus on the theoretical developments of mixture models and related topics, including Everitt and Hand (1981), Titterington, Smith, and Makov (1985), McLachlan and Basford (1988), McLachlan and Peel (2000), Mengersen, Robert, and Titterington (2011), and McNicholas (2016). These traditional monographs provide an excellent account on statistical theory and inference concerning mixture models (also with Bayesian approaches or topics in discriminant analysis). However, a more direct connection with real-life examples to promote the application of mixture models, especially in the fields of medical and health sciences, is not the focal point in these monographs. This book attempts to fill this gap, with emphasis on methodological challenges coming from real world problems in medical and health sciences. The monograph by Schlattmann (2009) addresses the issues relating to medical applications of mixture models, such as modelling of count data, disease mapping, and heterogeneity in meta-analysis. In this book we will focus on advanced mixture models recently developed to tackle problems associated with the analysis of multilevel data, longitudinal data, and multivariate dependent data.

1.4.1 Overview of Book

The book is divided into three parts. The first part deals with the applications of mixture models in the analysis of independent data. The second part focuses on mixture models with random-effects components for the analysis of dependent data, and the third part presents some miscellaneous topics about missing data and "Big Data" analyses. In the first part, Chapter 2 describes mixtures of normal and multivariate normal distributions for the analysis of continuous data. For continuous non-normal data, Chapter 3 introduces mixtures of gamma distributions. In Chapter 4, we present mixtures of generalized linear models for count or categorical data. Mixture models for survival data are described in Chapter 5. In the second part, we go beyond the independence assumptions underlying standard mixture data as presented in Chapters 2 to 5. Such an extension of the mixture model facilitates the advanced development of mixture models with random-effects components to account for correlation among data. In Chapter 6, we introduce the fundamental concepts of random-effects models and the approaches for parameter estimation. Chapter 7 focuses on the analyses of multilevel and repeated-measured data, Chapter 8 describes mixture models for correlated multivariate continuous data. Finally, Chapter 9 in the third part presents how missing data can be handled in the applications of mixture models and Chapter 10 discusses cluster analysis of "Big Data" using mixture models.

1.4.2 Sample Size Considerations for Mixture Models

Sample size calculations form an important part of the study design in randomized clinical trials and observational cohort studies to ensure validity, reliability, as well as scientific and ethical integrity of the study. While many papers and review articles have been published with a focus on sample size and power determinations, including Button et al. (2013), Charan and Biswas (2013), and Thabane et al. (2010), few studies have been considered for sample size calculations in a mixture modelling framework. This is due to the complexity in understanding specific design options and preliminary information on parameter values such as (1) the number of components, (2) the mixing proportions, (3) model assumptions, and (4) the degree of separation between components as characterized, for example, by (a) component mean differences, (b) component variances, and (c) skewness of the assumed component distributions, as well as (5) the nature and variability of the measurements for estimating sample sizes before the study starts. Moreover, sample size planning for mixture models is specific to the particular type of mixture analysis of interest and the goal of the study; see, for example, Maxwell, Kelley, and Rausch (2008).

The following paragraphs summarize sample size determinations within the mixture modelling framework for related topics described in this book, notably, in assessing the number of components, the intervention effect in clinical trials, differential expression in analyses of gene expression data, and the standard errors of parameter estimates. For example, Munoz and Acuna (1999) adopted Monte Carlo simulations to assess the sample size requirements for a simple mixture of two univariate normal components with equal variances for the determination of the number of components g using the likelihood-ratio test (LRT) method of McLachlan (1987) (see Section 1.2.3) in relevant biological applications such as those concerning morphological traits. Their results suggest that a sample size over $n = 300$ or $n = 550$ will have more than 80% power for two-component normal mixtures with a Mahalanobis distance of $\Delta = 2.5$ or $\Delta = 2$, respectively, where the Mahalanobis distance is defined as

$$\Delta = \sqrt{\delta(\mu_1, \mu_2, \sigma^2)} = (\mu_1 - \mu_2)/\sigma, \tag{1.22}$$

quantifying the separation between the two univariate normal components with means μ_1 and μ_2, respectively, and equal variances σ^2. In other contexts, sample size formula for the mixture-based LRT approach with case-control sampling design of association studies has been provided by Kim et al. (2008), where the underlying continuous quantitative measure is modelled by a four-component mixture of univariate normal distributions. Alternatively, Lubke and Neale (2006) considered penalized likelihood approach, such as AIC and BIC, for model selection of mixture models with continuous latent variables (see Chapter 2) and showed that the accuracy of model selection using penalized likelihood approaches depends on the separation between components and the sample size, whereas unequal mixing proportions do not seem to have a major impact on model selection accuracy. In the context of model selection using the posterior predictive approach, the effect of sample size on assessing the number of components of a mixture of multivariate generalized Bernoulli distributions (see Chapters 4 and 10) has been studied by Rubin and Stern (1998) with a psychological data set concerning infant temperaments. With a simulation-based approach for determining the sample size using posterior predictive distributions of discrepancies selected to diagnose model fit, Rubin and Stern (1998) showed that the sample size of the data set was inadequate to reject the two-component solution and a much larger sample size is required.

In a randomized clinical trial setting, Nie, Chu, and Cole (2006) derived the sample size required for assessing the intervention effect using the LRT approach, where a zero-inflated log-normal mixture model with left censoring due to the presence of detection limits is fitted to the treatment and control groups separately. A zero-inflated log-normal model assumes that the non-negative continuous response variable has a probability mass at zero and a continuous log-normal density for values greater than zero (see Chapters 4 and 7). The sample size calculation is based on the assumption of equal proportions of true zeros in the treatment and control groups and the LRT is used for testing the difference in means for the non-zero part between the two groups. Alternatively, Wang, Zhang, and Lu (2012) considered a proportional hazards cure model in the analysis of time-to-event data in clinical trials. The cure model takes the form of mixture models, where a significant proportion of patients is cured (or being a long-term survivor) and the survival times of uncured patients are modelled under the proportional hazards approach; see Chapters 5 and 6. The sample size formula derived by Cai et al. (2014), available in an R-package "NPHMC", can be used to test the treatment effects in the short-term survival and/or the cure fraction under different accrual scenarios.

In the analysis of gene-expression data, one of the main objectives is to detect genes that are differentially expressed in a known number of tissue classes (Ng et al., 2015). Mixture models have been applied to estimate the distribution of effects sizes in two-class comparative studies and identify differentially expressed genes; see, for example, Lee and Whitmore (2002). In this context, Jorstad, Midelfart, and Bones (2008) present a closed-form algorithm for estimating sample sizes that control the false discovery rate (FDR); see also Pawitan et al. (2005). Alternatively, Wei, Li, and Bumgarner (2004) calculated the sample sizes required to detect a 1.5-, 2-, and 4-fold changes in expression level (defined as the log ratio between two classes of tissues; see Section 8.3) as a function of FDR, power and percentage of genes that have a standard deviation below a given percentile. And Müller et al. (2004) obtained an optimal sample size for massive multiple testing of differential gene expression using a Bayesian decision theoretic approach via Monte Carlo simulations and loss functions that combine the competing goals of controlling false-positive and false-negative decisions. Alternatively for microarray experiments with multiple measurements (for all genes in each tissue sample), Pan, Lin, and Le (2002) determine the number of replicates of microarrays required to detect changes in gene expressions. However, these methods assume that expression values are independent across genes, which may not be realistic in practice (see Chapter 8).

Sample size is also important for quantifying the uncertainty around statistics of interest in mixture modelling; see Section 1.3.4 for calculation of standard errors. When the underlying population distribution is not normal and the sample size is small, methods based upon the asymptotic normality framework may induce errors in the estimation of standard errors. In this context, Zheng and Frey (2004) evaluated the effect of sample sizes ($n = 25$, 50, and 100) on the bootstrap resampling approach for calculating the standard errors of parameter estimates via synthetic data sets generated from selected two-component mixtures of log-normal distributions. They found that the bootstrap resampling method tends to be more stable for larger sample sizes and the stability is further improved when the components are well separated. In general, the results are more robust when components in a mixture model are of comparable weight.

Consideration of sample size requirements for mixture model analyses is not straight forward. Caution is needed as there may be a great amount of uncertainty in selecting appropriate parameter values (such as the means), the number of components in the mixture model, and the validity of model assumptions such as homoscedasticity in component densities. If the number of components and the assumed component distribution form are known and the approximate degree of separation can be determined based on some prior information or studies, then estimating the sample size for mixture models should be more feasible practically. However, when irregular conditions occur, the above methods of sample size calculation may become invalid. In general, the sample size should be large enough that the estimates will be sufficiently precise and the difference of interest is likely to be detected. There are two heuristic approaches that may be used to estimate the sample size requirement for mixture model analyses. Firstly, sample size calculations assuming a single population for the particular study design and research goal may be used as a guide, bearing in mind the adjustment required for the effects of the mixing proportions, the number of components and their separation on the sample size. The basic rule of thumb may also be applicable, such as increasing the sample size when the correlation between data becomes larger (as there is less independent information available). The second approach is based on a Monte Carlo simulation method to decide empirically on the sample size; see, for example, Muthén and Muthén (2002) and Müller et al. (2004). This approach allows for variations in the specification of the mixture model and finds the sample size that yields the required power. In all situations, the interpretation of the significance of parameter estimates should be treated with caution, especially for those significant estimates with a small effect size (false positive) or those estimates with a large effect size but that do not achieve significance (false negative).

1.5 Computing Packages for Mixture Models

Developing computing packages for mixture models is critical to promote the use of mixture modelling for analyzing real world data, as software for mixture models is currently limited in traditional statistical programs, such as SPSS, SAS, and STATA. In the Structural Equation Modelling (SEM) package in IBM SPSS Amos, mixture modelling is allowed to assign entities to clusters without regard for a dependent variable. The SEM package adopts Bayesian method instead of the ML approach via the EM algorithm for parameter estimation; see Arbuckle (2013) for more details. In SAS, Jones, Nagin, and Roeder (2001) have developed a new procedure "TRAJ" to analyze longitudinal data by fitting a semi-parametric mixture model, where component densities include censored normal, Poisson, zero-inflated Poisson, and Bernoulli distributions. More recently, the "FMM" procedure, experimental in SAS/STAT 9.3, estimates the parameters in univariate finite mixture models using both ML and Bayesian approaches and produces statistics to evaluate parameters and model fit (SAS Institute Inc., 2011). In STATA (StataCorp LP, College Station, Texas), the FMM module fits a finite mixture regression model using the ML approach. A STATA plugin for estimating group-based trajectory models (also known as growth mixture models) is also available; see, for example, Jones and Nagin (2013).

In other computing platform, MPlus Version 8 allows the fitting of mixture regression models with cross-sectional data as well as longitudinal data (including growth mixture models); see Muthén and Muthén (1998-2015). In addition, the NAG C Library (Mark 24) comes with the new functionality (G03GA) of fitting a mixture of normal distributions by maximizing the log-likelihood function for a given covariance structure, number of components, and (optionally) the initial membership probabilities. This new functionality is also in Mark 24 of the NAG Toolbox for MATLAB.

This book uses computing programs in R and Fortran for implementing mixture modelling procedures described in each chapter, which can be obtained via the book's online website (`https://www.crcpress.com/9781482236750`). Some mixture software in other computing platforms can be found in the appendix of McLachlan and Peel (2000).

1.5.1 R Programs

R is a free software environment for statistical computing and graphics. It complies and runs on a wide variety of UNIX platforms, Windows and Mac OS. Currently, the version of R for Windows is 3.4.2, which can be downloaded from the Comprehensive R Archive Network (CRAN) through the website `https://cran.r-project.org/bin/windows/base/old/3.4.2/`.

The R software environment provides several packages for fitting mixture models. For example, the "MCLUST" package in R (Fraley and Raftery, 2006) adopts an eigen-decomposition of the component covariance matrices that leads to a variety of mixtures of normal distributions. The EM algorithm is adopted in the parameter estimation process, whereas the BIC is used for the model selection. It also has the option to include an additional Poisson component in the mixture model for background noise. Recently, version 5 of the package adds new covariance structures, initialization strategies for the EM algorithm, model selection criteria, and bootstrap-based inference; see Scrucca et al. (2016) for more details. In a similar manner to the MCLUST package, Andrews and McNicholas (2012) developed the "tEIGEN" family of mixture models that is a t-analogue of the MCLUST family. Alternatively, the "FPC" package developed by Hennig (2010) fits mixtures of linear regression models and merges components of a normal mixture model. The package can also handle mixed continuous/categorical variables. The "CAMAN" package (developed based on C.A.MAN (Computer-Assisted Mixture Analysis) of Böhning,

Schlattmann, and Lindsay (1998)) can fit normal mixtures as well as semi-parametric mixtures, especially for heterogeneous data in pharmacokinetics or meta-analysis. There are also R packages for the clustering and the discriminant analysis of high-dimensional data (Berg'e, Bouveyron, and Girard, 2012) and for fitting discrete mixtures of regression models (Leisch, 2004). More recently, there are "MIXTOOLS" v1.1.0 developed by Young et al. (2015) for fitting mixtures of multivariate normal, multinomial, and gamma distributions, "movMF" v.0.2-1 by Hornik and Grün (2014) for fitting mixtures of von Mises-Fisher distributions, and "clustMD" by McParland and Gormley (2016) for clustering of mixed (continuous and categorical) data using a parsimonious mixture of latent Gaussian variable models. A list of R packages that can be used for fitting mixture models is also available in the CRAN Task View established by Leisch and Gruen (2017).

The R program codes for Example 1.1 in Section 1.1.1, including the mixture modelling, quantile regression, and subgroup regression, are provided in Figure 1.5.

1.5.2 Fortran Programs

The Fortran codes described in this book can be compiled to executable programs run in Windows using Gfortran, which is free Fortran 95/2003/2008 compiler for the GNU Compiler Collection (GCC); see the homepage for Gfortran `https://gcc.gnu.org/fortran/` for more information. Some common options for installation of Gfortran on Windows include free software such as MinGW (Minimalist GNU for Windows) and Cygwin. In this book, the examples that involved using Fortran programs were performed using Cygwin, which offers a UNIX-like environment for Windows and would be convenient for readers who had experiences with UNIX functions and utilities in a command-line environment. The current version of Cygwin for Windows is 2.9.0, which can be downloaded from the Cygwin website `http://www.cygwin.com`, offering up-to-date builds of GCC and Gfortran. Installation of gcc-core, gcc-debuginfo, gcc-fortran, and gcc-g++ packages is required in the procedure of "Selection packages to install" in Cygwin for compiling fortran codes using Gfortran; see the instruction in the Cygwin website.

The EMMIX program of McLachlan and Peel (2000) for fitting mixture models of (multivariate) normal or t distributions is available in both R and Fortran environments. The MCLUST of Fraley et al. (2012) and the C.A.MAN of Böhning, Schlattmann, and Lindsay (1998) are also available in Fortran environment as well. Moreover, there are Fortran programs for fitting a mixture of univariate normal distributions to binned and truncated data ("MGT" program of Jones and McLachlan (1990)), mixtures of discrete, normal, Poisson and von Mises distributions using the Minimum Message Length (MML) approach ("SNOB" program of Wallace and Dowe (1994)), a mixture of two Gompertz distributions to censored survival data (McLachlan et al., 1997), and mixture models to mixed continuous and categorical variables ("MULTIMIX" of Hunt and Jorgensen (1999)).

Main Analysis for Example 1.1

```
setwd("c:/Example1.1/")              # set the work directory

data <- matrix(scan("hsb2_sci.txt"),ncol=1,byrow=T)       # read the science scores

library(mvtnorm)                              # install package "mvtnorm" and call it

require(EMMIX)                                # install the EMMIX program and call it

set.seed(386)                                 # set a seed for generating random numbers

# initialization using 10 k-means and 10 random starts
# assume g=2 components of univariate normal with unequal standard deviations
# convergence criterion (log likelihood < 1e-10 of the previous log likelihood value)
initobj <- init.mix(data, g=2, distr="mvn", ncov=3, nkmeans=10, nrandom=10, nhclust=F)
obj <- EMMIX(data, g=2, distr="mvn", ncov=3, clust=NULL, init=initobj, itmax=1000,
       epsilon=1e-10, debug=TRUE)

attributes(obj)
write.csv(cbind(obj$tau,obj$clust),"cluster_2gp.csv")     # save the clustering results


# quantile regression (Figure 1.2(a))
library(quantreg)                             # install package "quantreg" and call it

hsb2 <- read.table("hsb2.txt",header=T,sep="\t")           # read the "hsb2" data

# obtain quantile regression lines for taus = 10th, 25th, 50th, 75th, and 90th quantiles
fit1 <- rq(science ~ read + write + math + socst + female, tau=0.1, data=hsb2)
summary(fit1)
fit2 <- rq(science ~ read + write + math + socst + female, tau=0.25, data=hsb2)
summary(fit2)
fit3 <- rq(science ~ read + write + math + socst + female, tau=0.5, data=hsb2)
summary(fit3)
fit4 <- rq(science ~ read + write + math + socst + female, tau=0.75, data=hsb2)
summary(fit4)
fit5 <- rq(science ~ read + write + math + socst + female, tau=0.9, data=hsb2)
summary(fit5)

# subgroup regression (Figure 1.2(b))

hsb2c <- read.table("hsb2_2gp.txt",header=T,sep="\t")  # hsb2 data with cluster identity

# obtain linear regression lines for cluster = 1 and 2
c1 <- lm(science ~ read + write + math + socst + female, data=subset(hsb2c,cluster==1))
summary(c1)
c2 <- lm(science ~ read + write + math + socst + female, data=subset(hsb2c,cluster==2))
summary(c2)
```

FIGURE 1.5
The R program codes for Example 1.1.

2

Mixture of Normal Distributions for Continuous Data

2.1 Introduction

Many variables of interest in medical and health sciences can be treated as continuous data. When clusters in the data are essentially elliptical or any non-normal features in the data are attributed to some underlying group structure, then a mixture of elliptically symmetric component distributions (such as normal or t-distributions) could be used to explore the group structure of the data. One attractive feature of mixture of normal or t-distributions is that the implied clustering is invariant under affine transformations (such as any linear transformation of data).

Mixture of normal distributions, like its single population analog, is widely used to model continuous data. Let $y = (y_1^T, \ldots, y_n^T)^T$ be the observed data with probability density function (p.d.f.) $f(y;\Psi)$, where n is the number of observations. With a mixture of normal distributions, the density of y_j is given by

$$f(y_j;\Psi) = \sum_{h=1}^{g} \pi_h \phi(y_j;\mu_h,\Sigma_h) \qquad (j = 1,\ldots,n), \tag{2.1}$$

where π_h $(h = 1,\ldots,g)$ are the mixing proportions of g components that sum to one, and

$$\phi(y_j;\mu_h,\Sigma_h) = (2\pi)^{-\frac{p}{2}} |\Sigma_h|^{-\frac{1}{2}} \exp\{-\tfrac{1}{2}(y_j - \mu_h)^T \Sigma_h^{-1} (y_j - \mu_h)^T\} \tag{2.2}$$

denotes a p-dimensional multivariate normal distribution with mean μ_h and covariance matrix Σ_h. The vector Ψ of unknown parameters consists of the mixing proportions $\pi_1, \ldots, \pi_{g-1}$, the elements of the component means μ_h, and the distinct elements of the component-covariance matrices Σ_h.

In contrast to models of continuous data using a single normal distribution, a normal mixture model needs to consider how the component-covariance matrices are specified; that is, whether they are unequal (heteroscedastic) or they are restricted to being the same (homoscedastic), that is, $\Sigma_h = \Sigma$ for all $h = 1,\ldots,g$. For multivariate continuous data, consideration should also be given whether to adopt for Σ_h a diagonal matrix (independence among variates) or an unrestricted matrix (unequal correlation among variates).

For applications where the tails of the normal distribution are shorter than appropriate or the parameter estimates are affected by atypical observations (outliers), the fitting of mixtures of multivariate t-distributions provides a more robust approach to the fitting of normal mixture models; see McLachlan and Peel (2000, Chapter 7). The p-dimensional multivariate t component density with location parameter vector μ_h, positive-definite matrix Σ_h, and ν_h degrees of freedom is given by

$$t_p(y_j;\mu_h,\Sigma_h,\nu_h) = \frac{\Gamma(\frac{\nu_h+p}{2})|\Sigma_h|^{-1/2}}{(\pi\nu_h)^{\frac{1}{2}p}\Gamma(\frac{\nu_h}{2})\{1+\delta(y_j,\mu_h;\Sigma_h)/\nu_h\}^{\frac{1}{2}(\nu_h+p)}}, \tag{2.3}$$

where $\Gamma(\cdot)$ denotes the gamma function and

$$\delta(y_j,\mu_h;\Sigma_h) = (y_j-\mu_h)^T\Sigma_h^{-1}(y_j-\mu_h)$$

denotes the Mahalanobis squared distance between y_j and μ_h (with Σ_h as the covariance matrix). When ν_h tends to infinity, Y_j becomes marginally multivariate normal with mean μ_h and covariance matrix Σ_h. Therefore, the parameter ν_h can be seen as a robustness tuning parameter, which can be inferred from the data by computing its ML estimate; see McLachlan and Peel (2000, Chapter 7). The application of a mixture of multivariate t-distributions for clustering cluster-specific contrasts of mixed effects for identifying differentially expressed genes is described in Chapter 8.

2.2 E- and M-steps

Fitting the mixture model (2.1) can be implemented using a maximum likelihood (ML) approach. An estimate $\widehat{\Psi}$ is obtained by solving the log likelihood equation, $\partial \log L(\Psi)/\partial\Psi = 0$, iteratively via the EM algorithm (Ng, 2013). From (2.1), the log likelihood for Ψ is given by

$$\log L(\Psi) = \sum_{j=1}^{n} \log\{\sum_{h=1}^{g} \pi_h\phi(y_j;\mu_h,\Sigma_h)\}. \tag{2.4}$$

Within the EM framework, each y_j is conceptualized to have arisen from one of the g components of the mixture model (2.1). We denote $z_1,\ldots,z_n$ the unobservable component-indicator vectors, where the hth element z_{hj} of z_j is taken to be one or zero accordingly to the jth observation y_j does or does not come from the hth component; see, for example, Ng (2013). As described in Section 1.3.2, the EM algorithm approaches the computation of roots of the (incomplete-data) log likelihood (2.4) indirectly by proceeding iteratively in terms of the log likelihood, $\log L_c(\Psi)$, of the complete data that contain observed data y and missing data $z = (z_1^T,\ldots,z_n^T)^T$, where

$$\log L_c(\Psi) = \sum_{h=1}^{g}\sum_{j=1}^{n} z_{hj}\{\log\pi_h + \log\phi(y_j;\mu_h,\Sigma_h)\}. \tag{2.5}$$

On the $(k+1)$th iteration of the EM algorithm, the E-step computes the Q-function, which is the expectation of the complete-data log likelihood conditioned on observed y using the current fit for Ψ. That is,

$$Q(\Psi;\Psi^{(k)}) = \sum_{h=1}^{g}\sum_{j=1}^{n} \tau_h(y_j;\Psi^{(k)})\{\log \pi_h + \log \phi(y_j;\mu_h,\Sigma_h)\}, \tag{2.6}$$

where $E_{\Psi^{(k)}}$ denotes expectation using the current parameter vector $\Psi^{(k)}$ and

$$\tau_h(y_j;\Psi^{(k)}) = \frac{\pi_h^{(k)}\phi(y_j;\mu_h^{(k)},\Sigma_h^{(k)})}{\sum_{l=1}^{g}\pi_l^{(k)}\phi(y_j;\mu_l^{(k)},\Sigma_l^{(k)})} \tag{2.7}$$

is the estimate of the posterior probability that the jth observation y_j belongs to the hth component of the mixture model on the basis of $\Psi^{(k)}$ $(h = 1,\ldots,g;\ j = 1,\ldots,n)$. For mixtures with normal component densities, it is computationally advantageous to work in the E-step in terms of the sufficient statistics (Ng and McLachlan, 2003b) given by

$$\begin{aligned} T_{h1}^{(k)} &= \sum_{j=1}^{n} \tau_h(y_j;\Psi^{(k)}) \\ T_{h2}^{(k)} &= \sum_{j=1}^{n} \tau_h(y_j;\Psi^{(k)})y_j \\ T_{h3}^{(k)} &= \sum_{j=1}^{n} \tau_h(y_j;\Psi^{(k)})y_jy_j^T. \end{aligned} \tag{2.8}$$

The M-step updates the estimate of Ψ to $\Psi^{(k+1)}$ that maximizes the Q-function with respect to Ψ over the parameter space. For mixtures with normal component densities, the M-step exists in closed form. On the basis of the sufficient statistics in (2.8), the updates of the parameters in the M-step are given by

$$\begin{aligned} \pi_h^{(k+1)} &= T_{h1}^{(k)}/n \\ \mu_h^{(k+1)} &= T_{h2}^{(k)}/T_{h1}^{(k)} \\ \Sigma_h^{(k+1)} &= \{T_{h3}^{(k)} - T_{h1}^{(k)^{-1}} T_{h2}^{(k)} T_{h2}^{(k)^T}\}/T_{h1}^{(k)}; \end{aligned} \tag{2.9}$$

see McLachlan and Peel (2000, Chapter 3) and Ng and McLachlan (2003b). For homoscedastic normal components, where $\Sigma_h = \Sigma$ for all $h = 1,\ldots,g$ and Σ is unspecified, the updated estimate of the common component-covariance matrix Σ is given by

$$\Sigma^{(k+1)} = \sum_{h=1}^{g} T_{h1}^{(k)}\Sigma_h^{(k+1)}/n, \tag{2.10}$$

where $\Sigma_h^{(k+1)}$ is given by (2.9), and the updates of π_h and μ_h are as above in the heteroscedastic case (2.9).

The E- and M-steps are alternated repeatedly until convergence of the EM sequence of iterations, which may be determined by using a suitable stopping rule like the difference $L(\Psi^{(k+1)}) - L(\Psi^{(k)})$ changes by an arbitrarily small amount in the case of convergence of the sequence of likelihood values $\{L(\Psi^{(k)})\}$.

In the case of unrestricted component-covariance matrices Σ_h, $L(\Psi)$ is unbounded because each data point gives rise to a singularity on the edge of the parameter space. As discussed in Section 1.2.2, consideration has to be given to the problem of relatively large (spurious) local maxima that occur as a consequence of a fitted component having a very small (but non-zero) generalized variance (the determinant of the covariance matrix); see McLachlan and Peel (2000, Section 3.10) for more details. Such a component corresponds to a cluster containing a few data points either relatively close together or almost lying in a lower dimensional sub-space in the case of multivariate data. In practice, a lower bound of π_h (such as 1%) may be set to avoid clustering solutions that involve a component with too few data points; see also Section 1.2.3 about the determination of the number of components.

For a mixture of multivariate t-distributions, the conditional expectations in the E-step can be expressed in closed form; see McLachlan and Peel (2000, Chapter 7). On the M-step, the solutions for updating parameters π_h, μ_h, Σ_h $(h = 1, \ldots, g)$ exist in closed form. Only the update for the degrees of freedom ν_h needs to be computed iteratively, which can be obtained via an Expectation-Conditional Maximization (ECM) algorithm as discussed in McLachlan and Peel (2000) and Ng (2013).

2.3 Diagnostic Procedures

In contrast to the case of single populations, testing of normality and homoscedasticity for mixture models is not straightforward. This is due to the lack of classified data and also that the component membership of each data point is practically unknown. One way to handle this situation is to proceed with testing of normality and homoscedasticity on the basis of the classified data implied by the fitted mixture model with the ML approach; see, for example, Hawkins, Muller, and ten Krooden (1982) and McLachlan and Basford (1988, Sections 2.7 and 3.2). With this approach, the test statistic proposed by Hawkins (1981) can be applied to assess simultaneously the normality and homoscedasticity for each implied group of the data. The test considers the Mahalanobis squared distance between the mean vector of a group and each data point of this group, which is distributed according to an F distribution under the null hypothesis of normality and homoscedasticity. The test of Hawkins (1981) adopts the Anderson-Darling statistic for assessing the distribution of the areas to the right of the observed Mahalanobis squared distance value under the F distribution corresponding to the null hypothesis of normality and homoscedasticity,

$$H_0 : y \sim N_p(\mu_h, \Sigma), \tag{2.11}$$

on the basis of the classified p-dimensional data y_{hj} in G_h ($h = 1,\ldots,g$; $j = 1,\ldots,n_h$), where Σ is a common covariance matrix and n_h is the number of entities assigned to the component G_h ($h = 1,\ldots,g$). That is, $n_h = \sum_{j=1}^{n} \hat{z}_{hj}$, where $\hat{z}_{hj} = 1$ if the estimated posterior probability for the hth component $\tau_h(y_j;\hat{\Psi})$ is the largest among the g components; see also Section 1.2. Denote

$$\bar{y}_h = \sum_{j=1}^{n_h} y_{hj}/n_h, \tag{2.12}$$

$$S_h = \sum_{j=1}^{n_h} (y_{hj} - \bar{y}_h)(y_{hj} - \bar{y}_h)^T/(n_h - 1), \tag{2.13}$$

and

$$S = \sum_{h=1}^{g} (n_h - 1)S_h/(n - g) \tag{2.14}$$

the sample mean vector and sample covariance matrix for the hth component, and the overall pooled covariance matrix, respectively, the Hawkins (1981) test showed that

$$F_{hj} = \frac{\nu n_h \delta(y_{hj}, \bar{y}_h; S)}{p[(\nu + p)(n_h - 1) - n_h \delta(y_{hj}, \bar{y}_h; S)]} \tag{2.15}$$

follows an F distribution with p and $\nu = n - g - p$ degrees of freedom ($h = 1,\ldots,g$; $j = 1,\ldots,n_h$) under the null hypothesis (2.11), where $\delta(y_{hj}, \bar{y}_h; S)$ is the Mahalanobis squared distance between y_{hj} and $\bar{y}_h$ with respect to the covariance matrix S; see McLachlan and Basford (1988, Sections 2.5 to 2.7) and Jamshidian and Jalal (2010). Denote a_{hj} the probability that an F-distributed random variable with degrees of freedom p and ν exceeds F_{hj}, it follows that, under H_0 in (2.11), a_{hj} ($h = 1,\ldots,g$; $j = 1,\ldots,n_h$) are independently and identically distributed as a uniform distribution on the range (0, 1). Hawkins (1981) suggested the use of the Anderson-Darling statistic (with its first two components) for assessing the uniformity of a_{hj}, as this statistic is particularly sensitive to fit in the tails of the distribution. In particular, the first component of the Anderson-Darling statistic measures the tendency of a_{hj} to have an excess of values near either 0 or 1, as represented, respectively, by a large positive or a large negative value for the first component. As described in McLachlan and Basford (1988, Section 2.5), non-normality is indicated if any values of the Anderson-Darling statistic and its components are significant (greater than 2.54 in magnitude) and homoscedasticity is rejected if the values of the first component of the Anderson-Darling statistic are large in magnitude, but of opposite sign. The R package "MissMech" of Jamshidian, Jalal, and Jansen (2014) can be used for testing normality and homoscedasticity for several groups of completely observed multivariate data using the Hawkins (1981) test; see also Jamshidian and Jalal (2010).

If it is reasonable to assume the component population G_h is heteroscedastic, multivariate normality can be assessed under the null hypothesis

$$H_0 : y \sim N_p(\mu_h, \Sigma_h) \text{ in } G_h (h = 1,\ldots,g). \tag{2.16}$$

Under (2.16) assuming normality and heteroscedasticity, F_{hj} in (2.15) is computed by replacing $\delta(y_{hj}, \bar{y}_h; S)$ and ν by $\delta(y_{hj}, \bar{y}_h; S_h)$ and $\nu_h = n_h - p - 1$, respectively. Consequently, a_{hj} is defined to be the probability that an F-distributed random variable exceeds F_{hj} with degrees of freedom p and ν_h; see McLachlan and Basford (1988, Section 2.5).

For univariate data, the overall goodness-of-fit of a mixture model can be visualized by comparing the density of the fitted mixture model with a histogram of the data. With an arbitrary split of the data into bins, formal assessment of model fit can be obtained using chi-square goodness-of-fit test. Alternatively, Aitkin (1997) suggested to compare fitted and empirical cumulative density functions (CDFs), as there is no loss of information in the CDFs. For the validity of the chi-square test, we choose as many bins as possible without allowing the expected number in any bin to be smaller than 5.

For multivariate data, the use of chi-square goodness-of-fit test for comparing fitted and empirical density distributions is plausible, providing a sensible split of the multivariate data into bins is available.

2.4 Example 2.1: Univariate Normal Mixtures

We consider an adolescent depression data set available from Chapter 20 of the monograph "Case Studies in Biometry" by Lange et al. (1994). An aim of the original study (Garrison et al., 1992) was to investigate the frequency of major depressive disorder and dysthymia in 12-14 year olds, where the Center for Epidemiological Studies-Depression Scale (CES-D) of Radloff (1977) was used as the primary measure of adolescent depression. The data set, "ch20.dat", which is available in the FTP archive of the Department of Statistics, University of Oxford at `http://www.stats.ox.ac.uk/pub/datasets/csb/`, consists of $n = 3,189$ students of seventh to ninth grades enrolled in four public middle schools and two high schools in a single southeastern US school district. Besides the CES-D total score, other socio-demographic characteristics collected in the data set are race (white or black), gender, guardian status (live with both natural parents or not), as well as a family cohesion score based on the Family Adaptability and Cohesion Evaluation Scale; see Garrison et al. (1992) for more information.

A histogram of the CES-D total scores is presented in Figure 2.1, showing a positively (right) skewed distribution of scores (range: 0 to 54; mean: 15.56; SD: 9.68; skewness: 0.93). Assuming there is only a single population, the association between the CES-D scores and children's demographic characteristics could be studied using generalized linear (regression) models with a non-normal distribution. Alternatively, if the skewness feature in the CES-D total scores is attributed to some underlying group structure, then fitting the scores with a mixture of univariate normal distributions will reveal the group structure. Assuming heteroscedasticity, initial values of $\Psi^{(0)}$ for the EM algorithm were determined from ten random starts and ten k-means starts. We considered the number of components from g=1 to g=6. Based

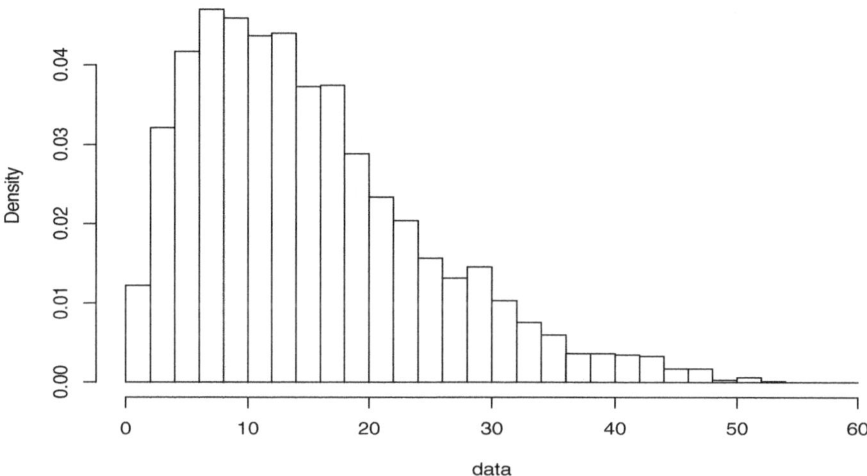

FIGURE 2.1
Histogram of CES-D scores (Example 2.1).

on model selection via BIC displayed in Table 2.1, four groups of young children were identified. The estimated parameters of the four-component mixture of normal distributions are given in Table 2.2.

The clustering result identifies four groups of young children corresponding to different CES-D depression scores. About 24.5% of students (Cluster 1) had a low level of depression (mean CES-D score = 5.96). The largest cluster (Cluster 2) represents a group of students (44.4%) who had a medium level of depression (mean CES-D score = 13.40). About 28.3% of students (Cluster 3) had a high level of depression (mean CES-D score = 24.60), whereas about 2.8% of students had a very high depression level with a mean CES-D score of 42.58. Additional information on the association between CES-D scores and children's characteristics were provided in Table 2.3. The table shows discrepancies within the four clusters defined on the basis of CES-D scores, with reference to the overall sample. In particular, the group of children with a very high mean CES-D score (Cluster 4) has a significantly higher proportion of females (85.5% versus 50.7%) and children not living with their parents (69.9% versus 48.2%), as well as having a lower cohesion score (mean: 42.23 versus 56.24).

The grouping of children's CES-D scores also provides additional information on the association between the CES-D score and child characteristics. Table 2.4 presents the impact of each characteristic on the CES-D score with the assumption of a single population as a whole (using a log normal distribution), along with the results on the

TABLE 2.1
Model selection for the adolescent depression data (Example 2.1).

No. of components (g)	Log likelihood	AIC	BIC	ICL	p-value[a]
1	-11762.7	23529	23542	23542*	0.00
2	-11479.4	22969	22999	24219	0.00
3	-11412.5	22841	22890	24709	0.00
4	-11400.3	22823	22889*	24596	0.03
5	-11394.7	22817*	22902	24942	0.10*
6	-11393.4	22821	22924	24744	–

* Number of components selected based on the criterion.
[a] p-value using bootstrap likelihood ratio statistic for g versus $g+1$.

TABLE 2.2
Estimates and standard errors (in parentheses) for the four-component mixture of normal distributions (Example 2.1).

	Cluster 1	Cluster 2	Cluster 3	Cluster 4
Mixing proportion π_h	0.245 (0.041)	0.444 (0.028)	0.283 (0.042)	0.028
Mean μ_h	5.96 (0.46)	13.40 (0.88)	24.60 (1.40)	42.58 (1.95)
Variance σ_h^2	7.2 (1.1)	21.2 (1.9)	48.8 (6.4)	19.9 (9.0)
Patients classified	849	1514	743	83
Range of CES-D scores	0-8	9-20	21-39	40-54

TABLE 2.3
Differentiation between the four clusters of children (Example 2.1).

Characteristic	Cluster 1 low CES-D	Cluster 2 medium CES-D	Cluster 3 high CES-D	Cluster 4 very high CES-D	Overall
Race					
White	765 (90.1%)	1244 (82.2%)	605 (81.4%)	69 (83.1%)	2683 (84.1%)
Black	84 (9.9%)	270 (17.8%)	138 (18.6%)	14 (16.9%)	506 (15.9%)
Gender					
Male	475 (55.9%)	777 (51.3%)	308 (41.5%)	12 (14.5%)	1572 (49.3%)
Female	374 (44.1%)	737 (48.7%)	435 (58.5%)	71 (85.5%)	1617 (50.7%)
Living with parents					
Yes	523 (61.6%)	758 (50.1%)	345 (46.4%)	25 (30.1%)	1651 (51.8%)
No	326 (38.4%)	756 (49.9%)	398 (53.6%)	58 (69.9%)	1538 (48.2%)
Cohesion score	63.66 (10.4)	56.46 (10.9)	48.87 (11.9)	42.23 (14.1)	56.24 (12.5)

Note: Data are frequency (percentage) for categorical variables or mean (standard deviation) for cohesion score.
Significant differences among the four clusters for all characteristics.

TABLE 2.4
Differentiation between the four clusters of children (Example 2.1).

Characteristic	Log normal	Multinomial logistic (relative to Cluster 1)		
		Cluster 2 medium CES-D	Cluster 3 high CES-D	Cluster 4 very high CES-D
Race (Black)	0.034 (0.157)	1.84 (< 0.001)	1.98 (< 0.001)	1.64 (0.139)
Gender (Female)	0.174 (< 0.001)	1.30 (0.004)	1.96 (< 0.001)	8.48 (< 0.001)
Not living with parents	0.033 (0.080)	1.19 (0.062)	1.10 (0.396)	2.08 (0.006)
Cohesion score	-0.021 (< 0.001)	0.94 (< 0.001)	0.89 (< 0.001)	0.85 (< 0.001)

Note: Data are estimated coefficients, with p-values given in parentheses.

CES-D scores for each CES-D cluster relative to Cluster 1 (low CES-D scores) using multinomial logistic regression. It can be observed that, when a single population is assumed, the CES-D score is significantly associated only with gender and cohesion scores. However, the results using multinomial logistic regression on the CES-D clusters indicate that children with black ethnicity tend to have moderately increased CES-D scores (Clusters 2 or 3). In addition, children not living with their parents are significantly more likely to have a very high CES-D score (Cluster 4).

Figure 2.2 plots the observed CES-D total scores together with the fitted mixture distributions. Goodness-of-fit can be studied by splitting the data into 31 bins arbitrarily. The three-component normal mixture model provided a good fit to the data (chi-square goodness-of-fit statistic: 10.53, p-value = 0.333).

With the mixture model (2.1) presented in Section 2.1, it is assumed that the mixing proportions π_h $(h = 1,\ldots,g)$ do not depend on covariates. Without covariates, the mixing proportions representing the prior probabilities of membership in each cluster are the same for all children. The implied clustering based on the posterior probabilities (2.7) thus represents a grouping of children entirely on the basis of their CES-D scores. If the mixing proportions are formulated to relate π_h with covariates (such as children's demographic characteristics), then the prior probabilities of cluster membership are not the same for all children. The effects of covariates will be incorporated into the mixture model via π_h. The clusters formed will then represent an "adjusted" clustering of children that takes into account children's characteristics. Another way to include covariates in the mixture model is through the component-specific means μ_h $(h = 1,\ldots,g)$ where $\mu_h = W\boldsymbol{\beta}_h$ are linked to the covariate vector x in W with unknown regression coefficient parameters denoted by $\boldsymbol{\beta}_h$ $(h = 1,\ldots,g)$; see also Equation (1.1). As discussed in Section 1.1.1, this mixture regression model when applied to Example 2.1 will address heterogeneity in regression coefficients $\boldsymbol{\beta}_h$ but not in the CES-D scores among students.

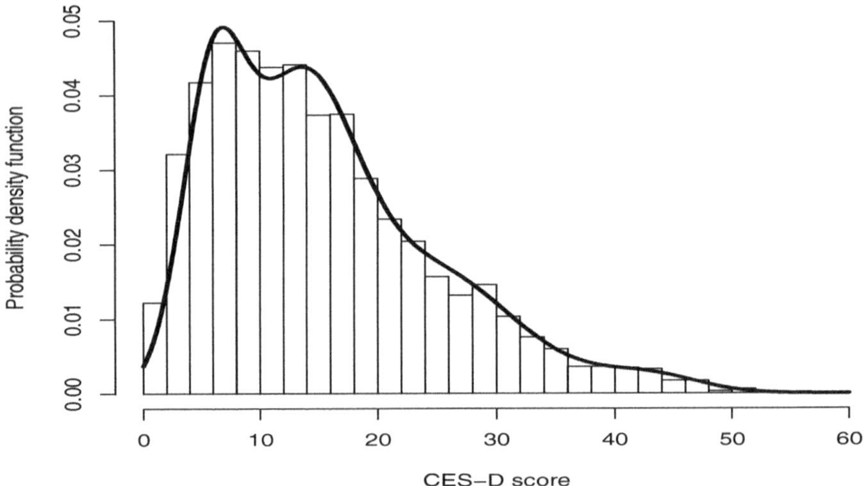

FIGURE 2.2
Empirical distribution of CES-D scores and fitted four-component mixture of normal distributions in solid line (Example 2.1).

2.5 Example 2.2: Multivariate Normal Mixtures

We consider a thyroid gland data set "new-thyroid.dat" available from the UCI Repository of machine learning databases (Bache and Lichman, 2013). The data comprised the measurement of five thyroid gland laboratory test results for 215 patients. The aim of the original study is to predict the functional state of the thyroid gland on the basis of the five laboratory measurements: T3-resin uptake (a percentage); total serum thyroxine (T4) as measured by the isotopic displacement method; total serum triiodothyronine (T3) as measured by radioimmuno assay; basal thyroid-stimulating hormone (TSH) as measured by radioimmuno assay; and the maximal difference of TSH after injection of 200 μg of thyrotropin-releasing hormone (TRH) as compared to the basal value. Based on complete medical records (including anamnesis, scan, etc.), three diagnostic classes are assigned to each patient which correspond to euthyroidism (normal thyroid function), hyperthyroidism (over-active thyroid function), and hypothyroidism (under-active thyroid function).

Descriptive statistics of the five laboratory measurements, along with the pairwise correlation coefficients, are given in Table 2.5. Visual inspection of normality of the five measurements using normal Q-Q plot are displayed in Figure 2.3. It can be

TABLE 2.5
Descriptive statistics of the five laboratory measurements for the thyroid function data (Example 2.2).

	T3-resin	Serum T4	Serum T3	Basal TSH	Diff. TSH
Mean	109.6	9.80	2.05	2.88	4.20
S.D.	13.1	4.70	1.42	6.12	8.07
Correlation coefficient:					
T3-resin	1.0	-0.494	-0.537	0.290	0.296
Serum T4	-0.494	1.0	0.719	-0.423	-0.410
Serum T3	-0.537	0.719	1.0	-0.242	-0.227
Basal TSH	0.290	-0.423	-0.242	1.0	0.498
Diff. TSH	0.296	-0.410	-0.227	0.498	1.0

TABLE 2.6
Model selection for the thyroid function data (Example 2.2).

No. of components (g)	Log likelihood	AIC	BIC	ICL	p-value[a]
1	-3140.5	6321	6388	6388	0.00
2	-2457.5	4997	5135	5136	0.00
3	-2238.4	4601	4810*	4814*	0.36*
4	-2198.0	4562	4842	4847	0.57
5	-2167.4	4543*	4893	4899	–

* Number of components selected based on the criterion.
[a] p-value using bootstrap likelihood ratio statistic for g versus $g+1$.

seen that departure from normality appears for T3-resin uptake (left skewed) as well as total serum T4 and T3, basal TSH, and the maximal difference of TSH (right skewed), which may possibly be due to group structure of the data.

Based on the correlation coefficients displayed in Table 2.5, unrestricted component-covariance matrices may be more appropriate. We thus fitted a mixture of normal distributions to the data with unequal and unrestricted covariance matrices from g=1 to g=5. Initial values of $\Psi^{(0)}$ for the EM algorithm were determined from ten random starts and ten k-means starts. Based on the results for model selection (Table 2.6), three groups of patients were identified. The estimated parameters of the three-component mixture of normal distributions are given in Table 2.7. The test of Hawkins (1981) was applied to the resulting clusters as if they represent a correct grouping of the data with respect to Clusters 1, 2, and 3. The results of this method for assessing normality and homoscedasticity are provided in Table 2.8. Against the null hypothesis (2.11) of homoscedastic multivariate normal components, Table 2.8(a) shows the difference in signs and significance of the first and second components of the Anderson-Darling statistic for the individual component populations, suggesting a very strong indication of heteroscedasticity. Table 2.8(b) shows the results

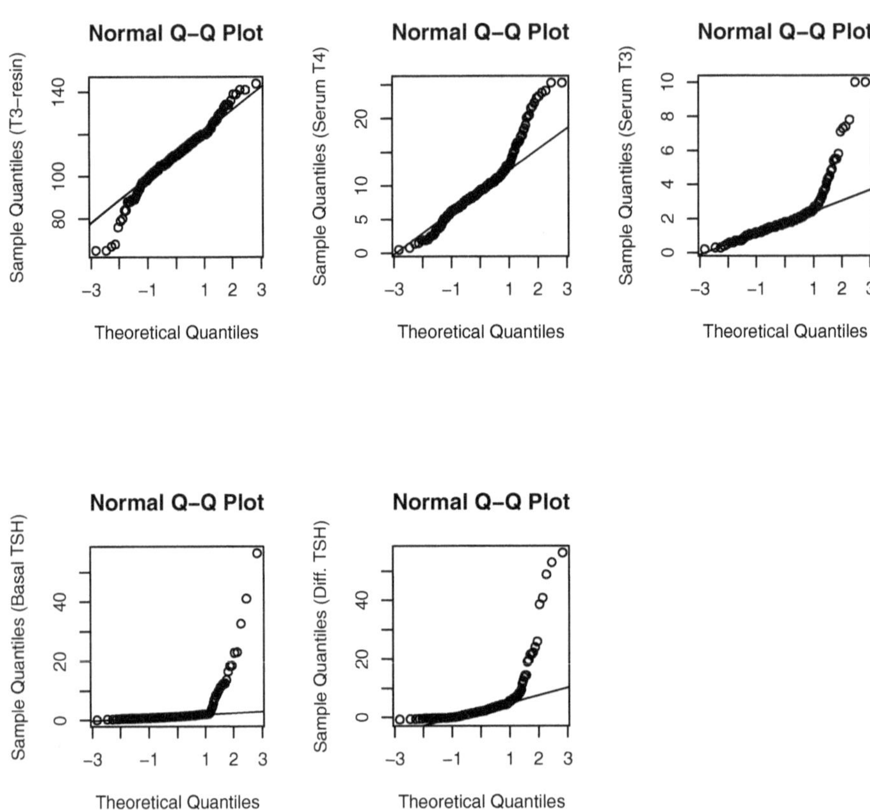

FIGURE 2.3
Normality check for the five laboratory measurements (T3-resin, Serum T4, Serum T3, Basal TSH, and Diff. TSH) of the thyroid function data (Example 2.2).

TABLE 2.7
Estimates and standard errors (in parentheses) for the three-component mixture of normal distributions (Example 2.2).

	Cluster 1	**Cluster 2**	**Cluster 3**
Mixing proportion π_h	0.693 (0.032)	0.177 (0.026)	0.130
Mean vector			
μ_{h1}	110.5 (0.638)	95.95 (2.741)	123.1 (1.968)
μ_{h2}	9.061 (0.159)	17.12 (0.846)	3.835 (0.389)
μ_{h3}	1.708 (0.043)	4.116 (0.338)	1.065 (0.087)
μ_{h4}	1.305 (0.041)	0.983 (0.071)	13.82 (2.107)
μ_{h5}	2.549 (0.125)	-0.030 (0.052)	18.70 (2.997)
Patients classified	149	38	28

of Hawkins (1981) test for the null hypothesis (2.16) assuming multivariate normality and heteroscedasticity. The nonsignificance of the Anderson-Darling statistics and their components suggest that it is reasonable to take the component distributions to be multivariate normal, but with unequal covariance matrices for the three components.

The data set has true classification of the patients into three diagnostic classes of normal, over-active, and under-active functions of the thyroid gland. The distribution table of the three clusters based on the mixture model compared with these true classifications is provided in Table 2.9. There are nine misclassifications (4.2%). The five laboratory measurements of these nine patients are presented in Table 2.10. Among the five patients belonging to Category 1 (normal function) who were misclassified, three were misclassified into Cluster 2 because of a negative difference in TSH and two were misclassified into Cluster 3 (one because of a higher Basal TSH reading and one because of a higher value in the difference of TSH). Four patients belonging to Category 3 (under-active function) were misclassified into Cluster 1 because of a lower Basal TSH reading.

2.6 Extensions of the Normal Mixture Model

Due to their computational tractability, normal mixture models are the most widely used mixture model-based approach for cluster analysis in applications in a wide variety of scientific fields; see, for example, McLachlan and Peel (2000), Boldea and Magnus (2009), Fahey et al. (2012), and Ng and McLachlan (2003b). Because of this, much attention in mixture modelling is given to extending normal mixture models for handling data with which mixtures of normal component densities are not adequate to fit. In this section, we briefly describe advanced developments of the normal mixture model within the mixture modelling literature. This includes the mixture of skew normal or skew-t distributions, the mixture of normal inverse Gaussian distributions,

TABLE 2.8

Results of Hawkins (1981) test for normality and homoscedasticity (Example 2.2) – applied to data in clustered form.

	Anderson-Darling statistic	Components of Anderson-Darling statistic First	Second
(a) Assuming homoscedasticity:			
Cluster 1	168.8	-15.2	-13.4
Cluster 2	14.3	3.52	-4.09
Cluster 3	35.4	5.46	-4.56
Totality	96.8	-9.2	-14.6
(b) Assuming heteroscedasticity:			
Cluster 1	0.36	-0.32	-0.63
Cluster 2	0.63	-0.35	-1.49
Cluster 3	0.48	-0.24	-1.12
Totality	0.85	-0.50	-1.55

TABLE 2.9

Distribution of three true diagnostic classes of thyroid functions over the three clusters obtained by the mixture of normal distributions (Example 2.2).

Cluster	Class 1 (normal)	Class 2 (over-active)	Class 3 (under-active)	Total
1	145	0	4	149
2	3	35	0	38
3	2	0	26	28
Total	150	35	30	215

TABLE 2.10

The five laboratory measurements of the nine misclassified patients (Example 2.2).

ID	Class	Cluster	T3-resin	Serum T4	Serum T3	Basal TSH	Diff. TSH
14	1	3	112	8.1	1.9	3.7	2.0
49	1	2	100	9.5	2.5	1.3	-0.2
50	1	3	118	8.1	1.9	1.5	13.7
51	1	2	100	11.3	2.5	0.7	-0.3
132	1	2	103	9.5	2.9	1.4	-0.1
188	3	1	108	3.5	0.6	1.7	1.4
189	3	1	120	3.0	2.5	1.2	4.5
213	3	1	103	5.1	1.4	1.2	5.0
215	3	1	102	5.3	1.4	1.3	6.7

the mixture of log-concave densities, the mixture of factor analyzers, the mixture of linear mixed-effects models, the mixture of sparse regression models, and the mixture of matrix normal distributions.

With applications involving highly asymmetric continuous data, such as studies with selective or truncated samples (Arnold and Beaver, 2002), the use of normal (or t) mixture models tends to overfit, where additional components are required to capture the skewness and asymmetry in the data, as illustrated by Wang, Ng, and McLachlan (2009). However, including such spurious and irrelevant components may lead to computational problems and invalid inferences being made (Ng, 2013). Mixtures of multivariate skew normal or skew t distributions have been proposed to fit asymmetric data in various applied problems; see, for example, Pyne et al. (2009), Wang, Ng, and McLachlan (2009), Vrbik and McNicholas (2012), Lee and McLachlan (2013), and Pyne et al. (2014). Let δ be a p-dimensional vector of skew parameters and suppose that conditional on w,

$$\begin{pmatrix} U_0 \\ U \end{pmatrix} \sim N\left(\begin{pmatrix} \mu \\ 0 \end{pmatrix}, \begin{pmatrix} \Sigma & 0 \\ 0 & 1 \end{pmatrix} \frac{1}{w}\right), \tag{2.17}$$

where the random variable $W \sim \text{gamma}(\nu/2, \nu/2)$; see Wang, Ng, and McLachlan (2009) and Ng (2013). Thus $Y = \delta|U| + U_0$ defines a p-dimensional multivariate skew normal or skew t distribution, whose density function is given by, respectively,

$$f(y;\mu,\Sigma,\delta) = 2\phi_p(y;\mu,\Omega)\Phi\left(\frac{\xi}{\sigma}\right) \tag{2.18}$$

or

$$f(y;\mu,\Sigma,\delta,\nu) = 2t_p(y;\mu,\Omega,\nu)T_{p+\nu}\left(\frac{\xi}{\sigma}\sqrt{\frac{\nu+p}{\nu+\eta}}\right). \tag{2.19}$$

In (2.18) and (2.19), $\Omega = \Sigma + \delta\delta^T$, $\xi = \delta^T\Omega^{-1}(y-\mu)$, $\sigma^2 = (1-\delta^T\Omega^{-1}\delta)$, $\phi_p(y;\mu,\Omega)$ is the p-dimensional normal distribution with mean μ and scale matrix Ω, and $\Phi(\cdot)$ denotes the cumulative distribution function corresponding to the skewing part. For (2.19), $\eta = (y-\mu)^T\Omega^{-1}(y-\mu)$, $t_p(y;\mu,\Omega,\nu)$ is the p-dimensional t distribution with location parameter μ, scale matrix Ω and degrees of freedom ν, and $T_{p+\nu}(\cdot)$ is the cumulative distribution function of a univariate (central) t random variable with degrees of freedom $(p+\nu)$. The mixture of multivariate skew normal or skew t distributions is defined by their component densities being specified with (2.18) and (2.19), respectively. These mixture models are referred to as a "restricted" multivariate skew normal or skew t mixture model, and the E- and M-steps involving the skew parameters δ_h $(h = 1,\ldots,g)$ can be expressed in closed form; see Wang, Ng, and McLachlan (2009), Vrbik and McNicholas (2012), Ng (2013), and Lee and McLachlan (2014). An "unrestricted" version of the multivariate skew t mixture model has been proposed by Lee and McLachlan (2014), where U in (2.17) is replaced by a p-dimensional variable vector U such that each element of the skew parameter vector δ is allowed to have a different random coefficient (the corresponding element in $|U|$) rather than a same scalar coefficient U. Due to the complicated expressions involved in the E-step with unrestricted multivariate skew t mixtures, both the E- and M-steps are modified to allow the data to be split into blocks for a block implementation of the EM algorithm, as illustrated in Lee and McLachlan (2016). Key applications with these models include the analysis of flow cytometry data (Pyne et al., 2009; Pyne et al., 2014; Wang, Ng, and McLachlan, 2009), financial risk analysis and image segmentation (Lee and McLachlan, 2016).

More recently, a mixture of multivariate normal inverse Gaussian (MNIG) distributions provided an alternative extension of a normal mixture model to a more flexible family of mixture distributions, which may be skewed and have fatter tails than a normal distribution, such as the distribution of financial returns data considered in O'Hagan et al. (2016). With this MNIG model, a positive scalar quantity u_{hj}, termed the mixing component, is assumed to be inverse Gaussian distributed $u_{hj} \sim IG(1, \gamma_h)$ such that the p-dimensional observation vector y_j can be assumed to be multivariate normally distributed conditioned on u_{hj} as $y_j | u_{hj} \sim N(\mu_h + u_{hj}\delta_h, u_{hj}\Sigma_h)$. The concept of the MNIG approach is to partition the variability into intrinsic (Σ_h) and mixing (u_{hj}) components, where the role of u_{hj}'s interaction with Σ_h is to control the "thickness" of the tails of the distribution and the role of δ_h is to capture the skewness of the data through the shifting of the mean μ_h with compensation for u_{hj}; see O'Hagan et al. (2016) for the details. In another context, mixture models with log-concave component densities have been developed by Hu, Wu, and Yao (2016) to relax the strong parametric assumption regarding the component densities in mixture models. Examples of log-concave densities are normal, Laplace, logistic, as well as gamma and beta with certain parameter constraints. Using an EM-type algorithm, it has been shown that the log-concave ML estimator for mixture models with log-concave densities exists and is consistent under fairly general conditions.

In applications concerning high-dimensional data, the mixture of factor analyzers model has been proposed to reduce the number of dimensions and concurrently perform clustering (McLachlan, Peel, and Bean, 2003). Key applications as a dimensionality reduction tool include the cluster analysis of gene-expression data, such as in Baek, McLachlan, and Flack (2010), where the number of genes p is much greater than the number of tissue samples n, as well as visualization of high-dimensional data and face recognition (Yang, Kriegman, and Ahuja, 2002). With mixtures of factor analyzers, each component-covariance matrix Σ_h is specified to have the form $\Sigma_h = B_h B_h^T + D_h$ $(h = 1, \ldots, g)$, which is formulated under the assumption that conditional on its membership of the hth component of the mixture,

$$Y_j = \mu_h + B_h U_{hj} + \varepsilon_{hj} \qquad (j = 1, \ldots, n), \tag{2.20}$$

where U_{hj} is a q-dimensional ($q < p$) vector of latent or unobservable factors and B_h is a $p \times q$ matrix of factor loadings. By assuming U_{hj} to be i.i.d. as $N(0, I_q)$ and independently the measurement error vector ε_{hj} to be i.i.d. as $N(0, D_h)$, then conditional on $U_{hj} = u_{hj}$, the Y_j are independently distributed as $N(\mu_h + B_h u_{hj}, D_h)$, where D_h is a diagonal matrix $(h = 1, \ldots, g)$; see McLachlan, Peel, and Bean (2003). More recently, Baek, McLachlan, and Flack (2010) proposed imposing constraints on B_h and D_h $(h = 1, \ldots, g)$ to further reduce the number of parameters. The so-called mixture of common factor analyzers (MCFA) helps to speed up the fitting process especially when p is large.

The analysis of correlated multivariate data is another application area where model-based clustering could be extended beyond the normal mixture model approach. For example, high-throughput biological experiments are now being conducted with replication for capturing either biological or technical variability in expression levels to improve the quality of inferences, as discussed in Lee et al. (2000). Because of the induced correlation among replicated measurements of gene expression from each tissue sample, standard normal mixture models assuming independence of observations (as described in Section 2.1) may no longer be valid. Extensions of normal mixture models for clustering correlated multivariate gene-expression data have been proposed in Ng et al. (2006) and Wang, Ng, and McLachlan (2012) through the use of random linear mixed-effect (LMM) modelling. By treating the unobservable random effects as missing data, the EM algorithm can be implemented via a conditional mode as illustrated in Ng (2013). The applications of these mixtures of LMM models for the analysis of correlated multivariate data, multilevel or longitudinal data are described in more detail in Chapters 8 and 10.

In the context of the partial least squares (PLS) approach to regression modelling with latent variables, a mixture of sparse regression models has been developed to explore the heterogeneity in relationships between multiple dependent (endogenous) latent variables and independent (exogenous) latent variables; see Ng and McLachlan (2014b). With the PLS approach considered in Hahn et al. (2002) and Ringle, Wende, and Will (2010), it is assumed that the endogenous latent variables η_j $(j = 1,\ldots,n)$ come from a multivariate normal mixture model, such that the mean vector of the hth component for the jth individual is given by $\mu_{hj} = (I - B_h)\eta_j - \Gamma_h\xi_j$, where ξ_j denotes the vector of exogenous latent variables, and $\Sigma_h = \text{diag}(\sigma_h^2)$ is a diagonal matrix constructed from the vector σ_h^2 that represents the variance of the random residuals. By specifying the relationship between η_j and ξ_j (the so called "inner" model) in terms of simultaneous regression equations as

$$B\eta_j + \Gamma\xi_j = \zeta_j, \tag{2.21}$$

the multivariate normal component densities are then given by

$$\phi(\eta_j;\mu_{hj},\Sigma_h) = \frac{|B_h|}{\sqrt{(2\pi)^q|\Sigma_h|}} \exp\{-\tfrac{1}{2}(B_h\eta_j + \Gamma_h\xi_j)^T\Sigma_h^{-1}(B_h\eta_j + \Gamma_h\xi_j)\}. \tag{2.22}$$

In (2.21), B is a $q \times q$ matrix with q being the number of endogenous latent variables, Γ is a $q \times p$ matrix where p is the number of exogenous latent variables, and ζ_j is a random vector of residuals. The matrices B and Γ represent the (path) coefficients relating to η_j and ξ_j, respectively, in the inner model. The relationships between the latent variables and the manifest variables are specified in the "outer" model; see, for example, Hahn et al. (2002) and Ng and McLachlan (2014b). With the PLS approach, the variance of the endogenous variables explained by the exogenous variables is maximized. Key application areas are marketing research and epidemiology where the aim is to capture heterogeneity in endogenous latent variables among individuals. A systematic computational algorithm has been developed by Ng and McLachlan (2014b) to identify sparse coefficient parameters in the implementation of the mixture of regression models, due to collinearity between endogenous latent variables at the component level.

Developments to normal mixture models have also been made for the purposes of handling matrix data including spatial multivariate data, multivariate time series, and longitudinal vector measurements in a wide variety of experimental settings. With reference to (2.1) and (2.2), a mixture of matrix normal distributions is expressed as

$$f(y;\Psi) = \sum_{h=1}^{g} \pi_h \phi(y; M_h, \Sigma_{ph}, \Sigma_{qh}), \tag{2.23}$$

where y is presented in a matrix form of dimensions $p \times q$ and the component density is given by

$$\phi(y; M_h, \Sigma_{ph}, \Sigma_{qh}) = (2\pi)^{-\frac{pq}{2}} |\Sigma_{ph}|^{-\frac{p}{2}} |\Sigma_{qh}|^{-\frac{q}{2}} \exp\{-\tfrac{1}{2} tr \Sigma_{ph}^{-1} (y - M_h) \Sigma_{qh}^{-1} (y - M_h)^T\}, \tag{2.24}$$

and where M_h denotes the $p \times q$ matrix of means, Σ_{ph} and Σ_{qh} represent $p \times p$ and $q \times q$ covariance matrices associated with rows and columns of the data matrix, respectively; see Viroli (2011). More recently, Dogru, Bulut, and Arslan (2016) have proposed a mixture of matrix variate t distributions for applications where a longer tail is required than that of a normal distribution or the parameter estimates are affected by outliers. As illustrated in Viroli (2011), an equivalent definition of (2.24) can be specified as a special case of the pq-dimensional normal distribution $\phi^{(pq)}(vec(M_h), \Sigma_{ph} \otimes \Sigma_{qh})$, where $vec(\cdot)$ takes a matrix and stacks the columns into a single vector of length pq and $\otimes$ is the Kronecker product. From this perspective, some problems in the cluster analysis of matrix data may be formulated in a vector format such that normal mixture models and extensions can be further modified to deal with the problems. A possible way is to incorporate random effects in the normal mixture models as considered by Celeux, Martin, and Lavergne (2005) and Ng et al. (2006); see also, for example, the mixtures of LMMs described in Chapter 8.

2.7 R Programs for Fitting Mixtures of Normal Distributions

The R package "EMMIX" can be downloaded from the following webpage: `http://www.maths.uq.edu.au/~gjm/mix_soft/EMMIX_R/index.html`. It was developed to fit mixture models of (multivariate) normal distributions or t distributions by ML via the EM algorithm to continuous data; see McLachlan and Peel (2000) and McLachlan and Ng (2009). The latest version EMMIX 1.0.1 includes new features such as fitting highly asymmetric continuous data with restricted skew normal or t mixture models, as discussed in Section 2.6. The R program codes for Examples 2.1 and 2.2 are provided in Figures 2.4 and 2.5, respectively.

Some other software in R for fitting mixture models of multivariate normal distributions include the MCLUST package (Fraley and Raftery, 2006), the FPC package (Hennig, 2010), the CAMAN package (Böhning, Schlattmann, and Lindsay, 1998), the MIXTOOLS package (Young et al., 2015), and those listed in the CRAN Task View established by Leisch and Gruen (2017); see also Section 1.5.1. The "mixAK" package (Komárek, 2009), on the other hand, adopts a Bayesian approach to fit mixtures of multivariate normal distributions and mixtures of generalized linear mixed models.

Main Analysis for Example 2.1

```
setwd("c:/Example2.1/")             # set the work directory

data <- matrix(scan("ch20_cesd.txt"),ncol=1,byrow=T)        # read the CES-D scores

library(mvtnorm)                             # install package "mvtnorm" and call it

require(EMMIX)                               # install the EMMIX program and call it

set.seed(216)                                # set a seed for generating random numbers

# initialization using 10 k-means and 10 random starts
# assume g=4 components of multivariate normal with unequal standard deviations
# convergence criterion (log likelihood < 1e-8 of the previous log likelihood value)
initobj <- init.mix(data, g=4, distr="mvn", ncov=3, nkmeans=10, nrandom=10, nhclust=F)
obj <- EMMIX(data, g=4, distr="mvn", ncov=3, clust=NULL, init=initobj, itmax=1000,
       epsilon=1e-8, debug=TRUE)

attributes(obj)
write.csv(cbind(obj$tau,obj$clust),"cluster_4gp.csv")      # save the clustering results

# calculate standard errors using B=50 bootstrap replications and save the results
# a larger value of epsilon in the convergence criterion for standard error estimation
std <- bootstrap(data, n=3189, p=1, g=4, distr="mvn", ncov=3, popPAR=initobj, B=50,
       replace=TRUE, itmax=1000, epsilon=1e-6)
write.csv(std,"cluster_4gp_se.csv")
```

FIGURE 2.4
The R program codes for Example 2.1.

Main Analysis for Example 2.2

```
setwd("c:/Example2.2/")             # set the work directory

data <- matrix(scan("new-thyroid.txt"),ncol=5,byrow=T)       # read the thyroid data

library(mvtnorm)                            # install package "mvtnorm" and call it

require(EMMIX)                              # install the EMMIX program and call it

set.seed(256)                               # set a seed for generating random numbers

# initialization using 10 k-means and 10 random starts
# assume g=3 multivariate normal with unequal and unrestricted covariance matrices
# convergence criterion (log likelihood < 1e-6 of the previous log likelihood value)
initobj <- init.mix(data, g=3, distr="mvn", ncov=3, nkmeans=10, nrandom=10, nhclust=F)
obj <- EMMIX(data, g=3, distr="mvn", ncov=3, clust=NULL, init=initobj, itmax=1000,
       epsilon=1e-6, debug=TRUE)

attributes(obj)
write.csv(cbind(obj$tau,obj$clust),"cluster_3gp.csv")     # save the clustering results

# calculate standard errors using B=50 bootstrap replications and save the results
# a larger value of epsilon in the convergence criterion for standard error estimation
std <- bootstrap(data, n=215, p=5, g=3, distr="mvn", ncov=3, popPAR=initobj, B=50,
       replace=TRUE, itmax=1000, epsilon=1e-5)
write.csv(std,"cluster_3gp_se.csv")
```

FIGURE 2.5
The R program codes for Example 2.2.

3

Mixture of Gamma Distributions for Continuous Non-Normal Data

3.1 Introduction

Many random variables of interest in medical and health sciences are non-symmetrical (or skewed), such as healthcare cost and utilization data. In this chapter, we introduce mixture models with gamma component distributions for handling clusters in the data that show non-symmetrical characteristics. Several other non-symmetrical component distributions can also be considered, such as skewed normal or skewed t-distributions. As described in Chapter 2, non-normal features in the data may be attributed to some underlying group structure, thus it is essential to compare to mixture models with symmetric component distributions (such as normal or t-distributions) as well.

The gamma distribution family is flexible to model data exhibiting different degrees of skewness. A typical gamma density function involves two parameters: μ quantifies the mean and α represents the shape of the gamma distribution. In other terminology, $\sigma^2 = \mu^2/\alpha$ represents the variance of the data, while $\theta = \mu/\alpha = \sigma^2/\mu$ is known as the scale parameter. The gamma distribution is a continuous distribution defined on positive values of the random variable. Figure 3.1 displays gamma distributions with various shape (α) and scale (θ) parameters. It can be seen from Figure 3.1 that the gamma density function is unimodal and skewed when the shape parameter α is greater than 1. When α increases, the skewness decreases. For $\alpha < 1$, the gamma density function is exponentially shaped. The gamma distribution is useful for modelling positive and positively skewed random variables, including healthcare costs (Jones et al., 2016; Venturini, Dominici, and Parmigiani, 2008), hospital length of stay (Lee et al., 2002; Lee et al., 2007), insurance claims (Jeon and Kim, 2013), intensity of ultrasonic images (Destrempes et al., 2009), evolutionary rates across nucleotide and amino acid sites (Mayrose, Friedman, and Pupko, 2005), human reference intervals in clinical pathology (Concordet et al., 2009), and waiting times (Mohammadi, Salehi-Rad, and Wit, 2013); even though the data are sometimes recorded as an integer and can thus be considered a count variable. In this case, Poisson distribution may also be considered, but the gamma distribution offers more flexibility in modelling data in terms of the variance-to-mean ratio (or known as the index of dispersion) of data represented by θ, compared to Poisson distribution which has a constant index of dispersion of 1.

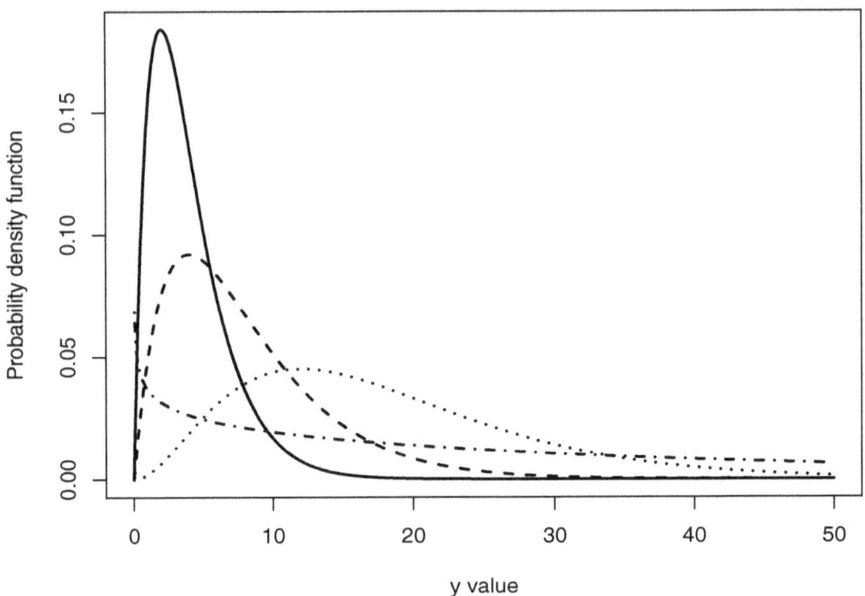

FIGURE 3.1
Gamma distributions with various shape (α) and scale (θ) parameters (Solid line: $\alpha = 2$, $\theta = 2$; Dashed line: $\alpha = 2$, $\theta = 4$; Dotted line: $\alpha = 3$, $\theta = 6$; Dash-dotted line: $\alpha = 0.8$, $\theta = 50$).

Let $y = (y_1, \ldots, y_n)^T$ be the observed data with probability density function (p.d.f.) $f(y; \Psi)$. With a mixture of gamma distributions, the density of y_j is given by

$$f(y_j; \Psi) = \sum_{h=1}^{g} \pi_h f_h(y_j; \mu_h, \alpha_h) = \sum_{h=1}^{g} \pi_h f_h(y_j; \psi_h) \qquad (j = 1, \ldots, n), \tag{3.1}$$

where π_h $(h = 1, \ldots, g)$ are the mixing proportions of g components that sum to one, $\psi_h = (\mu_h, \alpha_h)^T$ contains the unknown parameters for the hth component density, and

$$f_h(y_j; \psi_h) = \frac{(\alpha_h/\mu_h)^{\alpha_h} y_j^{\alpha_h - 1} \exp(-\alpha_h y_j/\mu_h)}{\Gamma(\alpha_h)} \tag{3.2}$$

denotes a gamma density function with $y_j > 0$, mean $\mu_h > 0$ and the shape parameter $\alpha_h > 0$ $(j = 1, \ldots, n;\ h = 1, \ldots, g)$, and $\Gamma(\alpha) = \int_0^\infty u^{\alpha-1} \exp(-u) du$ is the Gamma function. The vector Ψ of unknown parameters consists of the mixing proportions

$\pi_1, \ldots, \pi_{g-1}$, the elements of the component means μ_h and the component shape parameters α_h.

In many applications, there are characteristics of patients that may influence, for example, the healthcare costs or utilization outcomes. The underlying relationship between the outcome and its risk factors can be explored via a gamma mixture regression model, which is a special case of the class of generalized linear finite mixture models; see Jansen (1993). Let $x_{1j}, \ldots, x_{pj}$ be the p covariates of the jth individual $(j = 1, \ldots, n)$. There are two different approaches to include covariates in a mixture model: (a) to include covariates in the mixing proportions π_h via a logistic model such that the effects of the covariates on the prior probabilities of components belonging are modelled; see, for example, Lee et al. (2002); (b) to include covariates in the component densities $f_h(\cdot)$ such that the effects of the covariates on the mean outcome are modelled, as considered in Lee, Ng, and Yau (2001) and Lee et al. (2007). Specifically for the hth component, the model relates the mean for the jth individual μ_{hj} with the linear predictor η_{hj} of the individual's covariate vector x_j using a log link as

$$\begin{aligned} \log(\mu_{hj}) &= \eta_{hj} \\ &= x_{0j}\beta_{h0} + x_{1j}\beta_{h1} + \cdots + x_{pj}\beta_{hp} = x_j^T \boldsymbol{\beta}_h, \end{aligned} \tag{3.3}$$

where $x_{0j} = 1$ is included in $x_j = (x_{0j}, \ldots, x_{pj})^T$ with reference to β_{h0} for all $j = 1, \ldots, n$ and $\boldsymbol{\beta}_h = (\beta_{h0}, \ldots, \beta_{hp})^T$ contains the regression coefficients for the hth component $(h = 1, \ldots, g)$. A positive coefficient implies that an increase in the associated covariate value may increase the outcome, and vice versa for a negative coefficient. The log link function for relating the mean and the linear predictor (3.3) has been considered previously for single gamma regression models; see, for example, Chaudhuri and Mykland (1993) and Bedrick, Christensen, and Johnson (1996).

For gamma mixture regression models with the inclusion of covariates $x = (x_1^T, \ldots, x_n^T)^T$, (3.1) becomes

$$\begin{aligned} f(y_j, x_j; \Psi) &= \sum_{h=1}^{g} \pi_h f_h(y_j, x_j; \boldsymbol{\beta}_h, \alpha_h) \\ &= \sum_{h=1}^{g} \pi_h f_h(y_j, x_j; \psi_h) \qquad (j = 1, \ldots, n), \end{aligned} \tag{3.4}$$

with $\psi_h = (\mu_h, \alpha_h)^T$ replaced by $\psi_h = (\boldsymbol{\beta}_h^T, \alpha_h)^T$. As discussed in Section 1.1.1, the modelling of x_j via (3.3) in Model (3.4) attempts to address heterogeneous associations in the regression coefficients $\boldsymbol{\beta}_h$ between y and covariates x.

3.2 E- and M-steps

The fitting of the gamma mixture models (3.1) or (3.4) can be implemented using a maximum likelihood (ML) approach. An estimate $\widehat{\Psi}$ is obtained by solving the log likelihood equation, $\partial \log L(\Psi)/\partial\Psi = 0$, iteratively via the EM algorithm; see, for example, Ng (2013). With mixture model (3.4), the log likelihood for Ψ is given by

$$\log L(\Psi) = \sum_{j=1}^{n} \log\{\sum_{h=1}^{g} \pi_h f_h(y_j, x_j; \psi_h)\}, \tag{3.5}$$

where $f_h(y_j, x_j; \psi_h)$ is given by (3.2) for $h = 1, \ldots, g$, with $\log(\mu_{hj})$ specified as in (3.3). Within the EM framework, each y_j is conceptualized to have arisen from one of the g components of the mixture model (3.4). We denote $z_1, \ldots, z_n$ the unobservable component-indicator vectors, where the hth element z_{hj} of z_j is taken to be one or zero accordingly as the jth observation y_j does or does not come from the hth component; see, for example, Ng (2013). The EM algorithm approaches the computation of roots of the (incomplete-data) log likelihood (3.5) indirectly by proceeding iteratively in terms of the log likelihood, $\log L_c(\Psi)$, of the complete data that contain observed data y and missing data $z = (z_1^T, \ldots, z_n^T)^T$, where

$$\log L_c(\Psi) = \sum_{h=1}^{g} \sum_{j=1}^{n} z_{hj}\{\log \pi_h + \log f_h(y_j, x_j; \psi_h)\}. \tag{3.6}$$

On the $(k+1)$th iteration of the EM algorithm, the E-step computes the Q-function; the expectation of the complete-data log likelihood (3.6) conditioned on observed y using the current fit for Ψ. As the complete-data log likelihood is linear in the missing data z_{hj}, we simply have to calculate the current conditional expectation of the random variable Z_{hj} given the observed data y. That is,

$$\begin{aligned} Q(\Psi; \Psi^{(k)}) &= E_{\Psi^{(k)}}\{\log L_c(\Psi)|y\} \\ &= \sum_{h=1}^{g} \sum_{j=1}^{n} E_{\Psi^{(k)}}(Z_{hj}|y)\{\log \pi_h + \log f_h(y_j, x_j; \psi_h)\} \\ &= \sum_{h=1}^{g} \sum_{j=1}^{n} \tau_h(y_j; \Psi^{(k)})\{\log \pi_h + \log f_h(y_j, x_j; \psi_h)\}, \end{aligned} \tag{3.7}$$

where

$$\begin{aligned} \tau_h(y_j; \Psi^{(k)}) &= E_{\Psi^{(k)}}(Z_{hj}|y) \\ &= \mathrm{pr}_{\Psi^{(k)}}\{Z_{hj} = 1|y\} \\ &= \pi_h^{(k)} f_h(y_j, x_j; \psi_h^{(k)}) / \sum_{l=1}^{g} \pi_l^{(k)} f_l(y_j, x_j; \psi_l^{(k)}) \end{aligned} \tag{3.8}$$

is the posterior probability that the jth observation y_j belongs to the hth component of the mixture ($h = 1,\ldots,g$; $j = 1,\ldots,n$).

The M-step updates the estimate of Ψ to $\Psi^{(k+1)}$ that maximizes the Q-function with respect to Ψ over the parameter space. For mixtures with gamma component densities, the M-step involves solving non-linear equations. Denote $\tau_h(y_j;\Psi^{(k)})$ by $\tau_{hj}^{(k)}$, we have for gamma mixture regression models (3.4)

$$\begin{aligned}
\pi_h^{(k+1)} &= \sum_{j=1}^{n} \tau_{hj}^{(k)}/n;\\
\sum_{j=1}^{n} \tau_{hj}^{(k)} &\left[1 + \log\alpha_h - \beta_h^T x_j + \log y_j - \frac{y_j}{\exp(\beta_h^T x_j)} - \varphi(\alpha_h)\right] = 0;\\
\sum_{j=1}^{n} \tau_{hj}^{(k)} &\left[-1 + \frac{y_j}{\exp(\beta_h^T x_j)}\right]\alpha_h x_{lj} = 0 \qquad (l = 0,\ldots,p)
\end{aligned} \tag{3.9}$$

for $h = 1,\ldots,g$, where $\varphi(\alpha_h) = \Gamma'(\alpha_h)/\Gamma(\alpha_h)$ is the digamma function, which is the logarithmic derivative of the gamma function $\Gamma(\alpha_h)$.

For gamma mixture models without covariates, the last two sets of nonlinear equations are reduced to

$$\begin{aligned}
\mu_h^{(k+1)} &= \frac{\sum_{j=1}^{n} \tau_{hj}^{(k)} y_j}{\sum_{j=1}^{n} \tau_{hj}^{(k)}};\\
\sum_{j=1}^{n} \tau_{hj}^{(k)} &[1 + \log\alpha_h - \log\mu_h + \log y_j - y_j/\mu_h - \varphi(\alpha_h)] = 0,
\end{aligned} \tag{3.10}$$

where only the updates of α_h ($h = 1,\ldots,g$) are not in closed form.

The MINPACK routine HYBRD1 of Moré, Garbow, and Hillstrom (1980) in FORTRAN can be adopted to find a solution to the above systems of nonlinear equations. The E- and M-steps are alternated repeatedly until convergence of the EM sequence of iterations, which may be determined by using a suitable stopping rule like the difference $L(\Psi^{(k+1)}) - L(\Psi^{(k)})$ changes by an arbitrarily small amount in the case of convergence of the sequence of likelihood values $\{L(\Psi^{(k)})\}$.

The standard errors of the ML estimates of the parameters Ψ are not provided directly using the EM algorithm; see, for example, Ng (2013) for details. With an information-based approach, the asymptotic covariance matrix of the estimates of Ψ can be approximated by the inverse of the observed information matrix $I(\hat{\Psi};y)$, which involves the analytical evaluation of the second-order derivatives of the log likelihood for the gamma mixture model as illustrated in Lee, Ng, and Yau (2001) and Lee et al. (2007). As described in Chapter 1, the standard errors of the ML estimates based on the observed information matrix are valid only when the number of observations n is very large such that the asymptotic theory of ML applies to mixture models. To this end, we adopt a nonparametric bootstrap resampling approach to obtain the standard errors of the ML estimates of Ψ where an empirical distribution function of Y is obtained by sampling with replacement from the observed data y. The procedure is given in Chapter 1. We adopt $B = 100$ bootstrap replications, which are generally sufficient for standard error estimation as discussed in Efron and Tibshirani (1993).

3.3 Diagnostic Procedures

The goodness-of-fit of a mixture of gamma distributions can be assessed using an approximation of the Pearson statistic or the deviance for a gamma distribution; see McCullagh and Nelder (1989, p.34). Here the Pearson statistic is given by

$$\chi^2 = \sum_{j=1}^{n} (y_j - \hat{\mu}_j)^2 / var(\hat{\mu}_j), \tag{3.11}$$

where $\hat{\mu}_j = \sum_{h=1}^{g} \hat{z}_{hj} \hat{\mu}_h$, or for regression models, $\hat{\mu}_j = \sum_{h=1}^{g} \hat{z}_{hj} \hat{\eta}_{hj}$ with $\hat{z}_{hj} = \hat{\tau}_{hj}$, is the estimated mean for the jth individual $(j = 1, \ldots, n)$, while the variance function $var(\hat{\mu}_j) = \hat{\mu}_j^2$. For a gamma distribution, the deviance is expressed as

$$D = 2 \sum_{j=1}^{n} \{ -\log(y_j / \hat{\mu}_j) + (y_j - \hat{\mu}_j) / \hat{\mu}_j \}. \tag{3.12}$$

The degrees of freedom for both Pearson statistic (3.11) and deviance (3.12) are $n - d(g)$, where $d(g)$ is the number of free parameters in the mixture model with g components.

Alternatively, the overall goodness-of-fit can be visualized by comparing the empirical distribution of the data and the fitted mixture of gamma distributions. Similar to the procedures described in Section 2.3, the diagnosis for each component can be conducted on the basis of the classified data implied by the fitted mixture model using a Q-Q gamma plot.

3.4 Example 3.1: Mixture of Gamma Regression Model

We consider the National Health and Nutrition Examination Survey (NHANES) 2013-2014 questionnaire data regarding "Smoking-Cigarette Use". The NHANES is a program of studies designed to assess the health and nutritional status of adults and children in the United States. Information about the NHANES can be obtained from `http://www.cdc.gov/nchs/nhanes/`. The NHANES interview includes demographic, socioeconomic, dietary, and health-related questions. In this example, the Public Use File "SMQ_H.XPT" (updated version of September 2016) containing the questionnaire data about smoking-cigarette use is available at `http://wwwn.cdc.gov/nchs/nhanes/search/DataPage.aspx?Component=Questionnaire&CycleBeginYear=2013`. We retrieved the XPT formatted data file in Stata (Stata SE 13.1; StatCorp, College Station, TX, USA)

TABLE 3.1
Model selection for the number of attempts to quit smoking cigarettes (Example 3.1).

No. of components (g)	Log likelihood	AIC	BIC
1	-1242.7	2489	2498
2	-1152.1	2314*	2336*
3	n.c.	–	–

n.c. - nonconvergency.
* Number of components selected based on the criterion.

using the import SAS XPORT command. The smoking-cigarette use data set provides data on history of use, age at initiation, and other related details such as quit attempts for adults aged 18 years or older. There are 578 respondents who reported the number of times they had tried to quit smoking during the past 12 months. Among them, 577 (99.8% of 578) provided the length of time they were able to stop smoking cigarettes in their last attempt. In this example, we intend to cluster the respondents on the basis of the number of quit attempts and the number of smoking cessation days in the last attempt. These clustering analyses illustrate how gamma mixture models can be used for variables that are recorded as an integer value.

We fitted the data of respondent's number of attempts to quit smoking cigarettes in the past 12 months ($n = 578$) with mixtures of gamma distributions for g=1 to g=3. Model selection based on the BIC is displayed in Table 3.1. We identified two groups of survey respondents for the number of quit attempts. The estimated parameters of the two-component mixture of gamma distributions are given in Table 3.2. The largest cluster (Cluster 1) represents a group of respondents (77.2%) who had fewer attempts to quit smoking cigarettes in the past 12 months (mean = 2 attempts) and about 22.8% of respondents (Cluster 2) had more quit attempts, with a mean of 7.7. The adequacy of these final models was evaluated by means of the goodness-of-fit statistics displayed in Table 3.2. The ratios of Pearson statistic and deviance over the degrees of freedom are, respectively, 0.18 and 0.22, indicating no evidence of lack of fit for the two-component gamma mixture model.

The empirical distribution of number of quit attempts and fitted gamma mixture distributions were provided in Figure 3.2. In general, the two-component gamma mixture model describes suitably the number of attempts to quit smoking cigarettes.

We also fitted the data set using mixtures of normal distributions. Model selection based on BIC indicated there are four clusters (with the mixing proportions all greater than 1%). The empirical distribution and fitted normal mixture distributions were provided in Figure 3.3.

Because the normal mixture models had more components than those of gamma mixtures, it is anticipated that the normal mixture models may fit the empirical distributions of the data regarding the number of attempts to quit smoking cigarettes better than the gamma mixture models. Comparing Figure 3.3 with Figure 3.2, the normal mixture models fit better only when the number of attempts is large. This is

TABLE 3.2
Estimates and standard errors (in parentheses) for the two-component mixture of Gamma distributions on the number of quit attempts (Example 3.1).

	Cluster 1	**Cluster 2**
Mixing proportion π_h	0.772 (0.090)	0.228
Mean μ_h	2.014 (0.210)	7.719 (1.767)
Alpha α_h	3.609 (0.926)	2.135 (1.005)
Respondents classified	468	110
Range of counts	1-4	5-20
Pearson statistic	105.7	
Deviance	124.5	
Degrees of freedom	573	

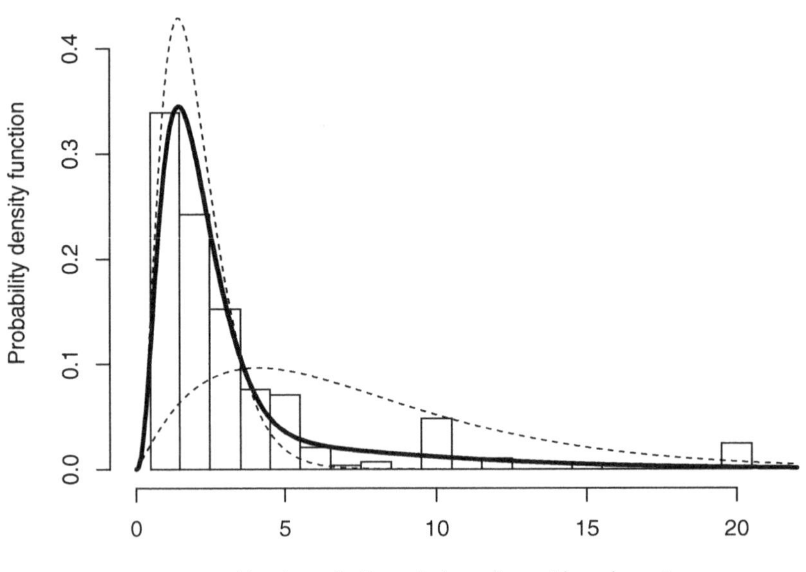

FIGURE 3.2
Empirical distribution of the number of times stopped smoking cigarettes and fitted gamma mixture distribution (Solid line) and its components (Dashed line).

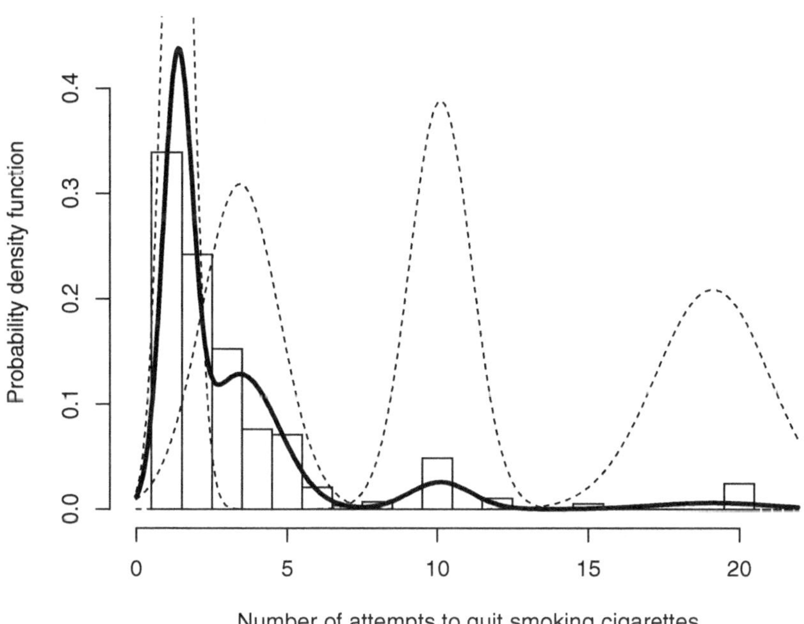

FIGURE 3.3
Empirical distribution of the number of times stopped smoking cigarettes and fitted normal mixture distribution (Solid line) and its components (Dashed line).

TABLE 3.3
Model selection for the number of days stopped smoking cigarettes (Example 3.1).

No. of components (g)	Log likelihood	AIC	BIC
1	-2550.0	5104	5113
2	-2442.9	4896	4918
3	-2424.2	4864	4899*
4	-2419.6	4861*	4909
5	n.c.	–	–

n.c. - nonconvergency.
* Number of components selected based on the criterion.

due to the two components of normal distributions having a mean close to 10 and 20 attempts, respectively. However, in practice, these two components will be sensitive to the data collection process which results in the presence of spikes in the probability density function. For example, more than 20 quit attempts were recorded as 20 in the questionnaire data about smoking-cigarette use in the NHANES, resulting in a spike in the probability density function at 20 quit attempts.

We then consider the number of days respondents stopped smoking in the last quit attempt. We fitted the data ($n = 577$; 3 respondents with 0 days were recoded as 0.5 day) with mixtures of gamma distributions for g=1 to g=5. Model selection based on the BIC is displayed in Table 3.3. We identified three groups of survey respondents for the number of days they had stopped smoking. The estimated parameters of the three-component mixture of gamma distributions are given in Table 3.4. The largest cluster (Cluster 1) represents a group of respondents (38.4%) who had stopped smoking for a very short duration in the last quit attempt (mean = 3 days). The next cluster (Cluster 2) represents a group of respondents (36.1%) who had stopped smoking for a medium duration (mean = 15 days), and about 25.5% of respondents (Cluster 3) stopped smoking for a long duration (mean = 123 days). The adequacy of these final models was evaluated by means of the goodness-of-fit statistics displayed in Table 3.4. The ratios of Pearson statistic and deviance over the degrees of freedom are, respectively, 0.25 and 0.36, indicating no evidence of lack of fit for the three-component gamma mixture model.

To assess the effects of risk factors on the number of days participants stopped smoking in their last attempt, the gamma mixture regression model (3.4) is fitted using respondents' smoking habits, demographic and family characteristics (available from the PUF "DEMO_H.XPT" in the NHANES website) via the log link model (3.3). These risk factors are potential determinants of the duration of smoking cessation, where the demographic and family factors are: gender, country of birth (USA born or not), education level (not completed high school, high school completed, college or above), marital status (married/living with partner, never married, widowed/divorced/separated), having children ≤ 5 in household (yes versus no), and the ratio of family income to poverty (range: 0 to 5 with a higher ratio indicating a

TABLE 3.4
Estimates and standard errors (in parentheses) for the three-component mixture of Gamma distributions on the number of days stopped smoking (Example 3.1).

	Cluster 1	**Cluster 2**	**Cluster 3**
Mixing proportion π_h	0.384 (0.098)	0.361 (0.078)	0.255
Mean μ_h	2.960 (0.695)	14.998 (4.272)	122.77 (18.58)
Alpha α_h	2.421 (0.758)	1.790 (1.177)	2.247 (1.523)
Respondents classified	232	211	134
Range of counts	0-6	7-40	42-357
Pearson statistic		142.5	
Deviance		202.6	
Degrees of freedom		569	

more wealthy family). Information about smoking habits is available from the PUF "SMQ_H.XPT", which includes the age respondents started smoking cigarettes regularly and the number of quit attempts during the past 12 months. For this mixture regression model, the number of quit attempts is quantified using the estimated posterior probability of belonging to the cluster of more quit attempts (Cluster 2 in Table 3.2), $\tau_2(y_j;\hat{\Psi})$.

There are $n = 543$ (94% of 577) respondents with complete information on the eight variables concerning smoking habits, demographic and family characteristics. The estimated parameters of the three-component mixture of gamma regression model are given in Table 3.5. The ratios of Pearson statistic and deviance over the degrees of freedom are, respectively, 0.19 and 0.27, indicating no problem in the overall goodness-of-fit for the three-component gamma mixture regression model.

The effect of risk factors on the mean days respondents stopped smoking in the last attempt are given by the exponential of the coefficients. With this mixture regression model, the mean duration of smoking cessation are 1.4 days (i.e., exp(0.371)), 17.7 days, and 167 days, respectively, for Clusters 1 to 3. From this perspective, we thus treat the three identified clusters as the short, medium, and long duration of smoking cessation in the last attempt (the ranges of duration for the three clusters are, respectively, 1-7 days, 4-45 days, and 30-357 days). As discussed in Section 1.1.1, clustering via a regression model in the component densities (see also Section 3.1 about mixture regression models) does not necessarily imply that all individual durations in the long duration cluster must be greater than those in the other two clusters.

From Table 3.5, it can be seen that the effect of risk factors differs between the three clusters; see also Table 3.4. In particular, the mean number of days respondents stopped smoking is reduced by about 3% (i.e., 1-exp(-0.029)) per each increased year of age they started to smoke regularly, for respondents who belong to the long duration group (Cluster 3). For respondents who belong to the short duration group, the mean number of days they stopped smoking is increased by about 60% (i.e., exp(0.470)) if they have more quit attempts and by about 24% (i.e., exp(0.216)) if they have a higher family income. Although not statistically significant, the results

TABLE 3.5
Three-component Gamma mixture regression results for the number of days stopped smoking in the last attempt (Example 3.1).

Variable	Mean (SD) or Count (%)	Cluster 1 coefficient (SE)	Cluster 2 coefficient (SE)	Cluster 3 coefficient (SE)
(Intercept)	—	0.371 (0.330)	2.871* (0.640)	5.118* (0.311)
Age smoke regularly	18.6 (5.9)	0.005 (0.014)	-0.019 (0.027)	-0.029* (0.011)
No. of quit attempts[a]	—	0.470* (0.217)	-0.216 (0.365)	-0.257 (0.303)
Gender				
Female	259 (47.7%)	-0.243 (0.161)	-0.016 (0.202)	-0.041 (0.136)
Male	284 (52.3%)		Reference	
Country of birth				
USA born	466 (85.8%)	-0.107 (0.215)	-0.336 (0.342)	-0.012 (0.215)
Others	77 (14.2%)		Reference	
Education level				
College	246 (45.3%)	-0.165 (0.221)	-0.337 (0.334)	0.176 (0.177)
completed HS	162 (29.8%)	-0.161 (0.228)	-0.194 (0.469)	0.012 (0.185)
Not completed HS	135 (24.9%)		Reference	
Marital status				
Widowed/divorced[b]	129 (23.8%)	0.443 (0.266)	-0.292 (0.344)	0.063 (0.174)
Never married	158 (29.1%)	0.271 (0.161)	0.295 (0.263)	-0.075 (0.199)
Married/living with partners	256 (47.1%)		Reference	
Children $\leq$ 5 in HH				
Yes	125 (23.0%)	0.218 (0.183)	0.162 (0.299)	0.205 (0.185)
No	418 (77.0%)		Reference	
Family income (ratio)	1.75 (1.5)	0.216* (0.043)	0.156 (0.083)	0.094 (0.055)
Pearson statistic			97.9	
Deviance			134.0	
Degrees of freedom			505	

* p-value < 0.05.
[a] No. of quit attempts is quantified using the estimated posterior probability of belonging to the cluster of more quit attempts.
[b] Including separated.

displayed in Table 3.5 show that female respondents born in the USA tend to have a shorter mean number of days of smoking cessation than males and those not born in the USA. On the other hand, respondents having children 5 years of age or younger in the household are more likely to have a longer mean number of days of smoking cessation.

3.5 Example 3.2: Mixture of Gamma Distributions for Clustering Cost Data

We consider the Medical Expenditure Panel Survey (MEPS) Household Component (HC) 2013 Public Use Files including Hospital Inpatient Stays (PUF no. HC-160D), Emergency Room Visit (PUF no. HC-160E), Outpatient Visit (PUF no. HC-160F), and Full Year Consolidated (PUF no. HC-163) data (Agency for Healthcare Research and Quality, 2015). The MEPS provides nationally representative estimates of health care use, expenditures, sources of payment, and health insurance coverage for the United States civilian non-institutionalized population; see, for example, Haas et al. (2005). Information on the MEPS can be obtained from `http://meps.ahrq.gov/mepsweb`. The four PUF data sets are available in the Household Component Event files at `http://meps.ahrq.gov/mepsweb/data_stats/download_data_files.jsp`, and provide detailed information on event-level hospital inpatient stays, emergency and outpatient visits as well as person-level health status, demographic and socio-economic characteristics. These data can be used to make estimates of health care utilizations and expenditures for the 2013 calender year. The SAS transport format data files "h160d.ssp", "h160e.ssp", and "h160f.ssp" contain 2,862 unique inpatient hospital stays, 7,510 emergency visits, and 13,145 outpatient visits reported by survey respondents, respectively. By collapsing these events for each respondent, there are 9,168 respondents overall.

Expenditures in MEPS are defined as the sum of payments for care received for each hospital stay, emergency and outpatient visit, including out-of-pocket payments and payments made by private insurance, Medicaid, Medicare and other sources; see Monheit, Wilson, and Arnett (1999) for details on expenditure definitions. Because more than half of hospital stays actually began with emergency room visits, we decided to combine the expenses for hospital stay and emergency visits. Thus we now have two subsets of data, namely the hospital/emergency expense and the expenditure for outpatient visits.

We fitted these two health care expenditure data subsets with mixtures of gamma distributions, separately, for g=1 to g=5. For the hospital/emergency expense data subset, we have $n = 6,008$ respondents (3,160 out of 9,168 respondents with zero expenses are excluded). The outpatient visits expenditure data subset contains $n = 4,303$ respondents (4,865 respondents with zero expenses excluded). Model selection based on various criteria is displayed in Table 3.6. We identified four groups of survey respondents for the expenditure in hospital/emergency. The estimated parameters of the four-component gamma mixture model are given in Table 3.7. About 17.4% of respondents (Cluster 1) had very low expenses (mean = $246) in hospital/emergency in 2013. Cluster 2 represents a group of respondents (31.2%) who had higher expenses in hospital/emergency (mean = $840) than those of Cluster 1. The largest sub-group of respondents is Cluster 3, who had a mean expense of $7,665 in hospital/emergency. Cluster 4 represents a group of respondents (11.4%) with the highest expense in hospital/emergency (mean = $32,644).

For the expenditure due to outpatient visits, the BIC indicated there are four clusters; see Table 3.6(b). However, the fourth component contains only 6 respondents ($\hat{\pi}_4 = 0.4\% < 1\%$), we thus instead consider the clustering solution based on three groups of respondents. The estimated parameters of this three-component gamma mixture model is provided in Table 3.8. Cluster 1 represents a group of respondents (41.6%) who had very low expenses in outpatient visits in 2013 (mean = $251). The largest sub-group of respondents is Cluster 2, who had a mean expense of $2,181 in outpatient visits. Cluster 3 represents a group of respondents (11.7%) with the highest expense in outpatient visits (mean = $10,098) in 2013.

The adequacy of these two mixture models was evaluated by means of the goodness-of-fit statistics. In contrast to count data in Example 3.1, the fitting of mixture models to cost data is not as good due to some extremely high costs in the expenditure data. From Table 3.7 for the hospital/emergency expense, the ratio of Pearson statistic over the degrees of freedom is 0.58, indicating no problem in the overall fit for the four-component gamma mixture model. However, the ratio of deviance over the degrees of freedom is 1.22, which is greater than one. For the outpatient expense data (Table 3.8), the ratio of Pearson statistic over the degrees of freedom is 0.70, indicating no evidence of lack of fit for the three-component gamma mixture model, whilst the ratio of deviance over the degrees of freedom is 1.08, which is slightly greater than one.

We also fitted the two health care expenditure data subsets using mixtures of normal distributions. Model selection based on BIC indicated there are eight and six clusters (with the mixing proportions all greater than 1%) for the hospital/emergency expense data subset and expenditure data of outpatient visits, respectively. The empirical distributions of expenditure data and fitted gamma mixture distributions and normal mixture distributions are provided in Figures 3.4 and 3.5, respectively. The normal mixture models had more components than those of gamma mixtures, and thus the normal mixture models fit the empirical distributions of expenditure data better than the gamma mixture models, as observed in Figures 3.4 and 3.5. However, the interpretation of this large number of components with the normal mixture models may not be straightforward.

To illustrate this further, we consider the two-dimensional clustering solutions of the 9,168 survey respondents in terms of expenditure for hospital/emergency and expenditure for outpatient visits; see Table 3.9 where zero expenditure is assigned to the very low expense group. We intend to explore the respondent's characteristics that differentiate four exclusive groups: $n = 2{,}965$ respondents with very low expenses in both data subsets; $n = 242$ respondents with high expenses in hospital

TABLE 3.6
Model selection for the two health care expenditure data subsets (Example 3.2).

No. of components (g)	Log likelihood	AIC	BIC
(a) Expenditure for hospital stay or emergency room visits			
1	-57151.7	114307	114321
2	-56318.3	112647	112680
3	-56161.9	112340	112393
4	-56142.3	112307	112380*
5	-56130.2	112288*	112382
(b) Expenditure for outpatient visits			
1	-36664.0	73332	73345
2	-36155.5	72321	72353
3	-36016.7	72049	72100
4	-35994.6	72011	72081*
5	-35983.4	71995*	72084

* Number of components selected based on the criterion.

TABLE 3.7
Estimates and standard errors (in parentheses) for the four-component mixture of Gamma distributions fitted on expenditure data (hospital stay or emergency room visits, $n = 6,008$) (Example 3.2).

	Cluster 1	Cluster 2	Cluster 3	Cluster 4
Mixing proportion π_h	0.174 (0.066)	0.312 (0.063)	0.400 (0.032)	0.114
Mean μ_h	246 (40)	840 (136)	7665 (476)	32644 (3360)
Alpha α_h	1.737 (0.426)	1.708 (0.834)	0.999 (0.120)	0.723 (0.031)
Respondents classified	1238	2127	2312	331
Range of expenses	3-310	311-1854	1857-28938	29021-321195
Pearson statistic		3499.0		
Deviance		7339.9		
Degrees of freedom		5997		

TABLE 3.8
Estimates and standard errors (in parentheses) for the three-component mixture of Gamma distributions fitted on expenditure data (outpatient visits, $n = 4,303$) (Example 3.2).

	Cluster 1	**Cluster 2**	**Cluster 3**
Mixing proportion π_h	0.416 (0.017)	0.467 (0.028)	0.117
Mean μ_h	251 (15)	2181 (224)	10098 (2228)
Alpha α_h	1.510 (0.063)	1.020 (0.054)	0.750 (0.090)
Respondents classified	2153	1905	245
Range of expenses	4-578	579-8496	8520-303957
Pearson statistic		3007.2	
Deviance		4640.4	
Degrees of freedom		4295	

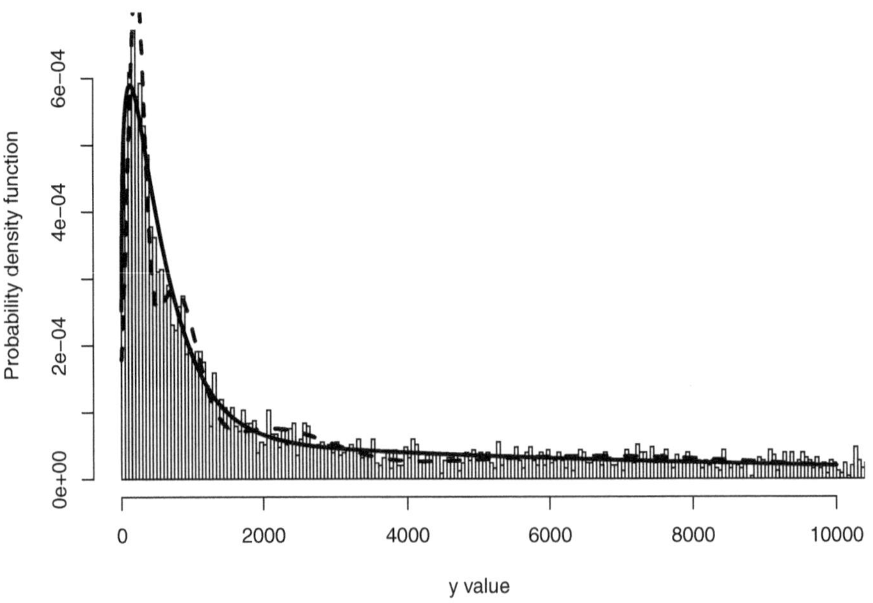

FIGURE 3.4
Empirical distribution of expenditure data of hospital stay or emergency room visits, fitted gamma mixture distribution (Solid line), and fitted normal mixture distribution (Dashed line).

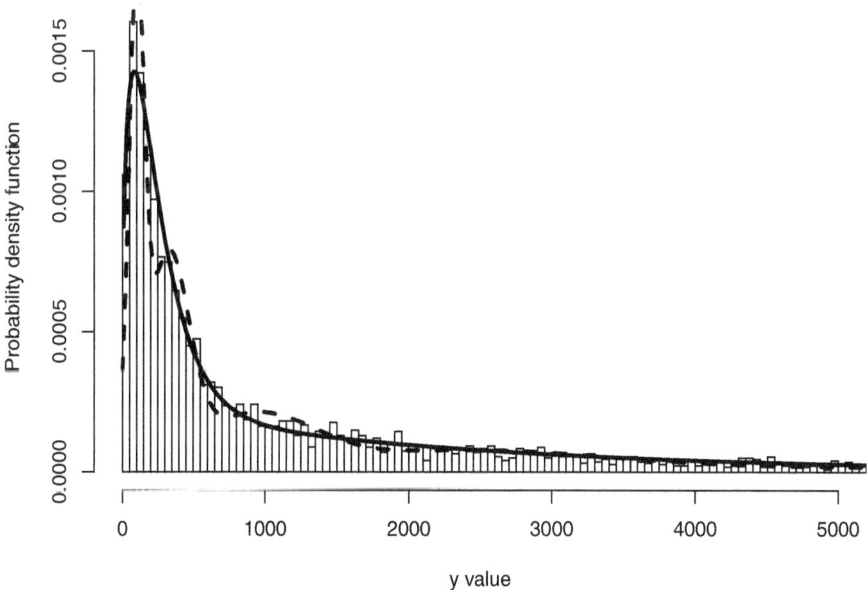

FIGURE 3.5
Empirical distribution of expenditure data of outpatient visits, fitted gamma mixture distribution (Solid line), and fitted normal mixture distribution (Dashed line).

TABLE 3.9
Sequential two-way clustering of survey respondents (Example 3.2).

Clustering based on expenditure in hospital stay or emergency room visits	**Clustering based on expenditure in outpatient visits**			
	Very low expense group	Medium expense group	High expense group	Total
Very low expense group	**2965** (42.2%)	1285 (18.3%)	**148** (2.1%)	4398
Low expense group	1887 (26.9%)	212 (3.0%)	28 (0.4%)	2127
Medium expense group	1924 (27.4%)	336 (4.8%)	**52** (0.7%)	2312
High expense group	**242** (3.4%)	**72** (1.0%)	**17** (0.2%)	331
Total	7018	1905	245	9168

Highlighted counts correspond to (1) Group 1, $n = 2{,}965$, very low expenses in both data subsets; (2) Group 2, $n = 242$, high expenses in hospital stay/emergency room visits but very low expenses in outpatient visits; (3) Group 3, $n = 148$, high expenses in outpatient visits but very low expenses in hospital stay/emergency room visits; and (4) Group 4, $n = 52 + 72 + 17 = 141$, high expenses in both of the two data subsets or either one (with medium expenses in the other).

stay/emergency room visits but very low expenses in outpatient visits; $n = 148$ respondents with high expenses on outpatient visits but very low expenses on hospital stay/emergency room visits; and $n = 141$ respondents with high expenses in both of the two data subsets or either one (with medium expenses in the other), all highlighted in bold in Table 3.9. The SAS transport format data files "h163.ssp" contain demographic, socioeconomic and health-related information of 36,940 respondents. As measurements of some variables regarding health conditions such as high blood pressure, diabetes and arthritis were restricted to respondents over 17 years of age, we thus consider only respondents who are equal to or older than 18 years of age in the following analyses.

Table 3.10 displays the discrepancies in respondent's characteristics among the four groups. In particular, respondents in Group 1 (Very Low on hospital/emergency and outpatient) are younger with lower family income compared to the average of the sample population. They are also more likely to be females with a lower proportion of them having health insurance. Group 2 (High on hospital/emergency; Very Low on outpatient) is characterized by older respondents in the West region with private or public health insurance. They appear to be older (mean age of 58.6) and the proportion of widowed/divorced/separated is higher. Respondents in Group 3 (Very Low on hospital/emergency; High on outpatient) are younger than those in Group 2 with a higher proportion of those being married and in the Midwest region. The mean family income of this group is the highest, also having a higher proportion being covered with private health insurance (71.6%). Group 4 (High on hospital/emergency and outpatient) is characterized by older respondents in the Northeast or Midwest regions and nearly all of them (97.7%) have private or public health insurance. From Table 3.10, it is observed that existence of health conditions increases medical expenditure. Specifically, high blood pressure and diabetes are associated with increased expenses in hospital stay or emergency room visits, whilst arthritis is associated with increased expenses in hospital/emergency as well as outpatient visits.

Finally, multinomial logistic regression analysis was performed to identify risk factors that differentiate respondents with high expenses in hospital/emergency and/or outpatient (Groups 2 to 4) and those with very low expenses in both hospital/emergency and outpatient (Group 1). Adjusted relative risk ratios (RRRs) with 95% CIs were presented in Table 3.11. Compared to respondents who have very low expenses in both hospital/emergency and outpatient (Group 1), respondents with high expenses in hospital stay or emergency room visits (Group 2) were more likely to be older (adjusted RRR = 1.016; 95% CI=1.006-1.027 for each increased year of age), male (adjusted RRR = 1.416; 95% CI=1.038-1.931), in the South region (adjusted RRR = 1.668; 95% CI=1.042-2.672) or the West region (adjusted RRR = 2.564; 95% CI=1.564-4.205), with high blood pressure (adjusted RRR = 1.490; 95% CI=1.033-2.149) or diabetes (adjusted RRR = 2.113; 95% CI=1.497-2.983). Respondents with high expenses in outpatient visits (Group 3) were characterized by those with arthritis (adjusted RRR = 1.648; 95% CI=1.099-2.471) and higher family incomes (adjusted RRR = 1.062; 95% CI=1.014-1.112) for each increased unit of percentage above the federal poverty line). Finally, respondents with high expenses in both hospital/emergency and outpatient (Group 4) were more likely to have high blood pressure (adjusted RRR = 1.913; 95% CI=1.237-2.956) or arthritis (adjusted RRR = 1.705; 95% CI=1.132-2.569). Having private health insurance is associated with all high medical expenditure groups with increasing adjusted RRRs from Group 2 (2.420; 95% CI=1.274-4.597) to Group 4 (8.989; 95% CI=2.770-29.17). Having public only health insurance is also associated with high expenditure in hospital stay or emergency room visits (Group 2: 2.128; 95% CI−1.114-4.066) and Group 4: 4.789; 95% CI=1.450-15.82).

3.6 Fortran Programs for Fitting Mixtures of Gamma Distributions

The Fortran programs "gammamix.f" (for gamma mixture models) and "gammamixregr.f" (for gamma mixture regression models) as well as external Fortran programs "psifn.f", "i1mach.f", "r1mach.f", "random.f", "hybrid.f", "chol.f", and "syminv.f" can be download from the book's online webpage. Gammamix.f fits mixture models of gamma distributions (Equations (3.1) and (3.2)) via the EM algorithm. Gammamixregr.f fits a gamma mixture regression model (3.4) with covariates in the linear predictor η_{hj} of the component means via a log link function (3.3) $(h = 1,\ldots,g;\ j = 1,\ldots,n)$.

The instructions to compile gammamix.f and gammamixregr.f with other external Fortran programs in Cygwin into executable programs GAMMAMIX and

TABLE 3.10
Characteristics of respondents (four groups with different expenditure levels; Example 3.2).

Variable	Group 1 V.Low-V.Low (n=2,184)	Group 2 High-V.Low (n=196)	Group 3 V.Low-High (n=134)	Group 4 High-High (n=129)	Overall (n=2,643)
Age* (year)	49.2 (17.9)	58.6 (19.0)	52.8 (14.4)	55.6 (16.2)	50.4 (18.0)
Gender*					
Male	782 (35.8%)	87 (44.4%)	56 (41.8%)	52 (40.3%)	977 (37.0%)
Female	1402 (64.2%)	109 (55.6%)	78 (58.2%)	77 (59.7%)	1666 (63.0%)
Race					
White	1440 (65.9%)	136 (69.4%)	94 (70.1%)	88 (68.2%)	1758 (66.5%)
Black	562 (25.7%)	49 (25.0%)	24 (17.9%)	31 (24.0%)	666 (25.2%)
Others[a]	182 (8.3%)	11 (5.6%)	16 (11.9%)	10 (7.8%)	219 (8.3%)
Country of birth					
USA	952 (82.4%)	94 (91.3%)	53 (84.1%)	55 (87.3%)	1154 (83.4%)
Others	203 (17.6%)	9 (8.7%)	10 (15.9%)	8 (12.7%)	230 (16.6%)
missing	1029	93	71	66	1259
Marital status*					
Married	941 (43.1%)	86 (43.9%)	77 (57.5%)	63 (48.8%)	1167 (44.2%)
Others[b]	636 (29.1%)	71 (36.2%)	37 (27.6%)	45 (34.9%)	789 (29.9%)
Never married	607 (27.8%)	39 (19.9%)	20 (14.9%)	21 (16.3%)	687 (26.0%)
Region*					
Northeast	450 (20.6%)	26 (13.3%)	31 (23.1%)	32 (24.8%)	539 (20.4%)
Midwest	510 (23.4%)	38 (19.4%)	38 (28.4%)	36 (27.9%)	622 (23.5%)
South	800 (36.6%)	75 (38.3%)	46 (34.3%)	40 (31.0%)	961 (36.4%)
West	424 (19.4%)	57 (29.1%)	19 (14.2%)	21 (16.3%)	521 (19.7%)
HBP*[c]					
Yes	959 (43.9%)	134 (68.4%)	58 (43.3%)	86 (66.7%)	1237 (46.8%)
No	1225 (56.1%)	62 (31.6%)	76 (56.7%)	43 (33.3%)	1406 (53.2%)
Diabetes*					
Yes	317 (14.5%)	70 (35.7%)	21 (15.7%)	35 (27.1%)	443 (16.8%)
No	1867 (85.5%)	126 (64.3%)	113 (84.3%)	94 (72.9%)	2200 (83.2%)
Arthritis*					
Yes	788 (36.1%)	105 (53.6%)	64 (47.8%)	72 (55.8%)	1029 (38.9%)
No	1396 (63.9%)	91 (46.4%)	70 (52.2%)	57 (44.2%)	1614 (61.1%)
Family size	2.7 (1.5)	2.5 (1.5)	2.6 (1.3)	2.5 (1.5)	2.7 (1.5)
missing	78	9	5	7	99
Family income*[d]	307.7 (322.7)	308.5 (328.6)	454.6 (331.9)	320.3 (287.0)	315.8 (323.4)
Health insurance*					
Any private	1078 (49.4%)	101 (51.5%)	96 (71.6%)	84 (65.1%)	1359 (51.4%)
Public only	724 (33.2%)	83 (42.3%)	28 (20.9%)	42 (32.6%)	877 (33.2%)
Uninsured	382 (17.5%)	12 (6.1%)	10 (7.5%)	3 (2.3%)	407 (15.4%)

Note: Only respondents $\geq$18 years old were considered. Data are mean (SD) or count (%).
Notation: V.Low-V.Low group: respondents with very low expenses in both data subsets; High-V.Low group: respondents with high expenses in hospital stay/emergency room visits but very low expenses in outpatient visits; V.Low-High group: respondents with very low expenses in hospital stay/emergency room visits but high expenses in outpatient visits; High-High group: respondents with high expenses in both of the two data subsets or either one (with medium expenses in the other); see Table 3.9.
* p-value < 0.05. Test the difference between groups using either chi-square tests or ANOVA. [a]Others: American Indian/Alaska Native/Asian/Native Hawaiian/Pacific Islander/Multiple races. [b]Others: Widowed/Divorced/Separated. [c]HBP – High blood pressure. [d]Family income as percentage above the federal poverty line based on family size and composition.

TABLE 3.11
Risk factors for high expenses in hospital/emergency and/or outpatient (Example 3.2).

Risk factor	Group 2 High-V.Low (n=196)	Group 3 V.Low-High (n=134)	Group 4 High-High (n=129)
Age (year)	1.016* (1.006-1.027)	1.002 (0.990-1.015)	1.000 (0.988-1.013)
Gender			
Male	1.416* (1.038-1.931)	1.348 (0.936-1.941)	1.288 (0.885-1.875)
Female	Reference	Reference	Reference
Region			
Northeast	Reference	Reference	Reference
Midwest	1.378 (0.815-2.333)	1.012 (0.614-1.668)	0.967 (0.584-1.601)
South	1.668* (1.042-2.672)	0.941 (0.582-1.523)	0.696 (0.427-1.136)
West	2.564* (1.564-4.205)	0.650 (0.359-1.177)	0.750 (0.422-1.334)
HBP[a]			
Yes	1.490* (1.033-2.149)	0.803 (0.533-1.208)	1.913* (1.237-2.956)
No	Reference	Reference	Reference
Diabetes			
Yes	2.113* (1.497-2.983)	1.114 (0.666-1.863)	1.528 (0.987-2.366)
No	Reference	Reference	Reference
Arthritis			
Yes	1.313 (0.940-1.833)	1.648* (1.099-2.471)	1.705* (1.132 2.569)
No	Reference	Reference	Reference
Family income[b]	0.963 (0.911-1.018)	1.062* (1.014-1.112)	0.950 (0.887-1.019)
Health insurance			
Any private	2.420* (1.274-4.597)	2.617* (1.312-5.223)	8.989* (2.770-29.17)
Public only	2.128* (1.114-4.066)	1.325 (0.622-2.825)	4.789* (1.450-15.82)
Uninsured	Reference	Reference	Reference

Note: Only respondents ≥18 years old were considered.

Data are adjusted relative risk ratio (95% CI).

Notation: V.Low-V.Low group: respondents with very low expenses in both data subsets; High-V.Low group: respondents with high expenses in hospital stay/emergency room visits but very low expenses in outpatient visits; V.Low-High group: respondents with very low expenses in hospital stay/emergency room visits but high expenses in outpatient visits; High-High group: respondents with high expenses in both of the two data subsets or either one (with medium expenses in the other); see Table 3.9.

* Significant at the 0.05 level.

[a] HBP – High blood pressure.

[b] Family income as percentage above the federal poverty line based on family size and composition (continuous value).

GAMMAMIXREGR, respectively, are provided in Figure 3.6. The program codes for fitting the data regarding quit attempts and the number of days respondents stopped smoking (Example 3.1) using GAMMAMIX are provided in Figures 3.7 and 3.8, respectively. The program code for the mixture regression analysis about the number of days stopped smoking (Example 3.1) using GAMMAMIXREGR is provided in Figure 3.9.

In Example 3.2, the program codes for fitting the cost data about hospital/emergency expenses and outpatient expenses using GAMMAMIX are provided in Figures 3.10 and 3.11, respectively.

Instruction to compile gammamix.f and gammamixregr.f in Cygwin

```
$ cd c:/Chapter3              # change to the work directory

$ gfortran -c psifn.f         # compile a Fortran file to obtain an .o file

$ gfortran -c i1mach.f

$ gfortran -c r1mach.f

$ gfortran -c random.f

$ gfortran -c hybrid.f

$ gfortran -c chol.f

$ gfortran -c syminv.f

$ gfortran -c gammamix.f           # compile the gammamix.f to obtain gammamix.o

# to obtain the executable program GAMMAMIX from the .o files
$ gfortran -o GAMMAMIX gammamix.o psifn.o i1mach.o r1mach.o random.o hybrid.o

$ gfortran -c gammamixregr.f       # compile the gammamixregr.f to obtain gammamixregr.o

# to obtain the executable program GAMMAMIXREGR from the .o files
$ gfortran -o GAMMAMIXREGR gammamixregr.o psifn.o i1mach.o r1mach.o random.o hybrid.o
chol.o syminv.o
```

FIGURE 3.6
Instruction to compile Fortran programs.

Main Analysis for Example 3.1 (fitting data about quit attempts)

```
$ cd c:/Chapter3              # change to the work directory

$ ./GAMMAMIX                  # call the GAMMAMIX program
 Adopt single or mixed gamma? (0/1)
1                                             # consider a gamma mixture model
 How many components?
2                                             # consider 2 components

# output file is "fort.25"


# calculate standard errors using B=100 bootstrap replications

$ ./GAMMAMIX                  # call the GAMMAMIX program
 Adopt single or mixed gamma? (0/1)
1                                             # consider a gamma mixture model
 How many components?
2                                             # consider 2 components

 Input the seeds for random number generation: input 3 integers between 1 and 30,000,
 e.g. 23 120 3411.
23 120 3411                                   # enter the seeds

 Do you want to print out the estimates
   for each bootstrap?     INPUT: 1(Yes); 0(No)
1                                             # to print out the estimates

# output file is "fort.25"
```

FIGURE 3.7
The program codes for Example 3.1 (quit attempts).

Main Analysis for Example 3.1 (fitting data about days stopped smoking)

```
$ cd c:/Chapter3            # change to the work directory

$ ./GAMMAMIX                # call the GAMMAMIX program
 Adopt single or mixed gamma? (0/1)
1                                          # consider a gamma mixture model
 How many components?
3                                          # consider 3 components

# output file is "fort.25"


# calculate standard errors using B=100 bootstrap replications

$ ./GAMMAMIX                # call the GAMMAMIX program
 Adopt single or mixed gamma? (0/1)
1                                          # consider a gamma mixture model
 How many components?
3                                          # consider 3 components

 Input the seeds for random number generation: input 3 integers between 1 and 30,000,
 e.g. 23 120 3411.
23 120 341                                 # enter the seeds

 Do you want to print out the estimates
   for each bootstrap?     INPUT: 1(Yes); 0(No)
1                                          # to print out the estimates

# output file is "fort.25"
```

FIGURE 3.8
The program codes for Example 3.1 (days stopped smoking).

Main Analysis for Example 3.1 (regression analysis on days stopped smoking)

```
$ cd c:/Chapter3              # change to the work directory

$ ./GAMMAMIXREGR              # call the GAMMAMIXREGR program
 Adopt single or mixed gamma distribution? (0/1)
1                                            # consider a gamma mixture model
 How many components?
3                                            # consider 3 components
 Do you want the output of tau? (0/1)
1                                            # to output the mixing proportions

# output file is "fort.25"


# calculate standard errors using B=100 bootstrap replications

$ ./GAMMAMIXREGR              # call the GAMMAMIXREGR program
 Adopt single or mixed gamma distribution? (0/1)
1                                            # consider a gamma mixture model
 How many components?
3                                            # consider 3 components
 Do you want the output of tau? (0/1)
0                                            # do not output the mixing proportions

 Input the seeds for random number generation: input 3 integers between 1 and 30,000,
 e.g. 23 120 3411.
23 120 3411                                  # enter the seeds

# output file is "fort.25"
```

FIGURE 3.9
The program codes for Example 3.1 (regression analysis on days stopped smoking).

Main Analysis for Example 3.2 (fitting data about hospital/emergency expenses)

```
$ cd c:/Chapter3              # change to the work directory

$ ./GAMMAMIX                  # call the GAMMAMIX program
 Adopt single or mixed gamma? (0/1)
1                                            # consider a gamma mixture model
 How many components?
4                                            # consider 4 components

# output file is "fort.25"


# calculate standard errors using B=100 bootstrap replications

$ ./GAMMAMIX                  # call the GAMMAMIX program
 Adopt single or mixed gamma? (0/1)
1                                            # consider a gamma mixture model
 How many components?
4                                            # consider 4 components

 Input the seeds for random number generation: input 3 integers between 1 and 30,000,
 e.g. 23 120 3411.
23 120 3411                                  # enter the seeds

 Do you want to print out the estimates
   for each bootstrap?     INPUT: 1(Yes); 0(No)
1                                            # to print out the estimates

# output file is "fort.25"
```

FIGURE 3.10
The program codes for Example 3.2 (hospital/emergency expenses).

Main Analysis for Example 3.2 (fitting data about outpatient expenses)

```
$ cd c:/Chapter3             # change to the work directory

$ ./GAMMAMIX                 # call the GAMMAMIX program
 Adopt single or mixed gamma? (0/1)
1                                            # consider a gamma mixture model
 How many components?
3                                            # consider 3 components

# output file is "fort.25"


# calculate standard errors using B=100 bootstrap replications

$ ./GAMMAMIX                 # call the GAMMAMIX program
 Adopt single or mixed gamma? (0/1)
1                                            # consider a gamma mixture model
 How many components?
3                                            # consider 3 components

 Input the seeds for random number generation: input 3 integers between 1 and 30,000,
 e.g. 23 120 3411.
23 120 34                                    # enter the seeds

 Do you want to print out the estimates
   for each bootstrap?     INPUT: 1(Yes); 0(No)
1                                            # to print out the estimates

# output file is "fort.25"
```

FIGURE 3.11
The program codes for Example 3.2 (outpatient expenses).

4

Mixture of Generalized Linear Models for Count or Categorical Data

4.1 Introduction

This chapter describes mixture of generalized linear models for handling count or categorical data. These types of response variables are common in a variety of medical and health science applications. For example, the number of hospital admissions (Axon et al., 2016), the number of occupational injuries (Lee, Wang, and Yau, 2001), or the number of illness spells (Böhning and Schlattmann, 1992) are count data corresponding to the number of times that a particular event of interest occurs within a study period, whereas the category of body mass index (Cameron et al., 2014) or the pattern of comorbidity in chronic diseases (Ng, Tawiah, and McLachlan, 2019) contribute to response variables in a categorical form. For analyzing count data arising from two or more latent sub-populations, Poisson mixture (regression) models can be adopted to analyze heterogeneity in count variables when a single Poisson distribution does not fit the empirical distribution of count data very well; see, for example, Böhning and Schlattmann (1992). Alternatively, mixture models of negative binomial distributions can be used to handle over-dispersion of count data, which is commonly encountered in medical and healthcare studies (Deb and Trivedi, 1997). In many such studies, over-dispersion of count data can be attributed to the presence of excess number of observations with zero counts, for example, corresponding to zero hospital admissions or zero disabled activities (Scuffham et al., 2019; Gill et al., 2015). In the literature, the zero-inflated Poisson (ZIP) and the zero-inflated negative binomial (ZINB) have received considerable attention to analyze count data with excess zeros; see, for example, Dietz and Böhning (2000), Lambert (1992), Lee, Wang, and Yau (2001), and Yau, Wang, and Lee (2003). In this context, it is therefore important to assess different models including determining the number of components, among mixtures of Poisson or negative binomial distributions, ZIP, or ZINB, and select the best model for the analysis of count data. There are various score tests proposed to compare these models, such as Ridout, Hinde, and Demétrio (2001), Lee, Wang, and Yau (2001), Lee, Xiang, and Fung (2004), and Yau, Wang, and Lee (2003). Within the mixture model framework, a zero-inflated model is a two-component mixture model containing a degenerate component of point mass at zero to account for excess zeros and a component of a Poisson (or a negative binomial) distribution. Of the related context in the economic literature, there are "two-part" models that handle zero-inflated count data by considering a logit (or a probit) model on the dichotomized response variable (zero versus positive counts) to estimate the probability of incurring positive counts (the first part). Separately and independently from the first part, the second part evaluates the sub-sample with positive responses to estimate the conditional mean counts using a zero-truncated count distribution; see Alfo and Maruotti (2010), Mihaylova et al. (2011), and the discussion in Scuffham et al. (2019).

Concerning the clustering of categorical data within the framework of a mixture of generalized linear models, mixture models with component densities belonging to the binomial family have been considered to tackle meta-analysis problems with a binary outcome concerning mortality after myocardial infarction and response to drug therapy (Aitkin, 1999); see also McLachlan and Peel (2000, Section 5.11). Generalization in mixture modelling to handle response variables with multiple categories has been given by Graham and Miller (2006), Li and Zhang (2008), and Rigouste, Cappé, and Yvon (2007), with applications to document classification and text clustering, and also by Topchy, Jain, and Punch (2005) in the context of consensus clustering of ensembles represented by categorical cluster labels.

4.2 Poisson Mixture Regression Model

Let Y_j be the non-negative count response variable with corresponding covariate vector x_j for $j = 1, \ldots, n$, where n denotes the number of observations. A g-component mixture of Poisson distributions is then given by

$$f(y_j; x_j) = \sum_{h=1}^{g} \pi_h f_h(y_j; \mu_{hj}), \tag{4.1}$$

where we view $f(y_j; x_j)$ as a density function in this case of discrete Y_j. In (4.1),

$$f_h(y_j; \mu_{hj}) = \exp(-\mu_{hj}) \mu_{hj}^{y_j} / y_j! \tag{4.2}$$

follows a Poisson distribution, which has a log function linking the mean μ_{hj} and the covariate vector x_j as

$$\log(\mu_{hj}) = \eta_{hj} = x_j^T \beta_h \qquad (h = 1, \ldots, g;\ j = 1, \ldots, n), \tag{4.3}$$

where β_h is the vector of regression coefficients (the first element accounts for an intercept term); see Xiang et al. (2005) and McLachlan and Peel (2000, Section 5.8). The vector of unknown parameters Ψ thus contains the mixing proportions $\pi_1, \ldots, \pi_{g-1}$ and the regression coefficients $(\beta_1^T, \ldots, \beta_g^T)^T$. From (4.1) and (4.2), it means that

$$E(Y_j) = \sum_{h=1}^{g} \pi_h \mu_{hj}$$

and

$$\text{var}(Y_j) = E(Y_j) + \sum_{h=1}^{g} \pi_h \mu_{hj}^2 - (E(Y_j))^2.$$

That is, unlike a Poisson distribution which has the mean equal to the variance, a mixture of Poisson distributions allows excess variation among $Y_1, \ldots, Y_n$ unless

$$\sum_{h=1}^{g} \pi_h \mu_{hj}^2 = (E(Y_j))^2 = (\sum_{h=1}^{g} \pi_h \mu_{hj})^2,$$

which is the case when $\mu_{1j} = \mu_{2j} = \cdots = \mu_{gj}$ for $j = 1, \ldots, n$.

In Poisson mixture model (4.1), the mixing proportions π_h can also be allowed to depend on covariates x_j. This situation will be considered in later sections concerning zero-inflated models.

Within the EM framework, the complete-data log likelihood is given by

$$\log L_c(\Psi) = \sum_{h=1}^{g} \sum_{j=1}^{n} z_{hj} \{\log \pi_h + \log f_h(y_j; \mu_{hj})\}, \tag{4.4}$$

where z_{hj} is one or zero accordingly as y_j does or does not belong to the hth component $(h = 1, \ldots, g)$. The E-step concerning the calculation of the Q-function is essentially the same as that given in Equation (3.7) in Section 3.2. The M-step updates the estimates for π_h and β_h that maximize the Q-function. In particular, closed-form equation is available for π_h $(h = 1, \ldots, g)$, as

$$\pi_h^{(k+1)} = \sum_{j=1}^{n} \tau_{hj}^{(k)} / n \tag{4.5}$$

for $h = 1, \ldots, g-1$ and $\pi_g^{(k+1)} = 1 - \sum_{h=1}^{g-1} \pi_h^{(k+1)}$. For β_h, the maximization involves solving the following non-linear equation:

$$\sum_{j=1}^{n} \tau_{hj}^{(k)} \{y_j - \exp(\eta_{hj})\} x_j = 0 \qquad (h = 1, \ldots, g). \tag{4.6}$$

The MINPACK routine HYBRD1 of Moré, Garbow, and Hillstrom (1980) can be used to find a solution to (4.6). In (4.5) and (4.6),

$$\tau_{hj}^{(k)} = \frac{\pi_h^{(k)} \exp(-\mu_{hj}) \mu_{hj}^{y_j} / y_j!}{\sum_{l=1}^{g} \pi_l^{(k)} \exp(-\mu_{lj}) \mu_{lj}^{y_j} / y_j!} \qquad (h = 1, \ldots, g). \tag{4.7}$$

is the current posterior probability of component membership for the jth observation $(j = 1, \ldots, n)$. For Poisson mixture models without covariates x_j, we have $f_h(y_j; \mu_h) = \exp(-\mu_h) \mu_h^{y_j} / (y_j!)$ and the M-step becomes in closed form as

$$\mu_h^{(k+1)} = \frac{\sum_{j=1}^{n} \tau_{hj}^{(k)} y_j}{\sum_{j=1}^{n} \tau_{hj}^{(k)}}. \tag{4.8}$$

The standard errors of the ML estimates of the parameters Ψ in Poisson mixture (regression) models can be obtained using a non-parametric bootstrap resampling approach, as described in Section 1.3.4.

4.3 Zero-Inflated Poisson Regression Model

Zero-inflated Poisson regression models have been widely used for analyzing count data with excess zeros; see, for example, Lambert (1992), Lee, Wang, and Yau (2001) and the references therein. For a Poisson distribution with mean μ, the expected number of zeros is $ne^{-\mu}$. In many real-world applications, there are many more observations with zero counts than expected. To account for the presence of excess zero counts, a ZIP regression model can be formulated as

$$f(y_j;x_j) = \pi(x_j)I_{[0]}(y_j;x_j) + (1-\pi(x_j))f_2(y_j;\mu_{2j}), \tag{4.9}$$

where $I_{[0]}(y_j;x_j)$ is a degenerate distribution with mass one at $y_j = 0$ and the second component (as given in (4.2)) is a Poisson distribution with mean $\mu_{2j} = \mu_j$; see McLachlan and Peel (2000, Section 5.9) for more details. It follows from (4.9) that a discrete count variable Y_j with realizations y_j and associated covariates x_j $(j = 1,\ldots,n)$ under a ZIP distribution can be expressed as:

$$\begin{aligned} P(Y_j = 0) &= \pi(x_j) + (1-\pi(x_j))\exp(-\mu_j), \\ P(Y_j = y_j) &= (1-\pi(x_j))\frac{\exp(-\mu_j)\mu_j^{y_j}}{y_j!}, \qquad y_j = 1,2,\ldots, \end{aligned} \tag{4.10}$$

where $0 < \pi(x_j) < 1$ is the mixing proportion that incorporates extra zeros than those permitted by single Poisson distribution with $\pi(x_j) = 0$ for all $j = 1,\ldots,n$. Under (4.10), we have

$$\begin{aligned} E(Y_j) &= (1-\pi(x_j))\mu_j \\ \text{var}(Y_j) &= (1-\pi(x_j))[\mu_j + \pi(x_j)\mu_j^2]. \end{aligned} \tag{4.11}$$

Thus the ZIP regression model (4.9) also allows extra variation than a Poisson distribution, as a Poisson mixture regression model described in Section 4.2. With the ZIP regression model (4.9), the mixing proportion is modelled by a logistic function as

$$\pi(x_j;\nu) = \frac{\exp(\nu^T x_j)}{1+\exp(\nu^T x_j)} = \frac{\exp(\xi_j)}{1+\exp(\xi_j)}, \tag{4.12}$$

where the first element of x_j is assumed to be one to account for an intercept term and $\xi_j = \nu^T x_j$ for $j = 1,\ldots,n$. With $\log(\mu_j) = \eta_j = x_j^T\beta$ $(j = 1,\ldots,n)$, the vector of unknown parameters is $\Psi = (\nu^T, \beta^T)^T$. It follows that the complete-data log likelihood becomes

$$\begin{aligned} \log L_c(\Psi) = &\sum_{y_j=0} \{z_j \log \pi(x_j;\nu) + (1-z_j)[\log(1-\pi(x_j;\nu)) - \exp(\eta_j)]\} + \\ &\sum_{y_j>0} \{\log(1-\pi(x_j;\nu)) - \exp(\eta_j) + y_j\eta_j - \log(y_j!)\}, \end{aligned} \tag{4.13}$$

where z_j is one or zero accordingly as y_j is or is not an excess zero. On the $(k+1)$th iteration of the E-step, the Q-function can be partitioned into two parts corresponding to the parameters in v and β, respectively, as:

$$
\begin{aligned}
Q_v &= \sum_{j=1}^{n} [\tau_j \xi_j - \log(1+\exp \xi_j)] \\
Q_\beta &= \sum_{j=1}^{n} (1-\tau_j)(y_j \eta_j - \exp(\eta_j) - \log(y_j!)),
\end{aligned} \tag{4.14}
$$

where

$$
\tau_j^{(k)} = \begin{cases} \frac{1}{1+\exp\{-x_j^T v^{(k)} - \exp[x_j^T \beta^{(k)}]\}} & \text{if } y_j = 0 \\ 0 & \text{if } y_j \geq 1 \end{cases} \tag{4.15}
$$

is the current estimated posterior probability of being an excess zero. The M-step updates the estimates for v and β that maximize the corresponding Q-functions. From (4.14), the M-step involves solving the following system of non-linear equations:

$$
\begin{aligned}
&\sum_{j=1}^{n} \{\tau_j^{(k)}(1-\pi(x_j)) - (1-\tau_j^{(k)})\pi(x_j)\} x_j = 0 \\
&\sum_{j=1}^{n} (1-\tau_j^{(k)})\{y_j - \exp(\eta_j)\} x_j = 0.
\end{aligned} \tag{4.16}
$$

The MINPACK routine HYBRD1 of Moré, Garbow, and Hillstrom (1980) can be used to solve (4.16).

The ZIP regression model (4.9) has been extended by Lee, Wang, and Yau (2001) to accommodate the extent of individual exposure risk (such as, the number of hours actually worked in analysis of injury counts due to manual handling for hospital orderlies) within the regression model. In this context, individual exposure information denoted by t_j $(j = 1, \ldots, n)$ can be accounted for in the Poisson part, where the conditional rate becomes $t_j \mu_j = t_j \exp \eta_j = t_j \exp(x_j^T \beta)$. That is, an exposure term $\log t_j$ can be treated as a covariate in the log link function; see (4.3). With this ZIP regression model with exposure adjustment, the EM algorithm is conducted according to (4.13) to (4.16), but with η_j and $x_j^T \beta$ replaced by $\log(t_j \mu_j)$ and $(x_j^T \beta + \log t_j)$, respectively; see Lee, Wang, and Yau (2001) for more details.

The standard errors of the ML estimates of the parameters Ψ in ZIP regression models can be obtained using a non-parametric bootstrap resampling approach, as described in Section 1.3.4.

4.4 Zero-Inflated Negative Binomial Regression Models

A ZIP regression model is useful to analyze count data with extra zeros relative to the Poisson distribution. However, in applications where the non-zero observations are over-dispersed, estimation of parameters may be seriously biased for a ZIP regression model. A way of dealing with over-dispersed Poisson counts is to adopt a negative binomial distribution to model the non-zero over-dispersed observations; that is, a ZINB model.

With reference to (4.9) for a ZIP model, the second component for a ZINB is now a two-parameter negative binomial distribution with mean μ_j and a scale parameter $(1/r)$:

$$f_2(y_j;\mu_j) = \begin{pmatrix} y_j + r - 1 \\ y_j \end{pmatrix} p_j^r (1-p_j)^{y_j}, \tag{4.17}$$

where $p_j = r/(r+\mu_j) = r/(r+\exp(\eta_j))$. The ZINB is then given by

$$\begin{aligned} P(Y_j = 0) &= \pi(x_j) + (1-\pi(x_j))p_j^r, \\ P(Y_j = y_j) &= (1-\pi(x_j)) \begin{pmatrix} y_j + r - 1 \\ y_j \end{pmatrix} p_j^r (1-p_j)^{y_j}, \quad y_j = 1,2,\ldots. \end{aligned} \tag{4.18}$$

When the scale parameter $(1/r) \to 0$ (or $r \to \infty$), the ZINB distribution (4.18) reduces to the ZIP distribution (4.10), as

$$\lim_{r\to\infty} f_2(y_j;\mu_j) = \frac{\exp(-\mu_j)\mu_j^{y_j}}{y_j!}.$$

With the ZINB regression model (4.18),

$$\begin{aligned} E(Y_j) &= (1-\pi(x_j))\mu_j \\ \text{var}(Y_j) &= (1-\pi(x_j))[\mu_j + \pi(x_j)\mu_j^2 + \mu_j^2/r], \end{aligned} \tag{4.19}$$

indicating that the ZINB regression model has larger variance than the ZIP regression model (4.10) unless $r \to \infty$, which reduces (4.19) to (4.11). Let the vector of unknown parameters be $\Psi = (v^T, \beta^T, r)^T$, the complete-data log likelihood for the ZINB regression model (4.18) is given by

$$\begin{aligned} \log L_c(\Psi) = &\sum_{y_j=0} \{z_j \log \pi(x_j;v) + (1-z_j)[\log(1-\pi(x_j;v)) - r\log(p_j)]\} + \\ &\sum_{y_j>0} \{\log(1-\pi(x_j;v)) + \log\frac{\Gamma(y_j+r)}{\Gamma(y_j+1)\Gamma(r)} + r\log(p_j) + y_j\log(1-p_j)\}, \end{aligned} \tag{4.20}$$

where z_j is one or zero accordingly as y_j is or is not an excess zero and $\Gamma(\cdot)$ denotes a gamma function. Similar to the case for the ZIP regression model, the Q-function can be partitioned into two parts corresponding to the parameters in v and (β,r), respectively, as:

$$\begin{aligned} Q_v &= \sum_{j=1}^{n} [\tau_j \xi_j - \log(1+\exp\xi_j)] \\ Q_{\beta,r} &= \sum_{j=1}^{n} (1-\tau_j)\{\log\frac{\Gamma(y_j+r)}{\Gamma(y_j+1)\Gamma(r)} + r\log(p_j) + y_j\log(1-p_j)\}, \end{aligned} \tag{4.21}$$

where

$$\tau_j^{(k)} = \begin{cases} \frac{1}{1+\exp\{-x_{ij}^T v^{(k)}\} p_j^r} & \text{if } y_j = 0 \\ 0 & \text{if } y_j \geq 1 \end{cases} \tag{4.22}$$

is the current estimated posterior probability of being an excess zero. The M-step updates the estimates for v, β, and r, respectively, by solving the following system of non-linear equations:

$$\sum_{j=1}^{n} \{\tau_j^{(k)}(1-\pi(x_j)) - (1-\tau_j^{(k)})\pi(x_j)\}x_j = 0$$

$$\sum_{j=1}^{n} (1-\tau_j^{(k)}) \frac{\{y_j - \exp(\eta_j)\}x_j}{(1+\exp(\eta_j)/r)} = 0$$

$$\sum_{j=1}^{n} (1-\tau_j^{(k)})\{\varphi(y_j+r) - \varphi(r) - \log(p_j)\} = 0, \tag{4.23}$$

where the digamma function $\varphi(r) = \Gamma'(r)/\Gamma(r)$ is the logarithmic derivative of the gamma function $\Gamma(r)$. The MINPACK routine HYBRD1 of Moré, Garbow, and Hillstrom (1980) can be adopted to solve (4.23). A non-parametric bootstrap resampling approach can be used to obtain the standard errors of the ML estimates of the parameters Ψ in ZINB regression models.

4.4.1 Example 4.1: Pancreas Disorder Length of Stay Data

We consider the pancreas disorder length of stay (LOS) data of Yau, Wang, and Lee (2003) for a group of 261 patients hospitalized between 1998 and 1999 in Western Australia. The disease encompassed acute pancreatitis, chronic pancreatitis, and other minor classifications, but malignant tumour was excluded. The data set comprised many same-day separations (zero LOS; 17%). Possible explanations for the excess of zero LOS might be due to advances in pharmaceuticals, medical technologies and clinical practice that enable more effective health services for improving patients' outcomes and leading to a decline in LOS. Yau, Wang, and Lee (2003) considered various models including a single Poisson distribution, a single negative binomial distribution, and a ZIP model. However, none of them gave an adequate fit to the empirical LOS distribution of Y_j. They then fitted a ZINB model and found a better fit, describing the LOS variations reasonably well; see Figure 4.1 for the fitted frequency distribution plot. Table 4.1 presents the estimates of key parameters and summary statistics for the above four different fitted models. It can be seen that single Poisson or NB distribution fails to account for extra zeros, whereas inadequate fit of a ZIP model indicates there is also simultaneous over-dispersion for the non-zero counts compared to a Poisson distribution. As indicated by Yau, Wang, and Lee (2003), a two-component Poisson mixture model gave a χ^2 goodness-of-fit statistic of 28.2, with a p-value <0.01. A three-component Poisson mixture model gave a larger log-likelihood value of -576.597 than the ZINB model. But model selection based on the BIC still picks the ZINB model, which has BIC=1176.8 compared to BIC=1181.0 for the three-component Poisson mixture model.

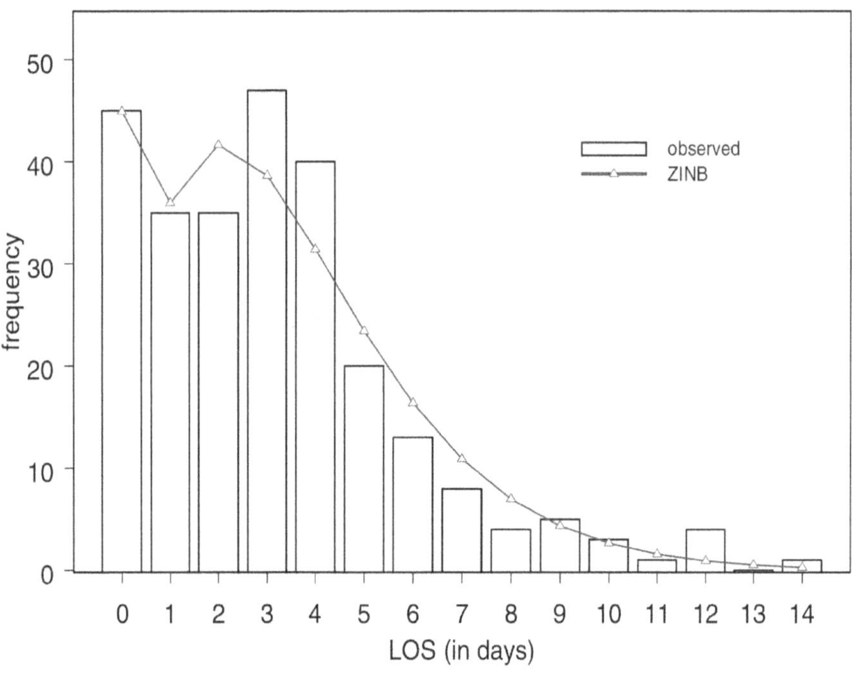

FIGURE 4.1
Empirical and fitted ZINB distributions of the pancreas disorder data (Example 4.1), Adapted from Yau, Wang, and Lee (2003).

To assess the effect of covariates on LOS, the following risk factors of clinical- and patient-related characteristics were considered: (1) Age (in years), (2) gender (male versus female), (3) marital status (single/others versus married), (4) of Aboriginal or Torres Strait Islander origin (yes versus no), (5) payment type (private/others versus public), (6) admission status (emergency versus elective), (7) treatment classification (general surgery versus GP/general medicine/gastroenterology), and (8) the number of diagnoses; see Yau, Wang, and Lee (2003) for more information. Among this cohort of 261 patients with pancreas disorder, 35% are female, 48% are married, and 32% are of Aboriginal or Torres Strait Islander origin. The average age is 36 years. Table 4.2 presents the results using the ZINB regression model. The estimate (standard error) of the scale parameter $(1/r)$ was 0.129 (0.038) and significantly greater than 0, indicating substantial over-dispersion in the non-zero counts and that a ZINB regression model is required.

From Table 4.2, it can be seen from the logistic part that the risk of having overnight or longer (non-zero) LOS was significantly higher for younger patients (adjusted OR = 1.237 for each year younger; 95% CI = 1.02 to 1.51) and patients with more diagnoses (adjusted OR = 16.38 for each additional diagnosis; 95%

TABLE 4.1
Estimates and summary statistics for four fitted models (Example 4.1). Adapted from Yau, Wang, and Lee (2003).

Parameter	single Poisson	single NB	ZIP	ZINB
mean (μ)	3.142 (0.110)	3.142 (0.170)	3.703 (0.191)	3.483 (0.246)
scale ($1/r$)	—	0.448 (0.075)	—	0.254 (0.019)
mixing proportion (π)	—	—	0.151 (0.011)	0.098 (0.027)
Log-likelihood	-640.737	-583.652	-600.399	-580.033
Deviance (d.f.)	626.5 (260)	506.9 (259)	545.9 (259)	499.7 (258)
χ^2 goodness-of-fit (d.f.)	136.0 (6)	16.7 (7)	31.4 (6)	9.5 (6)
p-value	<0.001	0.02	<0.001	0.148

TABLE 4.2
Estimates and standard errors (in parentheses) for ZINB regression model (Example 4.1). Adapted from Yau, Wang, and Lee (2003).

Parameter	Logistic part	NB part
Intercept	-2.285 (3.132)	0.483 (0.279)
Age (in years)	0.213* (0.100)	-0.004 (0.004)
Male	0.829 (1.685)	-0.102 (0.091)
Single/others	2.318 (2.061)	0.224* (0.092)
Aboriginality	-1.641 (2.157)	-0.138 (0.101)
Private/others	-1.122 (1.695)	0.253 (0.149)
Emergency	-15.958 (23.851)	0.583* (0.226)
General surgery	-3.358 (3.900)	0.226* (0.095)
No. of diagnoses	-2.796* (1.412)	0.082* (0.029)
Log-likelihood	-493.549	
Deviance	326.741	
Pearson statistic	263.132	
Degree of freedom	242	

* p-value < 0.05.

CI = 1.03 to 260.7). From the negative binomial part, it was observed that non-married patients (adjusted rate ratio of 1.251; 95% CI = 1.04 to 1.50) or those with more diagnoses (adjusted rate ratio = 1.085 for each additional diagnosis; 95% CI = 1.03 to 1.15) were positively associated with prolonged length of hospitalization. As expected, emergency admissions (adjusted rate ratio = 1.791; 95% CI = 1.15 to 2.79) or those requiring surgical treatments (adjusted rate ratio = 1.254; 95% CI = 1.04 to 1.51) also contributed to a longer LOS. As discussed in Yau, Wang, and Lee (2003), some regression coefficients and associated standard errors in the logistic part were large in magnitude because this cohort of patients were recruited from 36 hospitals and there was significant inter-hospital variation in the proportion of same-day separations. Also, the proportion of having an overnight or longer LOS for emergency-admitted patients was very high (95.4%), making it difficult to estimate the coefficient of this factor in the logistic part in relation to the odds of same-day stay. In general, the overall fit using the ZINB regression model to this pancreas disorder data set can be improved by accounting for the multilevel structure of the data (i.e., patients nested within hospitals) by using a ZINB mixed regression model to adjust for random hospital effects; see Chapter 7 and Yau, Wang, and Lee (2003) for more details.

4.5 Score Tests for Zero-Inflation in Count Models

As described in Section 4.1, it is important to assess different count models in order to identify the best model for the analysis of count data. In general, model selection methods such as BIC may be used to compare various count models in a form of penalized likelihood. Specific to applications where count data have a large proportion of zeros, score tests for zero-inflation (comparing a ZIP regression model versus a Poisson regression model) and for assessing a ZINB regression model versus a ZIP regression model have been proposed by Van den Broek (1995) and Ridout, Hinde, and Demétrio (2001), respectively. The score test of Van den Broek (1995) considered specific ZIP models that assume a constant proportion of excess zeros; i.e., $\pi(x_j) = \pi$ in (4.9). The test assesses whether an extra component of point mass at zero (the alternative hypothesis) is appropriate, where a fit of the Poisson regression model under only the null hypothesis is required. In particular, under the null hypothesis, the score test statistic is given by

$$S(\hat{\beta}) = \frac{[\Sigma_{j=1}^{n}(I_{(y_j=0)}/\exp(-\hat{\mu}_j) - 1)]^2}{\Sigma_{j=1}^{n}[1/\exp(-\hat{\mu}_j) - 1] - \hat{\lambda}^T X(X^T \Lambda X)^{-1} X^T \hat{\lambda}}, \tag{4.24}$$

where $I_{(y_j=0)}$ is an indicator function, $\lambda = (\mu_1, \mu_2, \ldots, \mu_n)^T$, $X = (x_1^T, x_2^T, \ldots, x_n^T)^T$, and $\Lambda = diag(\hat{\mu}_j)$; see Lee, Xiang, and Fung (2004) and the references therein for other tests for zero-inflation including the likelihood ratio test and the Wald test. For a ZIP regression model with accommodated individual exposure risk t_j $(j = 1, \ldots, n)$, the score test statistic (4.24) can be adapted to test for zero-inflation by replacing $\hat{\mu}_j$ by $\tilde{\mu}_j = t_j \exp(x_j \tilde{\beta})$, where $\tilde{\beta}$ is the estimate of β under the (null) Poisson regression model with offset $\log(t_j)$; see Section 4.3 and Lee, Wang, and Yau (2001) for more information.

For testing a ZIP against a ZINB regression model with the score test of Ridout, Hinde, and Demétrio (2001), the null ZIP model is a general model with both $\pi(x_j)$ and μ_j depending on x_j; see (4.10). To test the null hypothesis $H_0 : (1/r) = 0$ (i.e., a ZIP) against $H_1 : (1/r) > 0$, the score function proposed by Ridout, Hinde, and Demétrio (2001) is given by

$$U(\tilde{\beta}, \tilde{v}) = \tfrac{1}{2} \Sigma_{j=1}^{n} \{[(y_j - \tilde{\mu}_j)^2 - y_j] - I_{(y_j=0)} \tilde{\mu}_j^2 \tilde{\pi}(x_j)/\tilde{p}_{0,j}\}, \tag{4.25}$$

where parameters with tildes denote ML estimates under H_0, and $\tilde{p}_{0,j} = P(Y_j = 0)$ is from the ZIP model. The score statistic for testing H_0 takes the form

$$S(\tilde{\beta}, \tilde{v}) = U(\tilde{\beta}, \tilde{v}) \sqrt{\tilde{J}_{rr}}, \tag{4.26}$$

where the scalar $\sqrt{\tilde{J}_{rr}}$ is the upper left-hand element of the inverse information matrix evaluated at the ML estimates of the ZIP model. As described by Lee, Xiang, and Fung (2004), the score function $U(\tilde{\beta}, \tilde{v})$ has a limiting normal distribution with mean 0 and variance $(\tilde{J}_{rr})^{-1}$. It follows that the score test statistic $S(\tilde{\beta}, \tilde{v})$ has an asymptotic standard normal distribution under H_0.

However, the score test statistics in (4.24) and (4.26) may be influenced by outliers and extreme observations. For example in (4.24), $S(\hat{\beta})$ depends on both the number of observed zeros and the estimated $\hat{\mu}_j$ of the Poisson model, and is thus influenced by extreme counts. To this end, Lee, Xiang, and Fung (2004) illustrated the importance of sensitivity analysis of zero-inflated tests and developed sensitivity measures based on the local influence approach of Cook (1986).

Consider a score statistic S from observations $Y = (y_1, y_2, \ldots, y_n)^T$ and suppose a perturbation $\omega = (\omega_1, \omega_2, \ldots, \omega_n)^T$ is introduced into the data. Following Cook (1986), ω can be written as $\omega = \omega_0 + \alpha l$, where ω_0 represents the null point of no perturbation, α is the magnitude of perturbation, and $l = (l_1, l_2, \ldots, l_n)^T$ is a unit direction vector. Let $S(\omega)$ denote the score statistic under the model obtained based on perturbed data, the $(n+1) \times 1$ dimensional surface of interest is given by $(\omega^T, S(\omega))^T$. The slope on the path of the surface at $\alpha = 0$ is the derivative of $S(\omega)$ with respect to α evaluated at $\alpha = 0$, as

$$\left. \frac{\partial S(\omega)}{\partial \alpha} \right|_{\alpha=0} = \sum_{j=1}^{n} \left(\left. \frac{\partial S(\omega)}{\partial \omega_j} \right|_{\omega=\omega_0} \right) l_j, \tag{4.27}$$

which can be used to investigate the local behaviours of observations. The direction l_{max} that maximizes the slope is the main diagnostic quantifier, where $l_{max} = \dot{S}/\sqrt{\dot{S}^T \dot{S}}$, with $\dot{S} = \partial S(\omega)/\partial \omega_j|_{\omega=\omega_0}$; see Lee, Xiang, and Fung (2004) and Xiang and Lee (2005). Observations corresponding to large elements of l_{max} in their absolute values are potentially influential on the score statistic S and require attention.

For assessing the score test for ZINB versus ZIP, the method of Lee, Xiang, and Fung (2004) examines the sensitivity of the score function $U(\tilde{\beta},\tilde{v})$ given in (4.25) instead of the score statistic $S(\tilde{\beta},\tilde{v})$. The diagnostic quantity of interest is thus $l_{max} = \dot{U}(\tilde{\beta},\tilde{v})/||\dot{U}(\tilde{\beta},\tilde{v})||$, where $\dot{U}(\tilde{\beta},\tilde{v}) = \partial U(\tilde{\beta},\tilde{v}|\omega)/\partial\omega|_{\omega=\omega_0} = (\dot{U}_1,\ldots,\dot{U}_n)^T$. The form for $\dot{U}_j$ $(j=1,\ldots,n)$ varies under different perturbations ω. For perturbation of case weights, ω represents a vector of weights to modify the contribution of each case to the log-likelihood function of the ZINB regression model. By taking the first-order derivative of the log-likelihood function with respect to α, it follows that

$$\dot{U}_j = [(y_j-\tilde{\mu}_j)^2 - y_j] - I_{(y_j=0)}\tilde{\mu}_j^2\tilde{\pi}(x_j)/\tilde{p}_{0,j} \tag{4.28}$$

for $j=1,\ldots,n$. An index plot of l_{max} can be obtained to identify cases that are potentially influential on the score statistic. For additive perturbation of individual covariates, which can take place in either the Poisson part (the null model) or logistic part or both, the null perturbation ω_0 is given by $\omega_0 = (0,\ldots,0)^T$. Suppose a continuous covariate X_t of X is additively perturbed in the Poisson part as $X_{t\omega} = X_t + \delta_t\omega$, with $\delta_t = ||X_t||$; as well as in the logistic part as $X_{t\omega} = X_t + \rho_t\omega$, with $\rho_t = ||X_t||$. Let $\underline{x}_i$ denote the ith row of X, the corresponding perturbed equations become $\log(\mu_{i\omega}) = \underline{x}_i^T\beta + \delta_t\beta_t\omega_i$ and $\log(\pi_{i\omega}/(1-\pi_{i\omega})) = \underline{x}_i^T v + \rho_t v_t\omega_i$. Under such additive perturbations, it can be shown that

$$\begin{aligned}\dot{U}_j = \tilde{\mu}_j\delta_t\tilde{\beta}_t\left[-2(y_j-\tilde{\mu}_j) - I_{(y_j=0)}\frac{\tilde{\mu}_j\tilde{\pi}(x_j)}{\tilde{p}_{0,j}}\left(2+\frac{1-\tilde{\pi}(x_j)}{\tilde{p}_{0,j}}\tilde{\mu}_j\exp(-\tilde{\mu}_j)\right)\right] - \\ I_{(y_j=0)}\frac{\tilde{\mu}_j^2\tilde{\pi}(x_j)(1-\tilde{\pi}(x_j))}{\tilde{p}_{0,j}}\rho_t\tilde{v}_t\left[1-\frac{\tilde{\pi}(x_j)(1-\exp(-\tilde{\mu}_j)}{\tilde{p}_{0,j}}\right]\end{aligned} \tag{4.29}$$

for $j=1,\ldots,n$; see Lee, Xiang, and Fung (2004) for more details and for multiplicative perturbation of individual covariates.

4.5.1 Example 4.2: Revisit of the Pancreas Disorder LOS Data

A feature of the LOS data considered in Example 4.1 is the presence of extra zeros and the simultaneous over-dispersion (Lee, Xiang, and Fung, 2004). As shown in Table 4.1, neither single Poisson, single negative binomial, nor the ZIP models can provide an adequate fit to the LOS data. The score test statistic (4.26) proposed by Ridout, Hinde, and Demétrio (2001) (without covariates) equals 8.502 (p-value < 0.01) and rejects the null ZIP model. To assess the sensitivity of the score statistic $S(\tilde{\beta},\tilde{v})$ with covariates, Lee, Xiang, and Fung (2004) considered a ZINB regression model in Example 4.1, with an additional clinical risk factor concerning the number of surgical/treatment procedures and exclusion of the risk factors regarding payment type and treatment classification. The score statistic $S(\tilde{\beta},\tilde{v}) = 3.82$ was significant (p <0.001) with respect to the asymptotic standard normal distribution, providing strong evidence against the ZIP regression model with the above covariates. Figure 4.2 displays the direction of maximum slope under case weight perturbation. It is shown that patients 52, 83, and 202 may be influential on the score test. As discussed in Lee, Xiang, and Fung (2004), simultaneous removal of these three outliers reduces

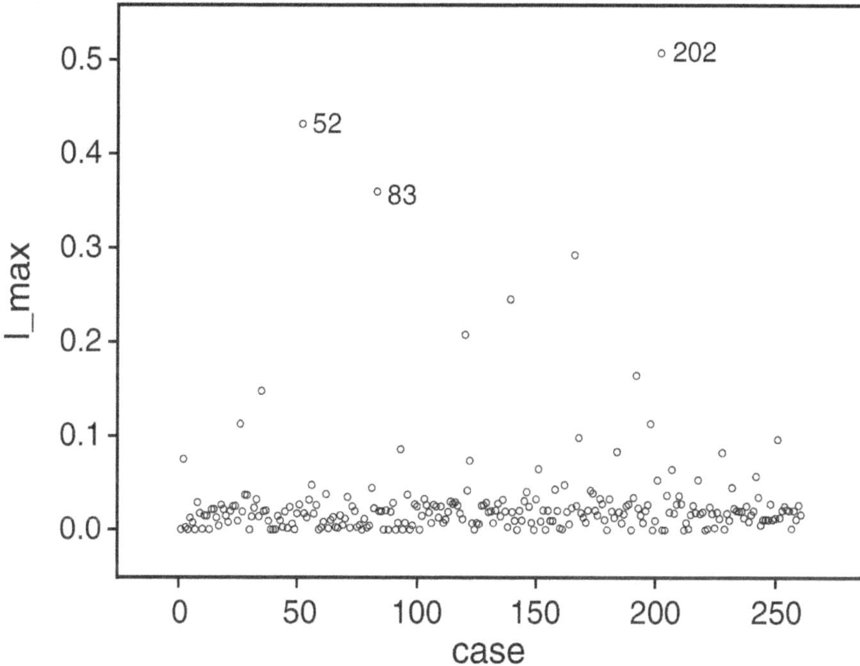

FIGURE 4.2
Direction of maximum slope under case weight perturbations for pancreas disorder LOS data (Example 4.2). Adapted from Lee, Xiang, and Fung (2004).

the score statistic $S(\tilde{\beta}, \tilde{v})$ to 1.39, which is not significant (p = 0.165). Indeed, it is shown that a ZIP regression model provides a reasonable fit to this reduced data set, where the Pearson statistic is 264.873, with degrees of freedom of 245 (p = 0.183). An inspection of the data reveals that Patient 202 corresponds to a 20-year-old female with the longest stay of 2 weeks, whereas patients 52 and 83 are both middle-aged males who incurred the second longest hospitalization of 12 days. Lee, Xiang, and Fung (2004) considered another sensitivity analysis for the covariate regarding the number of surgical/treatment procedures. The direction of maximum slope under additive perturbations of this covariate was plotted in Figure 4.3, which shows that Patient 202 and, to a lesser extent, Patient 129, are potentially influential on the score test statistic regarding the number of surgical/treatment procedures. Patient 129 had the largest number of five surgical/treatment procedures performed but incurred only a moderate stay of 8 days; see Lee, Xiang, and Fung (2004) for more information.

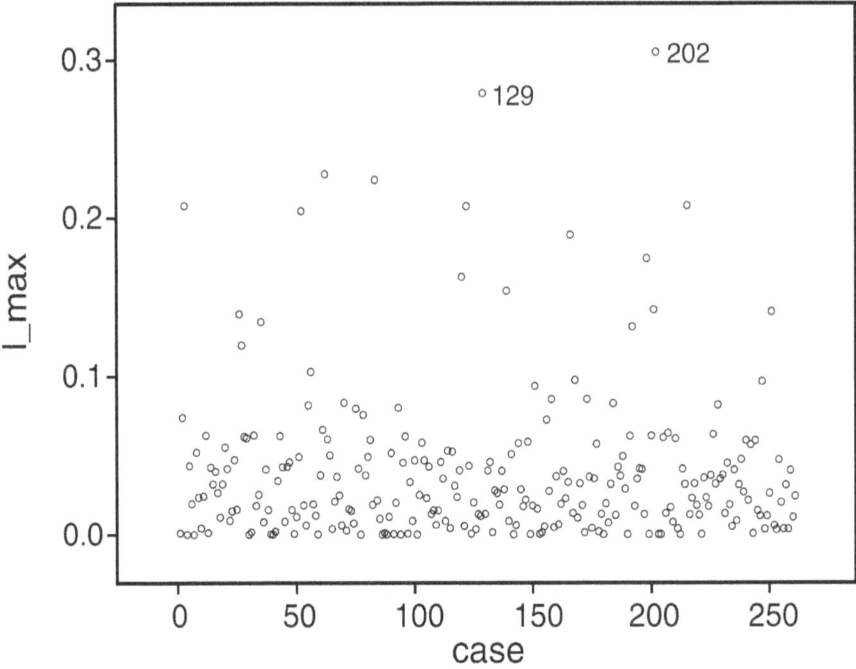

FIGURE 4.3
Direction of maximum slope under additive perturbations of number of procedures for pancreas disorder LOS data (Example 4.2). Adapted from Lee, Xiang, and Fung (2004).

4.6 Mixture of Generalized Bernoulli Distributions

Mixture of generalized Bernoulli distributions can be used to cluster multivariate categorical data. Let $y_1, \ldots, y_n$ be the observed p-dimensional categorical variable vectors from n subjects, with $y_j = (y_{1j}, \ldots, y_{pj})^T$ $(j = 1, \ldots, n)$. A multivariate generalized Bernoulli distribution consists of one draw on d_i labels with probabilities $\theta_{hi1}, \ldots, \theta_{hid_i}$ for each categorical variable $i = 1, \ldots, p$, and where $\theta_{hid_i} = 1 - \sum_{l=1}^{d_i-1} \theta_{hil}$. Assuming that the categorical variables $y_{1j}, \ldots, y_{pj}$ are independent of each other, the mixture of multivariate generalized Bernoulli distributions is given by

$$f(y_j; \Psi) = \sum_{h=1}^{g} \pi_h \prod_{i=1}^{p} \prod_{l=1}^{d_i} \theta_{hil}^{I(y_{ij}, l)} \qquad (j = 1, \ldots, n), \tag{4.30}$$

where $\pi_1, \ldots, \pi_g$ are the mixing proportions that sum to one and $I(y_{ij}, l)$ is an indicator function which is equal to one if $y_{ij} = l$ and is zero otherwise $(l = 1, \ldots, d_i)$.

Although the independence assumption of categorical variables in (4.30) may not be true, it often performs well in practice for handling multivariate categorical variables. This is because it usually requires fewer parameters to be estimated than more complicated alternative methods that try to model interactions between the categorical variables; see, for example, Ng (2015). With (4.30), the vector of unknown parameters Ψ contains the mixing proportions $\pi_1,\ldots,\pi_{g-1}$, and parameters $\theta_1,\ldots,\theta_g$ for component densities.

Mixture models with different types of Bernoulli components have been considered in various scientific fields. For example, univariate multinomial mixture models have been widely used in the context of text clustering; see Li and Zhang (2008) and Rigouste, Cappé, and Yvon (2007). Topchy, Jain, and Punch (2005) have considered a multivariate multinomial mixture model to obtain a consensus clustering of ensembles and Ambroise and Matias (2012) have adopted a mixture model of three-variate Bernoulli distributions to analyze random-graph networks. While it is well known that mixtures of univariate Bernoulli distributions are not identifiable (Ambroise and Matias, 2012), mixtures of multivariate generalized Bernoulli distributions are identifiable (Allman, Matias, and Rhodes, 2009) under two conditions: (a) the number of variates p is sufficiently large such that

$$p \geq 2\log_d g + 1, \tag{4.31}$$

where d is the minimum value of d_i $(i = 1,\ldots,p)$ and (b) the dimension of the parameter space for Ψ is smaller than that of the observed sample space. That is,

$$g\sum_{i=1}^{p}(d_i - 1) + g - 1 \leq S - 1, \tag{4.32}$$

where S is the number of distinct observed patterns of y_j, which has an upper bound of $\prod_{i=1}^{p} d_i$ corresponding to the dimension of the distribution space, as discussed in Ng (2015).

4.6.1 E- and M-steps

The unknown parameter vector Ψ can be estimated by the ML method via the EM algorithm, where the E- and M-steps are in closed form. On the $(k+1)$th iteration of the EM algorithm, the E-step calculates

$$\begin{aligned} E_{\Psi^{(k)}}(Z_{hj}|y) &= \tau_h(y_j;\Psi^{(k)}) \\ &= \frac{\pi_h^{(k)} f_h(y_j;\theta_h^{(k)})}{\sum_{r=1}^{g}\pi_r^{(k)} f_r(y_j;\theta_r^{(k)})} \\ &= \frac{\pi_h^{(k)}\prod_{i=1}^{p}\prod_{l=1}^{d_i}\theta_{hil}^{(k)\,I(y_{ij},l)}}{\sum_{r=1}^{g}\pi_r^{(k)}\prod_{i=1}^{p}\prod_{l=1}^{d_i}\theta_{ril}^{(k)\,I(y_{ij},l)}} \end{aligned} \tag{4.33}$$

for $h = 1,\ldots,g$; $j = 1,\ldots,n$, which are the current posterior probabilities of component membership. In the M-step, the estimates are updated to $\Psi^{(k+1)}$ as

$$\begin{aligned}
\pi_h^{(k+1)} &= \frac{\sum_{j=1}^{n} \tau_h(y_j;\Psi^{(k)})}{n}, \\
\theta_{hil}^{(k+1)} &= \frac{\sum_{j=1}^{n} I(y_{ij},l)\tau_h(y_j;\Psi^{(k)})}{\sum_{m=1}^{d_i}\sum_{j=1}^{n} I(y_{ij},m)\tau_h(y_j;\Psi^{(k)})} \qquad (i=1,\ldots,p;\ l=1,\ldots,d_i-1), \\
\theta_{hid_i}^{(k+1)} &= 1-\sum_{l=1}^{d_i-1}\theta_{hil}^{(k+1)} \qquad (4.34)
\end{aligned}$$

for $h = 1,\ldots,g$. The E- and M-steps are alternated repeatedly until convergence, which is indicated by the difference in the log likelihood values being less than 1×10^{-8} of the absolute value of the previous log likelihood. The EM algorithm is implemented from a variety of initial values for Ψ using the following scheme. As proposed by Ng (2015), initial values of $\pi_1,\ldots,\pi_{g-1}$ and $\theta_{hi1},\ldots,\theta_{hi(d_i-1)}$ for $h = 1,\ldots,g$ and $i = 1,\ldots,p$ can be generated randomly and independently from a uniform distribution $U(0,1)$, while keeping the constraint that $\sum_{h=1}^{g}\pi_h = 1$ and $\sum_{l=1}^{d_i}\theta_{hil} = 1$ for $i = 1,\ldots,p$ with all π_h and θ_{hil} being positive. The number of components in the mixture regression model is assessed using the BIC. The standard errors of the estimates of Ψ are obtained by the bootstrap resampling method with replacement, where the number of bootstrap replications is taken to be 100; see Ng (2015) for more information.

4.6.2 Cluster Analysis in Comorbidity Research

An example of applying mixture models as a clustering tool for categorical data can be seen in the analysis of morbidity data with the aims of quantifying heterogeneous comorbidity patterns of health conditions among individuals and identifying characteristics of individuals who are at risk of poorer health outcomes. The term "comorbidity" was introduced by Feinstein (1970) as the occurrence of other medical conditions in addition to an index condition of interest, but it has been generalized to represent co-occurrence of any two or more conditions (also referred to as multimorbidity) in the literature; see, for example, Ng, Holden, and Sun (2012) and van den Akker, Buntinx, and Knottnerus (1996).

Comorbidity is increasingly recognized as an important issue in medical care and health sciences, because of its association with poorer health-related outcomes, more complex clinical management, increases in health service utilization and associated costs, but a decrease in productivity (Caughey and Roughead, 2011; Holden et al., 2011; Valderas et al., 2009; Barnett et al., 2012). Comorbidity research has the potential to play an important role advancing evidence-based knowledge to address increasingly complex needs related to comorbidity, which is valuable for developing interventions and management strategies that are most effective and viable to better help the large number of individuals with multiple conditions (Jowsey et al., 2009; Smith et al., 2012).

As outlined in Ng (2015), there are two separate clustering tasks involved in comorbidity analysis. The first task corresponds to the identification of groups of comorbid conditions, with adjustment for co-occurrence of conditions by chance. Revealing comorbid condition groups beyond chance not only generates new hypotheses on potential shared biologic process but is also more relevant, having more substantial impact on treatment outcomes compared to conditions with minimal interaction (Valderas et al., 2009; Batstra, Bos, and Neeleman, 2002; Richardson and Doster, 2014). The second task is to cluster individuals according to their comorbidity patterns; the mixture of generalized Bernoulli distributions is applicable in this task of cluster analysis when the comorbidity patterns are presented as categorical variables.

Recently, Ng et al. (2018) conducted a systematic review of analytical methods used to identify groups of comorbid conditions. They revealed five analytical methods that use different approaches to form groups of comorbid conditions and concluded that the measure adopted to quantify co-occurrence between a pair of conditions is crucial in explaining the variation in the identified comorbid condition groups between these methods. Commonly used measures Jaccard or Yule Q coefficients (Foguet-Boreu et al., 2015; Vu, Finch, and Day, 2011; Marengoni et al., 2009) as well as tetrachroic correlation coefficients (Garin et al., 2016; Jackson et al., 2015; Prados-Torres et al., 2012) for comorbidity research do not explicitly adjust for co-occurrence by chance; see Ng et al. (2018) for more details. Alternatively, the asymmetric Somers' D statistic proposed by Ng, Holden, and Sun (2012) adjusts for comorbidity by chance and controls the false discovery rate (Benjamini and Hochberg, 1995), thus showing a higher power to detect non-random comorbidities compared to other common concordance statistics such as kappa, Kendall's Tau-b, gamma, and adjusted Rand index. Suppose that morbidity data are represented by a $n \times m$ binary matrix of one or zero, indicating the presence or absence of the ith health condition for the jth individual ($i = 1, \ldots, m$; $j = 1, \ldots, n$), where m and n are the numbers of health conditions and individuals, respectively. With the method of Ng, Holden, and Sun (2012), Ng (2015) described how an individual's comorbidity pattern can be summarized in terms of p categorical variables from the original m binary variables. Precisely, let p be the number of non-overlapping groups of comorbid conditions formed, then we can define

$$y_{ij} = \begin{cases} 0 & \text{if absence of conditions in the } i\text{th group} \\ 1 & \text{if presence of one condition in the } i\text{th group} \\ 2 & \text{if presence of more than one condition in the } i\text{th group} \end{cases} \tag{4.35}$$

for $i = 1, \ldots, p$. That is, individual morbidity data vector is reduced from m to p dimensions ($p < m$) that presents individual patterns of comorbidity with respect to the absence or presence of 1 (or >1) condition(s) in the ith group of comorbid conditions ($i = 1, \ldots, p$).

With reference to the formulation of mixture of generalized Bernoulli distributions in Section 4.6, $y_j = (y_{1j}, \ldots, y_{pj})^T$ ($j = 1, \ldots, n$) are p-dimensional categorical variable vectors, taking on d_i distinct labels defined in (4.35). Here, d_i equates to either two or three, depending on whether the number of comorbid conditions in the ith group is smaller than three or not. The mixture of generalized Bernoulli distributions (4.30) can be used to cluster individuals into, say g, components according to their comorbidity patterns defined by y_j $(j = 1, \ldots, n)$. In comorbidity research, knowledge concerning the characterized features of individuals that explain heterogeneity in comorbidity patterns is of great significance for its potential to identify individuals at risk of poorer health-related outcomes. Suppose $(x_{1j}, \ldots, x_{qj})$ is the q-dimensional covariate vector associated with the jth individual. The effect of the covariates on the probability of membership of the g components of comorbidity patterns can be modelled by incorporating x_j in the mixing proportions π_h via a logistic function. That is, (4.30) becomes a mixture regression model of multivariate generalized Bernoulli distributions as

$$f(y_j;\Psi) = \sum_{h=1}^{g} \pi_h(x_j;\alpha) \prod_{i=1}^{p} \prod_{l=1}^{d_i} \theta_{hil}^{I(y_{ij},l)} \qquad (j = 1, \ldots, n), \tag{4.36}$$

where

$$\pi_h(x_j;\alpha) = \frac{\exp(v_h^T x_j)}{1 + \sum_{l=1}^{g-1} \exp(v_l^T x_j)} \qquad (h = 1, \ldots, g-1) \tag{4.37}$$

and $\pi_g(x_j) = 1 - \sum_{l=1}^{g-1} \pi_l(x_j)$, and where the first element of x_j is assumed to be one to account for an intercept term $(j = 1, \ldots, n)$. In (4.37), α contains the elements in v_h, which are unknown constants representing the log odds ratio of component membership for each covariate x_j (Ng, Tawiah, and McLachlan, 2019). With this mixture regression model, the E-step (4.33) becomes

$$E_{\Psi^{(k)}}(Z_{hj}|y) = \tau_h(y_j, x_j; \Psi^{(k)}) = \frac{\pi_h(x_j;\alpha^{(k)}) \prod_{i=1}^{p} \prod_{l=1}^{d_i} \theta_{hil}^{(k)\,I(y_{ij},l)}}{\sum_{r=1}^{g} \pi_r(x_j;\alpha^{(k)}) \prod_{i=1}^{p} \prod_{l=1}^{d_i} \theta_{ril}^{(k)\,I(y_{ij},l)}} \tag{4.38}$$

for $h = 1, \ldots, g$ and $j = 1, \ldots, n$. The M-step for estimating $\theta_{hil}^{(k+1)}$ are still in closed form as

$$\theta_{hil}^{(k+1)} = \frac{\sum_{j=1}^{n} I(y_{ij}, l) \tau_h(y_j, x_j; \Psi^{(k)})}{\sum_{m=1}^{d_i} \sum_{j=1}^{n} I(y_{ij}, m) \tau_h(y_j, x_j; \Psi^{(k)})} \qquad (i = 1, \ldots, p;\ l = 1, \ldots, d_i - 1),$$

$$\theta_{hid_i}^{(k+1)} = 1 - \sum_{l=1}^{d_i - 1} \theta_{hil}^{(k+1)}. \tag{4.39}$$

However, the estimate of $\alpha^{(k+1)}$ is updated by solving the following system of non-linear equations

$$\sum_{j=1}^{n} \left(\tau_h(y_j, x_j; \Psi^{(k)}) - \frac{\exp(v_h^T x_j)}{1 + \sum_{l=1}^{g-1} \exp(v_l^T x_j)} \right) x_j = 0 \qquad (h = 1, \ldots, g-1), \tag{4.40}$$

with the MINPACK routine HYBRD1 (Moré, Garbow, and Hillstrom, 1980).

4.6.3 Example 4.3: Australian National Health Survey Data

We consider the morbidity data from the 2007-2008 Australian National Health Survey (NHS), which was conducted by the Australian Bureau of Statistics (ABS) from July 2007 to June 2008; see Australian Bureau of Statistics (2009). Information was collected about the prevalence of current long-term conditions (which were defined as medical conditions that were current at the time of the survey and that had lasted or expected to last for at least six months) from 20,788 Australians, with age ranges from the minimum age group of 0-4 up to the maximum age group of 85 or over. The NHS data in Confidentialised Unit Record Files (CURFs) are available on the ABS website at `http://www.abs.gov.au/ausstats/abs@.nsf/mf/4324.0`. In this example, we considered twenty-five conditions for adults aged 20-59 years (n=10,988). Using the method of Ng, Holden, and Sun (2012), 23 overlapping groups of comorbid conditions were obtained. Table 4.3 presents the overlapping groups sorted in the order of the strength of comorbidity beyond chance, which is defined as the average of Somers' D statistics for all pairs of conditions belonging to the group. With the false discovery rate controlled at $\alpha = 0.005$ level, the expected number of false positive among the significant pairs of conditions is less than 1; see Ng, Holden, and Sun (2012) for more details.

With the conversion procedure described in Ng, Tawiah, and McLachlan (2019), the non-overlapping groups of comorbid conditions were formed and were displayed in Figure 4.4 using the UCINET6 for Windows (Borgatti, Everett, and Freeman, 2002). It can be seen from Figure 4.4 that we had five ($p = 5$) non-overlapping groups of conditions, which were G1: {diabetes, cholesterol, angina, oedema, hypertension, disc disorder, rheumatoid, gout}, G2: {anaemia, anxiety, depression, mood disorder}; G3: {migraine, hayfever, sinusitis, bronchitis, asthma, dermatitis, psoriasis, back pain, food allergy}; G4: {hernia, osteroarthritis, osteroporosis}; and a singular group G5 (thyroid). The numbers of distinct labels for these $p = 5$ groups of comorbid conditions were therefore $d_i = 3$ for $i = 1$ to 4 and $d_5 = 2$. With reference to (4.31), mixture models of generalized Bernoulli distributions are identifiable when the number of components g is four or smaller.

We first fitted to the data matrix with n=10,988 and p=5 using a mixture model of generalized Bernoulli distributions without covariates (4.30) for $g = 1$ to $g = 4$. Model selection using the BIC indicated that there were three clusters. We then fitted the mixture regression model of multivariate generalised Bernoulli distributions (4.36) to the above data matrix with additional $q = 17$ covariates (including an intercept term in x_j). These demographic/lifestyle and health-related risk factors, represented by x_j in the mixing proportions (4.37), were as follows: gender, age, country of birth (English speaking countries or Other versus Australia as a reference), education (Does not have a non-school qualification versus Has a non-school qualification), employment status (Unemployed or Not in labour force versus Employed), marital status (Not married versus Married including De facto), smoking habit (Current or Past daily smoker versus Never smoked daily), exercise level in last 2 weeks (increasing from 0 to 4 for No exercise to High), consumption of fruit/vegetable (Did not meet recommended guidelines versus Recommended guidelines met),

TABLE 4.3
Overlapping groups of comorbid conditions for the NHS data (Example 4.3).

Group of comorbid conditions	Strength
Diabetes, cholesterol, angina, oedema, hypertension	0.262
Migraine, hayfever, sinusitis, bronchitis	0.226
Anxiety, depression, mood disorder	0.223
Hayfever, sinusitis, bronchitis, asthma, food allergy	0.220
Mood disorder, bronchitis, asthma	0.211
Diabetes, angina, hypertension, osteoarthritis	0.198
Angina, hypertension, gout	0.178
Hypertension, osteoarthritis, osteoporosis	0.173
Disc disorder, osteoarthritis, osteoporosis	0.167
Depression, migraine, sinusitis, bronchitis	0.163
Mood disorder, angina, oedema, asthma	0.156
Depression, mood disorder, migraine, bronchitis	0.153
Migraine, hayfever, bronchitis, back pain	0.151
Angina, disc disorder, osteoarthritis	0.142
Depression, mood disorder, angina	0.142
Oedema, sinusitis, asthma	0.136
Oedema, asthma, rheumatoid	0.134
Sinusitis, asthma, osteoporosis	0.128
Depression, migraine, bronchitis, back pain	0.128
Anaemia, migraine, sinusitis	0.126
Mood disorder, asthma, osteoporosis	0.124
Anaemia, mood disorder, migraine	0.120
Bronchitis, disc disorder, osteoarthritis	0.116

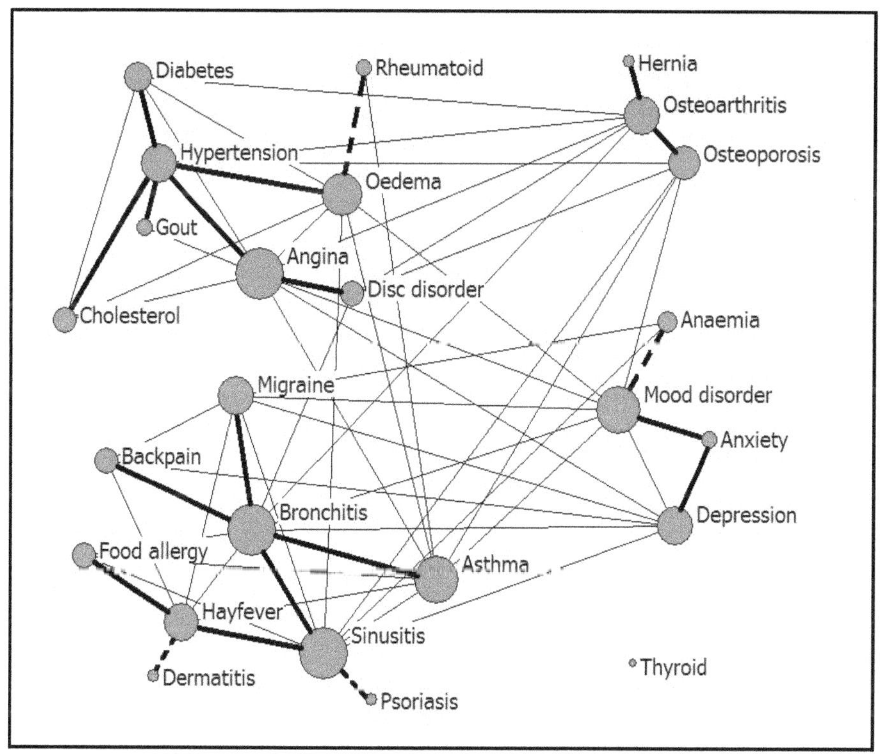

FIGURE 4.4
Significant non-random comorbidity between 25 medical conditions for the NHS data (Example 4.3) – Nodal size is proportional to the number of conditions that are significantly comorbid with the condition; bolded lines link the "closest" pairs of conditions (given a condition, its closest condition is the condition with which the pairwise Somers' D statistic is maximum and significant); some conditions link only to one condition of a group (these links are presented by bolded dashed lines).

private health insurance (Not covered versus Covered), main language spoken at home (Other language versus English only), residential area (Inner regional or Other areas versus Major cities) based on Australian Standard Geographical Classification (ASGC) remoteness area category, and housing tenure type (Renter or Other versus Owner).

Figure 4.5 displays the three clusters regarding the different patterns of probabilities of $y_{ij} = 1$ (the presence of one condition) and $y_{ij} = 2$ (the presence of more than one condition) in the $p = 5$ comorbidity groups, G1 to G5, described above. It can be seen that the majority of the participants (65.0%) belonged to Cluster 1, consisting of individuals with relatively low levels of comorbidity in all groups of comorbid conditions. Indeed, all three clusters showed a certain level of comorbidity in condition groups G3 (migraine, hayfever, sinusitis, bronchitis, asthma, dermatitis, psoriasis, back pain, food allergy). There were 25.3% of individuals belonging in Cluster 2, who had a relatively high level of comorbidity in condition groups G1 (diabetes, cholesterol, angina, oedema, hypertension, disc disorder, rheumatoid, gout). Finally, Cluster 3 (9.7%) consisted of individuals who had relatively high levels of comorbidity in G2 (anaemia, anxiety, depression, mood disorder) and G3.

Table 4.4 presents the adjusted odds ratios (ORs) of belonging to Cluster 2 (high level of comorbidity in G1) and Cluster 3 (high level of comorbidity in G2 and G3) relative to those in Cluster 1 for the thirteen risk factors. The standard errors of the ML estimates $\hat{\Psi}$ were obtained by 100 bootstrap replications. It can be seen from Table 4.4 that older individuals, being unemployed or not in labour force, or being a past or current daily smoker had increased chances of a high level of comorbidity with reference to Clusters 2 and 3. On the other hand, individuals who were born overseas, spoke other languages at home, or had a higher level of exercise were more likely to have a low level of comorbidity. The two clusters of individuals with high levels of comorbidity showed somewhat different risk factors between Clusters 2 and 3. High levels of comorbidity in cardiovascular and metabolic diseases (G1) were characterized by Cluster 2, appearing more frequently among men than women (adjusted OR: 2.152, 95% CI: 1.67, 2.77) who are living in inner regional areas (adjusted OR: 1.411, 95% CI: 1.08, 1.85) or other remote areas (adjusted OR: 1.436, 95% CI: 1.03, 2.00). Alternatively, the high level of comorbidity in mental health disorders and hayfever/allergy problems (characterized by Cluster 3) was more prominent among women (adjusted OR: 2.243, 95% CI: 1.72, 2.93), with higher education (having non-school qualification; adjusted OR: 1.429, 95% CI: 1.14, 1.79), not married (adjusted OR: 1.952, 95% CI: 1.51, 2.53), met recommended guidelines for consumption of fruit/vegetable (adjusted OR: 1.983, 95% CI: 1.08, 3.63), without private health insurance (adjusted OR: 1.327, 95% CI: 1.03, 1.71) nor housing ownership (adjusted OR: 1.805, 95% CI: 1.44, 2.27).

To study the impact of comorbidity on general wellbeing and health service use, we present in Table 4.5 health-related outcomes and key indicators of individuals across the three clusters. It can be observed from Table 4.5 that both Clusters 2 and 3 have a significantly higher proportion of: (1) a need for GP consultations in the last year; (2) self-assessed status of fair or poor health; (3) high or very high levels of psychological distress based on Kessler 10 classification (especially for Cluster 3, which

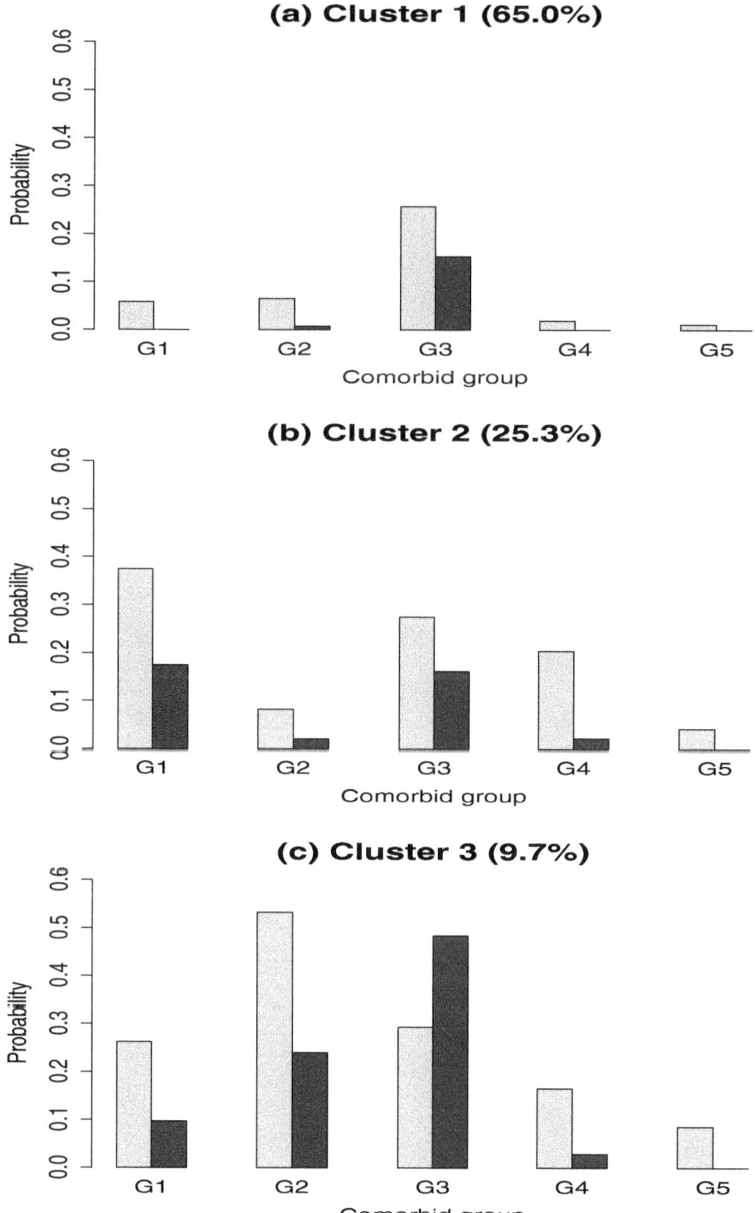

FIGURE 4.5
Probabilities of the presence of one condition (light grey bar) or more than one condition (dark grey bar) in the five comorbidity groups (G1 to G5) for NHS 2007-08 data (Example 4.3).

TABLE 4.4
Adjusted OR (95% CI) of belonging to high comorbidity clusters for NHS 2007-08 data (Example 4.3).

Risk factor	Adjusted OR relative to Cluster 1 (low comorbidity) Cluster 2 (25.3%)	Cluster 3 (9.7%)
Female	0.465* (0.36, 0.60)	2.243* (1.72, 2.93)
Age (per 5-year increment)	2.465* (2.29, 2.65)	1.484* (1.36, 1.62)
Country of birth		
Australia	Reference	Reference
English-speaking[a]	0.680* (0.51, 0.90)	0.595* (0.40, 0.89)
Other	0.926 (0.64, 1.33)	0.453* (0.27, 0.76)
No non-school qualification	0.921 (0.74, 1.15)	0.700* (0.56, 0.88)
Unemployed or not in labour force	1.587* (1.17, 2.15)	4.770* (3.80, 5.98)
Not married	1.000 (0.80, 1.25)	1.952* (1.51, 2.53)
Smoking status		
Never smoked daily	Reference	Reference
Current daily smoker	1.555* (1.14, 2.12)	1.700* (1.27, 2.27)
Past daily smoker	1.916* (1.51, 2.44)	1.986* (1.55, 2.55)
Higher exercise level	0.865* (0.80, 0.94)	0.865* (0.79, 0.95)
Low consumption of fruit/vegetable	1.138 (0.76, 1.70)	0.504* (0.28, 0.92)
No private health insurance	0.961 (0.75, 1.23)	1.327* (1.03, 1.71)
Other language spoken at home	0.559* (0.33, 0.95)	0.236* (0.11, 0.52)
Residential area		
Major city	Reference	Reference
Inner regional	1.411* (1.08, 1.85)	1.043 (0.79, 1.37)
Other areas	1.436* (1.03, 2.00)	0.698 (0.48, 1.01)
Housing tenure type not owned	0.974 (0.76, 1.26)	1.805* (1.44, 2.27)

* p-value < 0.05.
[a] English-speaking countries include Canada, Ireland, New Zealand, South Africa, United Kingdom, and United States of America.

TABLE 4.5
Health-related outcomes and key indicators among the three clusters of individuals for NHS 2007-08 data (Example 4.3).

	Frequency (percentage) or Mean (SD)[a]			
Indicator	**Cluster 1 (65.0%)**	**Cluster 2 (25.3%)**	**Cluster 3 (9.7%)**	**Total sample**
GP visit in last year				
No	5621 (73.8%)	1247 (50.2%)	368 (41.4%)	7236 (65.9%)
Yes	1997 (26.2%)	1235 (49.8%)	520 (58.6%)	3752 (34.1%)
Self-assessed health				
Excellent	1962 (25.8%)	294 (11.8%)	49 (5.5%)	2305 (21.0%)
Very good	3121 (41.0%)	793 (32.0%)	199 (22.4%)	4113 (37.4%)
Good	2015 (26.5%)	863 (27.2%)	292 (32.9%)	3170 (28.8%)
Fair	445 (5.8%)	368 (14.8%)	218 (24.5%)	1031 (9.4%)
Poor	75 (1.0%)	164 (6.6%)	130 (14.6%)	369 (3.4%)
Psychological distress				
Low level	5351 (70.3%)	1647 (66.4%)	168 (19.0%)	7166 (65.2%)
Moderate level	1649 (21.6%)	511 (20.6%)	237 (26.7%)	2397 (21.8%)
High level	493 (6.5%)	237 (9.6%)	269 (30.4%)	999 (9.1%)
Very high level	124 (1.6%)	86 (3.5%)	212 (23.9%)	422 (3.8%)
BMI group (self-reported)				
Under-weight	167 (2.6%)	21 (1.0%)	26 (3.6%)	214 (2.3%)
Normal weight	3058 (46.9%)	622 (28.4%)	269 (37.5%)	3949 (41.9%)
Overweight	2117 (32.5%)	847 (38.6%)	212 (29.5%)	3176 (33.7%)
Obese	1177 (18.1%)	703 (32.1%)	211 (29.4%)	2091 (22.2%)
BMI group (measured)				
Under-weight	124 (2.3%)	8 (0.5%)	15 (2.4%)	147 (1.9%)
Normal weight	2284 (43.0%)	414 (24.0%)	206 (33.6%)	2904 (38.0%)
Overweight	1855 (34.9%)	684 (39.7%)	191 (31.2%)	2730 (35.7%)
Obese	1053 (19.8%)	616 (35.8%)	201 (32.8%)	1870 (24.4%)
Household income	6.48 (2.64)	6.05 (2.91)	4.53 (2.89)	6.22 (2.78)

[a] Test for differences in frequencies among the three clusters using chi-square tests; test for differences in means using ANOVA (Significant differences among the three clusters for all indicators).

is associated with high comorbidity in G2 corresponding to mental health disorders); and (4) overweight or obese based on either self-reported or measured data (especially for Cluster 2, which is associated with high comorbidity in G1 corresponding to cardiovascular and metabolic diseases). In addition, it was found that Clusters 2 and 3 had generally lower household income based on the mean of household income deciles, compared to Cluster 1.

4.7 Computing Programs for Fitting Mixture of Generalized Linear Models

Computing programs for parameter estimation of Poisson mixture regression models (Section 4.2), ZIP (with exposure), or ZINB mixture regression models, along with the corresponding diagnostic and sensitivity analyses (Sections 4.3 to 4.5) were developed in S-plus macros and were available on request from the corresponding author of Xiang et al. (2005), Lee, Wang, and Yau (2001), Yau, Wang, and Lee (2003), and Lee, Xiang, and Fung (2004), respectively.

Computing programs for fitting mixture regression model of generalized Bernoulli distributions to comorbidity data in Section 4.6 are available at the book's online webpage. They include the Fortran program "mixgbdregr.f" and external Fortran programs "random.f" and "hybrid.f".

The mixgbdregr.f is compiled with random.f and hybrid.f in Cygwin to obtain the executable program MIXGBDREGR, which fits a mixture regression model of generalized Bernoulli distributions as given in (4.36) and (4.37) via the EM algorithm.

The program code for fitting the NHS 2007-08 morbidity data (Example 4.3) using MIXGBDREGR is provided in Figure 4.6.

Main Analysis for Example 4.3

```
$ cd c:/Chapter4                 # change to the work directory

$ gfortran -c random.f           # compile a Fortran file to obtain an .c file

$ gfortran -c hybrid.f           # compile a Fortran file to obtain an .c file

$ gfortran -c mixgbdregr.f       # compile the mixgbdregr.f to get mixgbdregr.o

# to obtain the executable program MIXGBDREGR from the .o files
$ gfortran -o MIXGBDREGR mixgbdregr.o random.o hybrid.o

$ ./MIXGBDREGR                   # call the MIXGBDREGR program
 Number of covariates? Input 0 if none
17                                         # consider nq=17

 Input the seeds for random number generation: input 3 integers between 1 and 30,000,
 e.g. 23 120 34.
23 120 34                                  # enter the seeds

 Number of bootstraps? Input 1 if not bootstrapping
1                                          # for original data

 Input the maximum number of iterations
500

# output file is "output"


# calculate standard errors using B=100 bootstrap replications

$ ./MIXGBDREGR                   # call the MIXGBDREGR program
 Number of covariates? Input 0 if none
17                                         # consider nq=17

 Input the seeds for random number generation: input 3 integers between 1 and 30,000,
 e.g. 23 120 34.
23 120 34                                  # enter the seeds

 Number of bootstraps? Input 1 if not bootstrapping
100                                        # for 100 bootstrap replications

 Do you want to print out the estimates
   for each bootstrap?     INPUT: 1(Yes); 0(No)
1                                          # to print out the estimates

 Input the maximum number of iterations
500

# output file is "output"
```

FIGURE 4.6
The program codes for Example 4.3.

5

Mixture Models for Survival Data

5.1 Introduction

Much research in the medical and health sciences focuses on the distribution of time to the occurrence of some event that represents failure. For example, failure can be the first onset of a particular disease, the recurrence of disease after treatment, or the death of an individual. Data collected in this type of research often contain censored observations, where their failure times are not yet observed at the end of the follow-up period and we know only that their failure times are greater than particular values. In this chapter, we describe how mixture models via the EM algorithm can be adopted for the survival analysis of censored failure time data.

In survival analysis, the focus is the distribution of failure time T. Let T have a p.d.f. $f(t)$, then the survival function $S(t)$ is defined as

$$S(t) = \text{pr}(T > t) = \int_t^\infty f(u)du, \tag{5.1}$$

where $S(t)$ is monotone, non-increasing, and equal to one at $t = 0$ and zero as time approaches infinity. The hazard function is defined as

$$h(t) = \lim_{\Delta t \to 0} \text{pr}\{t < T < t + \Delta t \mid T > t\}/\Delta t = f(t)/S(t), \tag{5.2}$$

which measures the instantaneous failure rate at time t. The only restriction on $h(t)$ is that it is non-negative, that is, $h(t) \geq 0$ for all t. Since

$$h(t) = -d\log S(t)/dt \quad \text{and} \quad S(t) = \exp\{-\int_0^t h(u)du\}, \tag{5.3}$$

$S(t)$ is derivable from $h(t)$ and vice versa. In most applications, we are interested in considering the effects of independent risk variables (covariates) on failure time T. In this case, the survival function and the hazard function are denoted by $S(t;x)$ and $h(t;x)$ respectively, where x is the covariate vector.

Cox's proportional hazards model (Cox, 1972) can be used to model covariate effects for survival data:

$$h(t;x) = h_0(t)\exp\{x^T\beta\}, \tag{5.4}$$

where β is the vector of regression coefficients and $h_0(t)$ is the baseline hazard function. The covariates are thus modelled as acting multiplicatively on the baseline hazard function. Moreover, model (5.4) implies that the hazard ratio for two individuals

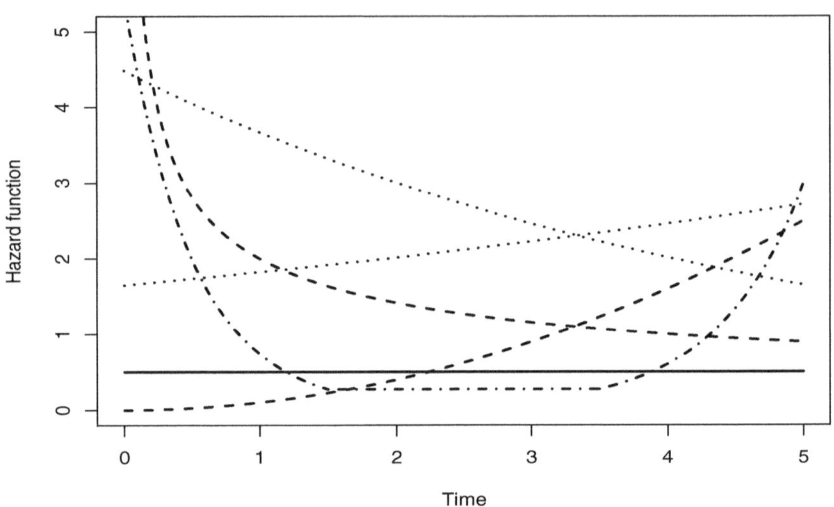

FIGURE 5.1
Several typical shapes of hazard functions: Exponential (solid line), Weibull with an increasing or decreasing hazard (dashed line), Gompertz with an increasing or decreasing hazard (dotted line), Bathtub shape (dashed-dotted line).

depends on the difference of their linear predictors $x^T\beta$ at any time, and is constant over time provided that the covariates do not change. In terms of survival functions, the assumption of proportional hazards (5.4) implies

$$S(t;x) = \{S_0(t)\}^{\exp(x^T\beta)}, \tag{5.5}$$

where

$$S_0(t) = \exp\{-\int_0^t h_0(u)du\} = \exp\{-H_0(t)\}$$

is the baseline survival function, and $H_0(t)$ denotes the cumulative hazard function. In applications, $h_0(t)$ may have a specified parametric lifetime distribution such as exponential and Weibull, or it may be left as an arbitrary non-negative function. Figure 5.1 displays several typical shapes of hazard functions. The exponential hazard function, $h_0(t) = \lambda$ ($\lambda > 0$), is the simplest with a constant hazard rate over time. The Weibull hazard function is given by

$$h_0(t) = \lambda\alpha t^{\alpha-1} \qquad (\lambda, \alpha > 0), \tag{5.6}$$

which is flexible in that it can represent either a monotonic increasing ($\alpha > 1$), a constant (i.e., an exponential hazard when $\alpha = 1$), or a monotonic decreasing hazard function ($\alpha < 1$). From (5.6), the logarithm of the hazard function is linear in $\log(t)$. Closely related to the Weibull distribution, the Gompertz hazard function is defined as

$$h_0(t) = \exp(\lambda + \alpha t) \tag{5.7}$$

and its logarithmic hazard function is linear in time t. The Gompertz distribution can represent either a monotonic increasing ($\alpha > 0$), a constant ($\alpha = 0$), or a monotonic decreasing hazard function ($\alpha < 0$). Survival models with monotonic increasing hazard rates may arise when there is natural aging, while monotonic decreasing hazard functions are, however, less common in practice. The final "bathtub" shape of the hazard function given in Figure 5.1 characterizes the risk of failure following a medical treatment. The hazard function is decomposed into three overlapping phases in time: an early phase immediately following treatment in which the risk is initially relatively high and drops gradually, a middle phase of constant risk, and finally a late phase in which the failure risk starts to increase gradually as the patient ages; see Blackstone, Naftel, and Turner (1986) and Ng et al. (2004). Alternatively, Glasser (1980) showed that a gamma mixture with a common scale parameter can have a bathtub-shaped hazard function.

Issues of censoring often complicate the analysis of survival data. For example, there will be individuals who do not fail at the end of the study, or individuals who withdraw from the study before it ends. Such observations are censored, as we know only that their failure times are greater than particular values. In this chapter, we make an assumption of non-informative censoring which means that knowledge of a censoring time for an individual provides no further information about his/her likelihood of survival at a future time. Let the observed failure-time data be denoted by

$$y = (t_1, x_1^T, D_1, \ldots, t_n, x_n^T, D_n)^T,$$

where t_j is the failure time or censoring time for the jth individual, and x_j is a vector of covariates associated with the jth individual. D_j is an indicator of the failure status where $D_j = 1$ means that failure occurs at time t_j for the jth individual, and $D_j = 0$ indicates that the jth individual had not yet failed by time t_j.

For inference about the regression coefficients β in (5.4), the baseline hazard $h_0(t)$ can be eliminated in the estimating equations derived from Cox's log-partial likelihood in Cox (1975) as

$$\log(\beta) = \sum_{j=1}^{n} D_j [x_j^T \beta - \log \sum_{l \in R_j} \exp(x_l^T \beta)], \tag{5.8}$$

where R_j is the risk set which is the set of individuals who have not failed or been censored just prior to t_j. The efficiency of the partial likelihood relative to a full likelihood based on a parametric form of $h_0(t)$ was studied in various papers including Efron (1977) and Oakes (1981). In brief, it was shown that the partial likelihood is asymptotically equivalent to the full likelihood, provided that parameters β are not far from zero and censoring does not depend on covariates; see also the two monographs by Kalbfleisch and Prentice (2011) and Lawless (2011) for computational details on the method of partial likelihood and the techniques used for estimating the model parameters.

5.2 Application of Mixture Models in Survival Analysis

Mixture models have been used to model survival data in a variety of situations, as illustrated in McLachlan and McGiffin (1994), Phillips, Coldman, and McBride (2002), and Ng et al. (2004). Following previous chapters on specifying mixture models, we suppose that the p.d.f. $f(t)$ of the failure time T has the mixture form as

$$f(t) = \sum_{h=1}^{g} \pi_h f_h(t), \tag{5.9}$$

where $f_h(t)$ $(h = 1,\ldots,g)$ denote the component densities and π_h are the mixing proportions that sum to one. Thus the survival function $S(t)$ corresponding to (5.9) also has the mixture form as

$$S(t) = \int_t^\infty f(u)du = \int_t^\infty \sum_{h=1}^{g} \pi_h f_h(u)du = \sum_{h=1}^{g} \pi_h S_h(t), \tag{5.10}$$

where $S_h(t) = \int_t^\infty f_h(u)du$ is the component survival function corresponding to the component density $f_h(t)$ for $h = 1,\ldots,g$.

From (5.2), it follows that the unconditional hazard function with the survival mixture model (5.10) is given by

$$h(t) = \frac{f(t)}{S(t)} = \sum_{h=1}^{g} \pi_h \frac{h_h(t)S_h(t)}{S(t)} = \sum_{h=1}^{g} \pi_h h_h^*(t), \tag{5.11}$$

which does not have the form of a mixture of component hazard functions with $h_h(t) = f_h(t)/S_h(t)$ $(h = 1,\ldots,g)$. Suppose that $h_h(t)$ has the proportional hazards form $h_h(t;x) = h_{h0}(t)\exp\{x^T\beta_h\}$, it can be seen from (5.11) that $h_h^*(t;x)$ does not have a proportional hazards form in general; see McLachlan and McGiffin (1994) for more information.

Mixture specification of the hazard (or the cumulative hazard) function has also been considered, such as by Blackstone, Naftel, and Turner (1986) and Rosen and Tanner (1999). Taking the bathtub-shaped hazard function in Figure 5.1, for example, the hazard function can be specified by a three-component mixture model as

$$h(t) = \pi_1 h_1(t) + \pi_2 h_2(t) + \pi_3 h_3(t),$$

with its three components corresponding to the three phases of risk of failure following treatment, as described in Section 5.1 for the bathtub-shaped hazard function.

The unconditional survival function is then given by

$$\begin{aligned} S(t) &= \exp\{-\int_0^t h(u)du\} \\ &= \exp\{-\int_0^t \sum_{h=1}^{3} \pi_h h_h(u)du\} \\ &= \prod_{h=1}^{3} S_h(t)^{\pi_h}, \end{aligned} \tag{5.12}$$

which does not have the form of a mixture of component survival functions, with $S_h(t) = \exp\{-\int_0^t h_h(u)du\}$ $(h = 1,\ldots,g)$. Similarly, the unconditional p.d.f.

$$f(t) = \sum_{h=1}^{3} \pi_h h_h(t)S(t) \tag{5.13}$$

does not have a mixture form. This means that, unlike mixture specification of the survival function in (5.10), $\pi_h(h = 1,\ldots,g)$ in (5.13) specified under a mixture of hazard functions does not have an interpretation of the mixing proportions.

In this chapter, we will consider the use of mixture models as a flexible way of modelling survival data in three situations. The first situation is where the mixture modelling approach is directly applicable and when the adoption of a single parametric family is inadequate for modelling the distribution of failure time. The second situation is when an individual is exposed to g distinct types of failures (competing risks), such that D_j can now take on $(g+1)$ possible values to indicate the failure type or censoring for each individual $(j = 1,\ldots,n)$. The third situation corresponds to studies where a subgroup of patients have recovered after treatment for some particular disease and the focus of these studies is on the proportion of "cured" patients.

5.3 Mixture Models of Parametric Survival Distributions

In the first situation, we consider the use of survival mixture models to analyse mortality data. The survival experience of patients after diagnosis or certain treatments may be described by sub-populations of patients with different risks of mortality. This heterogeneity in risk can be modelled using a mixture of parametric survival distributions, with each component corresponding to a sub-population of patients with a certain mortality risk. Let x be a vector of covariates associated with each patient, the survival function of time to death T is modelled as

$$S(t;x) = \sum_{h=1}^{g} \pi_h S_h(t;x), \tag{5.14}$$

where π_h denotes the prior probability of patients belonging to the hth component and $S_h(t;x)$ is the conditional survival function given that the patients belong to the

hth component ($h = 1,\dots,g$). With the concomitant information x in (5.14), effects of demographic and clinical characteristics on the survival of patients with heterogeneous mortality risks can be determined. We specify the parametric survival distributions under Cox's proportional hazards form (Cox, 1972), where the conditional hazard function for the hth component is given by

$$h_h(t_j;x_j) = h_{h0}(t_j)\exp\{x_j^T\beta_h\} \qquad (h = 1,\dots,g). \tag{5.15}$$

In (5.15), $h_{h0}(t_j)$ is the baseline hazard function as described in Section 5.1 and β_h is the vector of regression coefficients. A positive regression coefficient for a covariate implies an increased mortality risk, whereas a negative coefficient indicates a reduced risk of mortality. Under the formulation based on (5.14) and (5.15), the vector of unknown parameters is

$$\Psi = (\pi_1,\dots,\pi_{g-1},\beta_1^T,\dots,\beta_g^T,\psi_1^T,\dots,\psi_g^T)^T,$$

where ψ_h is the vector of unknown parameters in the baseline hazard function ($h = 1,\dots,g$).

5.3.1 The EM Algorithm for Mixtures of Parametric Survival Models

The unknown parameter vector Ψ can be estimated via ML using the EM algorithm. From (5.14) and (5.15), the log likelihood for Ψ is given by

$$\log L(\Psi) = \sum_{j=1}^{n}\left[D_j\log\{\sum_{h=1}^{g}\pi_h f_h(t_j;x_j)\} + (1-D_j)\log\{\sum_{h=1}^{g}\pi_h S_h(t_j;x_j)\}\right], \tag{5.16}$$

where

$$S_h(t_j;x_j) = S_{h0}(t_j)^{\exp\{x_j^T\beta_h\}} \qquad (h = 1,\dots,g). \tag{5.17}$$

On the $(k+1)$th iteration, the E-step of the EM algorithm involves the calculation of the Q-function

$$\begin{aligned} Q(\Psi,\Psi^{(k)}) = \sum_{j=1}^{n}\Big[D_j\sum_{h=1}^{g}\tau_h(t_j;x_j,\Psi^{(k)})\log\{\pi_h f_h(t_j;x_j,\beta_h,\psi_h)\} + \\ (1-D_j)\sum_{h=1}^{g}\tau_h(t_j;x_j,\Psi^{(k)})\log\{\pi_h S_h(t_j;x_j,\beta_h,\psi_h)\}\Big], \end{aligned} \tag{5.18}$$

where

$$\tau_h(t_j;x_j,\Psi^{(k)}) = \tau_{hj}^{(k)} = \frac{\pi_h^{(k)} S_h(t_j;x_j,\beta_h^{(k)},\psi_h^{(k)})}{\sum_{l=1}^{g}\pi_l^{(k)} S_l(t_j;x_j,\beta_l^{(k)},\psi_l^{(k)})} \tag{5.19}$$

is the posterior probability that the jth individual with censored survival time t_j belongs to the hth component ($h = 1,\dots,g$). The Q-function (5.18) can be decomposed as

$$Q(\Psi,\Psi^{(k)}) = Q_\pi^{(k)} + Q_1^{(k)} + \cdots + Q_g^{(k)},$$

with respect to the parameters π_h $(h=1,\ldots,g-1)$, (β_1^T, ψ_1^T), ..., and (β_g^T, ψ_g^T), respectively. That is,

$$Q_\pi^{(k)} = \sum_{j=1}^{n}\sum_{h=1}^{g} \tau_{hj}^{(k)} \log \pi_h \tag{5.20}$$

and

$$\begin{aligned} Q_h^{(k)} &= \sum_{j=1}^{n} \tau_{hj}^{(k)} \{D_j \log f_h(t_j; x_j, \beta_h, \psi_h) + (1-D_j) \log S_h(t_j; x_j, \beta_h, \psi_h)\} \\ &= \sum_{j=1}^{n} \tau_{hj}^{(k)} \{D_j \log h_h(t_j; x_j, \beta_h, \psi_h) + \log S_h(t_j; x_j, \beta_h, \psi_h)\} \end{aligned} \tag{5.21}$$

for $h=1,\ldots,g$, where $D_j=1$ and $D_j=0$ indicate death and censoring, respectively.

With the proportional hazards form in (5.15) and (5.17), the M-step obtained the updated estimate $\Psi^{(k+1)}$ by maximizing $Q_\pi^{(k)}$ and $Q_h^{(k)}$ for $h=1,\ldots,g$, respectively. The maximization has a closed-form equation for π, where

$$\pi_h^{(k+1)} = \sum_{j=1}^{n} \tau_{hj}^{(k)}/n \qquad (h=1,\ldots,g-1). \tag{5.22}$$

and for β_h and ψ_h $(h=1,\ldots,g)$, the maximization involves solving the following set of non-linear equations:

$$\begin{aligned} &\sum_{j=1}^{n} \tau_{hj}^{(k)} \{D_j + \exp(x_j^T \beta_h) \log S_{h0}(t_j)\} x_j = 0 \\ &\sum_{j=1}^{n} \tau_{hj}^{(k)} \{D_j \frac{\partial h_{h0}(t_j)}{\partial \psi_h} \frac{1}{h_{h0}(t_j)} + \exp(x_j^T \beta_h) \frac{\partial S_{h0}(t_j)}{\partial \psi_h} \frac{1}{S_{h0}(t_j)}\} = 0. \end{aligned} \tag{5.23}$$

The MINPACK routine HYBRD1 of Moré, Garbow, and Hillstrom (1980) can be used to find a solution to (5.23). Assuming a Weibull hazard function for $h_{h0}(t)$ $(h=1,\ldots,g)$ as in (5.6), the set of non-linear equations (5.23) becomes:

$$\begin{aligned} &\sum_{j=1}^{n} \tau_{hj}^{(k)} \{D_j - \exp(x_j^T \beta_h) \lambda_h t_j^{\alpha_h}\} x_j = 0 \\ &\sum_{j=1}^{n} \tau_{hj}^{(k)} \{\frac{D_j}{\lambda_h} - \exp(x_j^T \beta_h) t_j^{\alpha_h}\} = 0 \\ &\sum_{j=1}^{n} \tau_{hj}^{(k)} \{\frac{D_j}{\alpha_h} + D_j \log t_j - \exp(x_j^T \beta_h) \lambda_h t_j^{\alpha_h} \log t_j\} = 0. \end{aligned} \tag{5.24}$$

5.3.2 Example 5.1: Survival Mixture Modelling of Mortality Data

We consider the mortality data of Moertel et al. (1995) in one of the first successful trials of adjuvant chemotherapy for colon cancer. The data set is available at

TABLE 5.1
Model selection for the colon cancer data (Example 5.1).

No. of components (g)	Log likelihood	AIC	BIC
1	-3957.6	7939	7997
2	-3904.2	7858*	7979*
3	-3895.5	7867	8050

* Number of components selected based on the criterion.

`https://vincentarelbundock.github.io/Rdatasets/datasets.html` and consists of 929 patients with resected colon cancer that had regional nodal involvement (Stage C); see Moertel et al. (1990) for more details. Patients were stratified according to the invasion by the primary lesion and the interval since surgery and also according to the number of lymph nodes involved ($>$4 versus $\leq$4). They were then randomly assigned to three treatment arms, namely observation, therapy with levamisole alone, and therapy with levamisole plus fluorouracil (5-FU), as described in Moertel et al. (1990). The median follow-up time of this data set is 6.5 years (a minimum of 5 years) (Moertel et al., 1995). Besides time to death in days and treatment arms, we consider risk variables of patient's demographic and clinical characteristics including age in years; gender; obstruction of colon by tumour; adherence to nearby organs; differentiation of tumour (poor versus moderately well/well); depth of invasion; interval since surgery to registration (long: 21-35 days versus short: 7-20 days); number of lymph nodes involved ($>$4 versus $\leq$4). In general, patient characteristics were not significantly different among the three treatment arms, except that more men received levamisole alone and fewer received levamisole plus 5-FU (p=0.028); see Moertel et al. (1990) for more information. In this analysis, we consider $n = 906$ patients (98% of 929 patients) with completed information on all risk variables.

We fitted Weibull mixture regression models (5.14) to the data with g=1 to g=3. Model selection displayed in Table 5.1 indicates there are two components corresponding to patients with different risks of mortality. The estimated regression coefficients for a single Weibull model and a two-component Weibull mixture model were provided in Table 5.2. It can be seen that, with a single Weibull regression model, a marginally increasing hazard (α=1.064) was determined. With a Weibull mixture model, we identified two sub-populations of patients. The larger sub-population consists of 62.9% of patients with an increasing hazard (α_1=1.137), whereas the smaller sub-population consists of 37.1% of patients with a bigger increasing hazard of α_2=1.940. While the treatment effects of levamisole alone relative to the observation group were not significant in both the single Weibull and the Weibull mixture models, there is apparently a Simpson's paradox as the direction of treatment effect of levamisole alone reversed when the two sub-populations are combined. Compared to observation, treatment of levamisole plus 5-FU was found to have a significant effect on reducing the mortality risk in the single Weibull model. With the Weibull mixture model, it was significant only in the sub-population with a smaller hazard

TABLE 5.2
Estimates and standard errors (in parentheses) for fitting a single Weibull and a two-component Weibull mixture model (Example 5.1).

Variable	A single Weibull	Two-component Weibull mixture	
		first component	second component
Constant ($\log \lambda_h$)	-10.3* (0.40)	-12.7* (1.77)	-17.4* (2.45)
Age (years)	0.008 (0.004)	0.017 (0.017)	0.018* (0.009)
Male	0.034 (0.109)	0.260 (0.314)	-0.170 (0.198)
LEV vs. Obs	-0.027 (0.113)	0.116 (0.283)	0.112 (0.359)
LEV+5-FU vs. Obs	-0.397* (0.140)	-1.043* (0.492)	0.110 (0.255)
Obstruction	0.280* (0.126)	0.722* (0.284)	1.260* (0.346)
Adhesion	0.162 (0.147)	0.364 (0.357)	0.008 (0.419)
Differentiation	0.350* (0.133)	0.566 (0.319)	0.992* (0.475)
Invasion depth	0.419* (0.126)	0.372 (0.499)	1.223* (0.412)
time since surgery	0.233* (0.107)	0.263 (0.267)	0.147 (0.240)
more than 4 LNs	0.966* (0.118)	1.823* (0.360)	0.882* (0.318)
Mixing proportion π_h	1.000	0.629 (0.034)	0.371
Shape parameter α_h	1.064 (0.033)	1.137 (0.114)	1.940 (0.253)

* p-value < 0.05.

rate (the first component). Survival among all eligible patients in the original study in Moertel et al. (1995) is provided in Figure 5.2. It illustrates that therapy with levamisole alone showed no effect (p = 0.57), and therapy with levamisole plus 5-FU was found to have an overall survival advantage over the observation group (p = 0.0007), as reported in Moertel et al. (1995). However, the significance is attributed mainly to the difference appearing at a later time (or could be interpreted as the group of patients with a relatively smaller hazard in the first component). From the results with the two-component Weibull mixture model in Table 5.2, having more than 4 lymph nodes involved and obstruction of colon by tumour had prognostic significance on both components of patients. Moreover, lymph nodes >4 had a greater effect on increasing the mortality risk for patients with a lower hazard (adjusted HR of 6.2 versus 2.4), whereas obstruction had a greater effect on patients with a higher hazard (adjusted HR of 3.5 versus 2.1). Poor differentiation (adjusted HR of 2.7) and deeper invasion (adjusted HR of 3.4) also had prognostic significance on increased mortality risk for patients with a higher hazard (the second component).

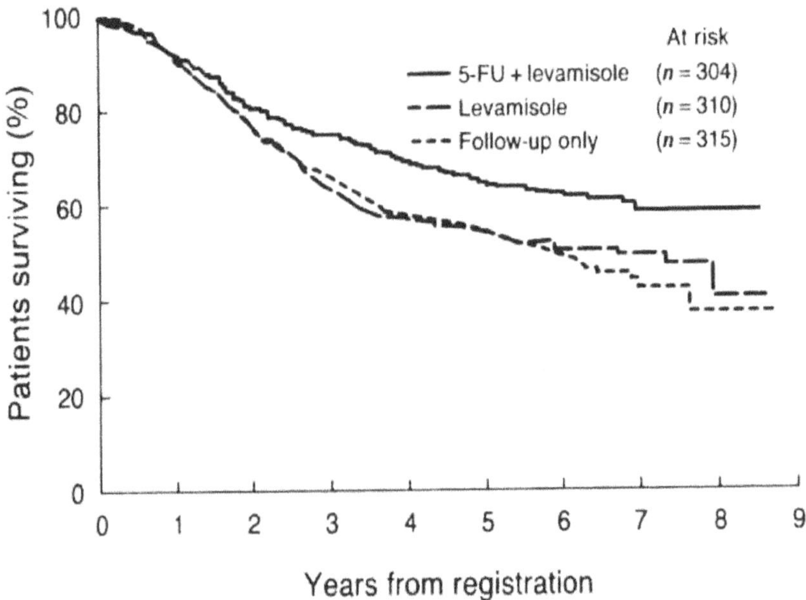

FIGURE 5.2
Survival by treatment arms for Example 5.1 colon cancer mortality data (Adapted from Moertel et al. (1995)).

5.4 Semi-Parametric Mixture Survival Models

The second situation concerns competing-risks problems, where an individual is exposed to g distinct types of failures. Let the observed failure-time data be

$$y = (t_1, x_1^T, D_1, \ldots, t_n, x_n^T, D_n)^T, \tag{5.25}$$

where t_j is the failure or censoring time for the jth individual, x_j is a vector of covariates associated with the jth individual, and $D_j = h$ indicates that the jth individual fails due to the hth type of failure and $D_j = 0$ represents a censored observation $(j = 1, \ldots, n)$. A mixture model approach can be considered for the regression analysis of such competing-risks data, where the survival function of the failure time is expressed as

$$S(t; x) = \sum_{h=1}^{g} \pi_h(x) S_h(t; x), \tag{5.26}$$

where $\pi_h(x)$ is the probability of failure from the hth type and $S_h(t; x)$ denotes the conditional survival function given failure is due to the hth type $(h = 1, \ldots, g)$. In the context of competing-risk analysis, the mixture model (5.26) assumes that an individual will fail from one of the g failure types, chosen by a stochastic mechanism at the outset, characterized by the marginal distribution $\pi_h(x_j)$ of each failure type given the characteristics of the individual x_j. That is, the sum of $\pi_h(x_j)$ is equal to one for all $j = 1, \ldots, n$ and the model (5.26) is formulated such that useful information is obtained about how the factors x influence the incidence of each failure type and how they affect the failure time among individuals who failed from each type.

Larson and Dinse (1985) were among the first to use Model (5.26) to handle competing-risks problems, where the component-hazard functions, $h_h(t; x)$ $(h = 1, \ldots, g)$ are assumed to follow a proportional hazards model, that is

$$h_h(t; x) = h_{h0}(t) \exp\{x^T \beta_h\} \quad (h = 1, \ldots, g). \tag{5.27}$$

In (5.27), β_h denotes a vector of regression coefficients and the baseline hazard functions, $h_{h0}(t)$ $(h = 1, \ldots, g)$, are taken to be piecewise constant. Common lifetime distributions can also be adopted to specify $h_{h0}(t)$, such as the Weibull distribution and the Gompertz distribution considered by Gelfand et al. (2000) and Gordon (1990b), respectively. The mixture modelling approach (5.26) provides an alternative method to the cause-specific hazard function approach of Prentice et al. (1978), where the estimation of the coefficients β_h is obtained separately for each failure type with failures from remaining types treated as censored observations; see also Kay (1986).

Here we describe an ECM-based semi-parametric mixture model approach proposed by Ng and McLachlan (2003a) that adopts a nonparametric specification for the baseline hazard functions in (5.27) in order to relax the parametric constraints. As illustrated by Ng and McLachlan (2003a) in a simulation study, the ECM-based semi-parametric approach is particularly useful when the true baseline hazard due to each cause does not follow any common lifetime distributions. In this situation, the semi-parametric approach consistently provides less biased estimates compared to a fully parametric approach when the baseline hazard functions were misspecified.

5.4.1 The ECM Algorithm

With the mixture survival model (5.26), the mixing proportions are assumed to have the logistic form as

$$\pi_h(x_j;\alpha) = \frac{\exp(v_h^T x_j)}{1+\sum_{l=1}^{g-1}\exp(v_l^T x_j)} \qquad (h=1,\dots,g-1), \tag{5.28}$$

where $\pi_g(x_j;\alpha) = 1-\sum_{l=1}^{g-1}\pi_l(x_j;\alpha)$ and the first element of x_j is assumed to be one to account for an intercept term. In (5.28), α contains the logistic coefficients in v_h $(h=1,\dots,g-1)$. This mixture modelling framework implies that a population may be split into g mutually exclusive components corresponding to each failure type. Let $\Psi = (\alpha^T, \beta_1^T, \dots, \beta_g^T)^T$ be the vector containing the unknown parameters for the regression and logistic coefficients according to (5.27) and (5.28), respectively. On the basis of the observed data y given by (5.25), the log likelihood function for Ψ under the mixture model (5.26) is expressed as

$$\begin{aligned}\log L(\Psi) \quad = \quad & \sum_{j=1}^{n}[\sum_{h=1}^{g} I(D_j,h)\log\{\pi_h(x_j;\,\alpha)f_h(t_j;x_j,\beta_h)\} \\ & \qquad +I(D_j,0)\log S(t_j;x_j,\Psi)],\end{aligned} \tag{5.29}$$

where $I(D_j,h)$ is an indicator function which is equal to 1 if $D_j = h$ and is 0 otherwise $(h=0,1,\dots,g)$ and $f_h(\cdot)$ is the p.d.f. for the hth component $(h=1,\dots,g)$. With reference to Section 1.3.2, the maximum likelihood estimation of Ψ can be posted as an incomplete-data problem by introducing an unobservable random vector z of zero-one indicator variables for each censored observation t_j, where $z_j = (z_{1j},\dots,z_{gj})^T$, and where $z_{hj} = 1$ or 0 according as the jth individual would have failed from hth type or not $(h=1,\dots,g)$. It follows that the Q-function on the $(k+1)$th iteration of the E-step is given by

$$\begin{aligned}Q(\Psi;\Psi^{(k)}) \quad = \quad & \sum_{j=1}^{n}\left[\sum_{h=1}^{g} I(D_j,h)\log\{\pi_h(x_j;\,\alpha)f_h(t_j;x_j,\beta_h)\}\right. \\ & \left. +\sum_{h=1}^{g} I(D_j,0)\tau_{hj}^{(k)}\log \pi_h(x_j;\,\alpha)S_h(t_j;x_j,\beta_h)\right],\end{aligned} \tag{5.30}$$

where $\Psi^{(k)}$ is the estimate of Ψ after the kth iteration and

$$\tau_{hj}^{(k)} = \frac{\pi_h(x_j;\,\alpha^{(k)})S_h(t_j;x_j,\beta_h^{(k)})}{\sum_{l=1}^{g}\pi_l(x_j;\,\alpha^{(k)})S_l(t_j;x_j,\beta_l^{(k)})} \tag{5.31}$$

is the posterior probability that the jth individual with censored survival time t_j would have failed due to failure type h $(h=1,\dots,g)$.

In the M-step, the Q-function (5.30) can be decomposed into

$$\begin{aligned}
& \sum_{j=1}^{n}\sum_{h=1}^{g}[I(D_j,h)\log\pi_h(x_j,\alpha)+I(D_j,0)\tau_{hj}^{(k)}\log\pi_h(x_j,\alpha)] \\
+ & \sum_{j=1}^{n}[I(D_j,1)\log f_1(t_j;x_j,\beta_1)+I(D_j,0)\tau_{1j}^{(k)}\log S_1(t_j;x_j,\beta_1)] \\
+ & \quad \vdots \\
+ & \sum_{j=1}^{n}[I(D_j,g)\log f_g(t_j;x_j,\beta_g)+I(D_j,0)\tau_{gj}^{(k)}\log S_g(t_j;x_j,\beta_g)] \qquad (5.32)\\
= & Q_0+Q_1+\ldots+Q_g,
\end{aligned}$$

respectively, such that the estimates of α and $\beta_1,\ldots,\beta_g$ can be updated separately by maximizing Q_0 and $Q_1,\ldots,Q_g$, respectively. Assuming proportional hazards (5.27) for the component-hazard functions, $h_h(t;x)$ $(h=1,\ldots,g)$, we have the hth component survival function

$$S_h(t;x)=S_{h0}(t)^{\exp\{x^T\beta_h\}} \qquad (5.33)$$

with the baseline survival function $S_{h0}(t)$ and

$$\begin{aligned}
Q_h = & \sum_{j=1}^{n}[-\{I(D_j,h)+I(D_j,0)\tau_{hj}^{(k)}\}H_{h0}(t_j)\exp(x_j^T\beta_h) \\
& +I(D_j,h)\{\log h_{h0}(t_j)+x_j^T\beta_h\}], \qquad (5.34)
\end{aligned}$$

where $H_{h0}(t)$ is the cumulative hazard function for the hth component $(h=1,\ldots,g)$. As noted by Ng and McLachlan (2003a), the full likelihood (5.29) is used here because a partial likelihood approach (Kalbfleisch and Prentice, 2011) cannot eliminate the baseline survival function (5.33) from the likelihood formed under the mixture model (5.26).

It follows from (5.32) that the updates $\alpha_h^{(k+1)}$ satisfy the equation

$$\sum_{j=1}^{n}[I(D_j,h)+I(D_j,0)\tau_{hj}^{(k)}-\pi_h(x_j,\alpha)]x_j=0. \qquad (5.35)$$

To maximize Q_h (5.34) with respect to β_h and $H_{h0}(t)$, an ECM algorithm, proposed by Meng and Rubin (1993), can be implemented to first obtain the update $H_{h0}^{(k+1)}(t)$ by maximization of Q_h with β_h fixed at $\beta_h^{(k)}$ and then calculating the update $\beta_h^{(k+1)}$ by maximization of Q_h with $H_{h0}(t)$ fixed at $H_{h0}^{(k+1)}(t)$ in the second CM-step; see, for example, Ng and McLachlan (2003a).

In the first CM-step concerning the update to $H_{h0}^{(k+1)}(t)$, the observed failure times are rearranged in increasing order and denote the m_h distinct failure times due to the hth failure type by $t_{(h1)}<\cdots<t_{(hm_h)}$ for $h=1,\ldots,g$. The baseline hazard function

$h_{h0}(t)$ is taken to be a step function with discontinuities at each observed failure time due to the hth cause. Adopting the approach proposed by Breslow (1974) that censored observations are considered as censored at the preceding uncensored failure time, it can be shown that, for fixed β_h $(h = 1,\ldots,g)$, Q_h in (5.34) is maximized with respect to $H_{h0}(t)$ at

$$H_{h0}^{(k+1)}(t_{(hl)}) = \sum_{j=1}^{l} \left(\frac{d_{hj}}{\sum_{r\in R(t_{(hj)})}[I(D_r,h)+I(D_r,0)\tau_{hr}^{(k)}]\exp(x_r^T\beta_h)} \right) \quad (5.36)$$

for $l = 1,\ldots,m_h$, where d_{hj} is the number of failures due to the hth type at time $t_{(hj)}$ and $R(t_{(hj)})$ is the risk set at time $t_{(hj)}$; see Ng and McLachlan (2003a). It follows from (5.36) that the updated estimates for the baseline survival functions are given by

$$S_{h0}^{(k+1)}(t_{(hl)}) = \exp\{-H_{h0}^{(k+1)}(t_{(hl)})\} \qquad (h = 1,\ldots,g;\, l = 1,\ldots,m_h). \quad (5.37)$$

However, the solution to the second CM-step for updating β_h does not exist in closed form. On differentiation of Q_h with respect to β_h for fixed $H_{h0}^{(k+1)}(t)$ obtained in (5.36), it follows that $\beta_h^{(k+1)}$ satisfies the equation

$$\sum_{j=1}^{n}[I(D_j,h) - \{I(D_j,h)+I(D_j,0)\tau_{hj}^{(k)}\}H_{h0}^{(k+1)}(t_j)\exp(x_j^T\beta_h)]x_j = 0. \quad (5.38)$$

The MINPACK routine HYBRD1 of Moré, Garbow, and Hillstrom (1980) can be adopted to find a solution for (5.38). Let $\hat{\alpha}$, $\hat{\beta}_h$, and $\hat{S}_{h0}(t)$ $(h = 1,\ldots,g)$ be the ML estimates of the corresponding parameters in Ψ, the estimated cumulative incidence function (Pepe, 1991) for the hth failure type is given by

$$\pi_h(x;\, \hat{\alpha})\{1 - \hat{S}_{h0}(t)^{\exp(x^T\hat{\beta}_h)}\} \quad (5.39)$$

and the conditional probability for the hth failure type within a specified time t given that failure due to other types does not occur during this period is estimated by

$$\frac{\pi_h(x;\, \hat{\alpha})\{1 - \hat{S}_{h0}(t)^{\exp(x^T\hat{\beta}_h)}\}}{\pi_h(x;\, \hat{\alpha}) + \sum_{l\neq h}^{g}\pi_l(x;\, \hat{\alpha})\hat{S}_{l0}(t)^{\exp(x^T\hat{\beta}_l)}}. \quad (5.40)$$

An alternative method to estimate the regression parameters β_h has been considered by Kuk (1992), who used a marginal likelihood approach to eliminate the baseline hazard functions $h_{h0}(t)$ as nuisance parameters during the analysis. In contrast to Kuk's method, the above ECM-based semi-parametric mixture model allows the nonparametric maximum likelihood estimates of the baseline survival functions $S_{h0}(t)$ $h = 1,\ldots,g$ in (5.37) to be used in the estimation of the logistic (α) and regression (β_h) parameters and does not require Monte Carlo approximation of the marginal likelihood, as illustrated by Ng and McLachlan (2003a).

The nonparametric bootstrap approach described in Section 1.3.4 can be used to obtain the standard errors of the ML estimates of Ψ, with the resampling scheme modified for the competing-risks problem. Let n_h $(h = 1,\ldots,g)$ be the number of failures due to hth type, and let n_{g+1} be the number of censored observations. With the modified scheme, the bootstrap data are obtained by sampling separately from each of the $(g+1)$ sets, corresponding to the hth failure type $(h = 1,\ldots,g)$ and the censored observations, with the sizes of these bootstrap sub-samples taken equal to n_h $(h = 1,\ldots,g)$ and n_{g+1} of the original data, respectively; see Golbeck (1992) for more information.

5.4.2 Example 5.2: Survival Analysis of Competing-Risks Data

We illustrate the ECM-based semi-parametric mixture survival model using the well known prostate cancer data of Kay (1986). The data set contains the survival times of 483 patients with prostate cancer who entered a clinical trial during 1967-1969 and who were randomly allocated to different levels of treatment with the drug diethylstilbestrol (DES). It is a subset of a full data set available from Andrews and Herzberg (1985). Complete information is available on eight categorical risk variables, namely drug treatment (RX=0: 0.0 or 0.2 mg; RX=1: 1.0 or 5.0 mg); age group (AG=0: $<$75 years; AG=1: 75-80 years; AG=2: $\geq$80 years); weight index (WT=0: $\geq$100 kg; WT=1: 80-99 kg; WT=2: $<$80 kg); performance rating (PF=0: normal; PF=1: limitation of activity); history of cardiovascular disease (HX=0: no; HX=1: yes); serum haemoglobin (HG=0: $\geq$12 g/100ml; HG=1: 9-12 g/100ml; HG=2: $<$9 g/100ml); size of primary lesion (SZ=0: $<30\ cm^2$; SZ=1: $\geq 30\ cm^2$), and Gleason stage/grade category (SG=0: $\leq$10; SG=1: $>$10). Three types of failure were considered: (a) death due to cancer, (b) death due to cardiovascular (CVD) disease, and (c) death due to other reasons; see Kay (1986) for more details.

With this data set, 149 patients died due to cancer, 139 patients died due to CVD disease, and 56 patients died from other causes. The number of censoring times is 139 (28.8% of 483 patients). The ECM-based semi-parametric mixture modelling approach with three components is fitted to the data, where standard errors of the maximum likelihood estimates are obtained by the nonparametric bootstrap approach with $B = 100$ replications; see Section 5.4.1. Table 5.3 displays the logistic coefficients and regression coefficients in the hazard functions. From Table 5.3, it is observed that higher DES dosage significantly increases the probability of death due to CVD, and thus significantly reduces the risk of death due to cancer. However, DES dosage does not have a significant effect on prolonging the time to death due to either CVD or cancer. Patients with a history of CVD (HX=1) have a higher probability of death due to CVD, compared to those patients without such a history. Patients with larger sized (SZ=1) or high-grade (SG=1) tumours have higher probability of death due to cancer or CVD and also shorter time to death given that death is due to the cause of cancer or CVD.

In Figure 5.3, we illustrate the effect of DES dosage on the cumulative incidence function of death due to cancer for patients with high-grade tumours (SZ=1, SG=1, and the other variables set to zero). For young (AG=0) patients with high-grade

TABLE 5.3
ECM-based semi-parametric mixture model (Example 5.2).

Variable	Logistic model		Three components		
	Cancer	CVD	Cancer	CVD	Other
Constant	−0.44 (0.59)	−0.44 (0.72)			
RX	0.56 (0.45)	1.27*(0.56)	−0.54 (0.31)	−0.18 (0.38)	0.59 (0.59)
AG	−0.52 (0.32)	−0.44 (0.32)	0.17 (0.25)	0.47*(0.23)	0.41 (0.45)
WT	−0.35 (0.27)	−0.35 (0.39)	0.41 (0.21)	−0.03 (0.27)	0.55 (0.48)
PF	−0.17 (0.59)	−0.17 (0.57)	0.10 (0.35)	0.52 (0.35)	0.81 (0.79)
HX	0.54 (0.41)	1.51*(0.50)	0.13 (0.28)	0.59 (0.43)	1.18 (0.89)
HG	0.50 (0.39)	0.20 (0.51)	0.36 (0.32)	−0.09 (0.41)	1.31 (0.90)
SZ	0.32 (0.57)	−1.20 (0.64)	0.52 (0.29)	1.18*(0.46)	1.05 (0.91)
SG	2.10*(0.56)	1.29*(0.60)	0.76*(0.37)	−0.26 (0.38)	1.41 (1.05)

Data are maximum likelihood estimates (standard errors).
* Significant at the 0.05 level.

tumours but no history of CVD, it can be seen from Figure 5.3 that the cumulative incidence of death due to cancer tends to increase much more rapidly in the low-dose DES group than in the high-dose group in the first two years. It can be seen from Table 5.3 that, aside from RX, HX, SZ, and SG, another significant risk factor for death due to CVD is AG (older patients have a significantly shorter time to death due to CVD).

We compared the ECM-based semi-parametric mixture model with a parametric three-component mixture of Weibull distributions, with results presented in Table 5.4. From the fully parametric approach (Table 5.4), it can be seen that higher DES dosage has a marginally significant effect on time to death due to cancer. In contrast to the result obtained by the ECM-based semi-parametric approach (Table 5.3), it is observed that patients with high-grade tumours (SG=1) have an increased chance of cancer death, but SG does not have a significant effect on time to death due to cancer. On the other hand, a larger size primary lesion (SZ=1) has a significant effect on the time to death due to cancer, which is different from the result obtained by the ECM-based semi-parametric approach.

Using the cause-specific hazard approach, Kay (1986) found that treatment with high dose DES significantly reduced the risk of cancer and other causes while increasing the risk of CVD. The HG, SZ, and SG variables were all significant for the time to death due to cancer. Besides the treatment, other significant risk factors for increased risk of CVD were AG and HX, whereas AG and WT were related to the failure from other causes. Using the cause-specific hazard approach, a separate model is fitted for each failure type, treating failure from other types as censored. Hence, the competing types of failure are not jointly estimated and estimation of the unconditional (marginal) probability $\pi_h(x)$ or the cumulative incidence function can only be accomplished by combining estimates of each failure type; see the discussion in Fine (1999), Tai et al. (2001), and Ng and McLachlan (2003a). In contrast, the mixture model-based approach (5.26) simultaneously estimates the logistic coefficients α in

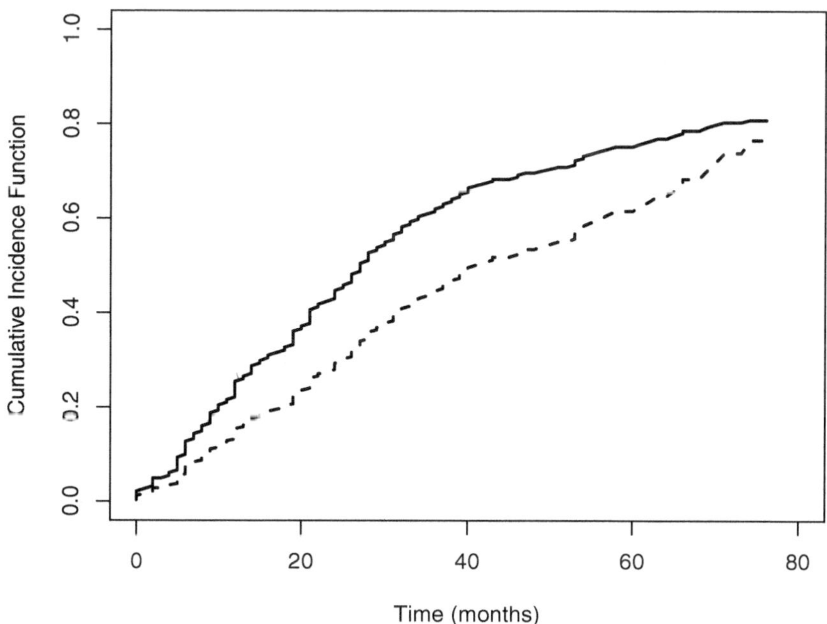

FIGURE 5.3
Cumulative incidence functions of death due to cancer for young patients (AG=0) with weight index $\geq$100, normal PF, no history of CVD, hemoglobin $\geq 12\ g/100ml$, and high-grade tumours (SZ=1, SG=1): Solid line (low-dose DES); Dashed line (high-dose DES).

TABLE 5.4
Three-component Weibull mixture model (Example 5.2).

Variable	Logistic model		Three components		
	Cancer	CVD	Cancer	CVD	Other
Constant	−0.32 (0.54)	−0.28 (0.60)			
RX	0.46 (0.42)	1.18*(0.43)	−0.41*(0.21)	−0.17 (0.17)	0.23 (0.42)
AG	−0.56 (0.36)	−0.47 (0.37)	0.15 (0.18)	0.34*(0.14)	0.16 (0.34)
WT	−0.32 (0.31)	−0.45 (0.35)	0.22 (0.18)	0.04 (0.14)	0.37 (0.30)
PF	−0.13 (0.52)	−0.01 (0.60)	0.18 (0.22)	0.29 (0.26)	0.69 (1.01)
HX	0.46 (0.49)	1.51*(0.49)	0.12 (0.23)	0.47 (0.27)	0.90 (0.56)
HG	0.45 (0.42)	0.22 (0.45)	0.31 (0.18)	−0.05 (0.24)	1.05 (0.66)
SZ	0.33 (0.56)	−1.00 (0.60)	0.53*(0.18)	0.74*(0.32)	0.62 (0.64)
SG	1.99*(0.49)	0.90*(0.52)	0.46 (0.27)	0.02 (0.23)	0.78 (0.65)

Data are maximum likelihood estimates (standard errors).
* Significant at the 0.05 level.

(5.28) and the regression coefficients β_h $(h = 1,\ldots,g)$ in (5.27), thus providing information with regard to how the risk factors influence the incidence of each failure type as well as how they affect the failure time for individuals failed due to each type.

It is noted that the mixture model-based approach assumes that the risk of failures due to g types are independent of each other. That is, there is no constraint in the corresponding component-survival functions $S_h(t_j)$ $(h = 1,\ldots,g)$. In situations where the risks are not independent, a constrained mixture model may be required, for example, in modelling the time to a re-replacement operation or death after the initial replacement of the aortic valve in the heart by a xenograft valve (Ng et al., 1999). It is necessary to impose constraints on the component survival functions because biologic valves have a finite working life and have to be replaced if the patient were to live for a sufficiently long enough time after the initial replacement operation.

5.5 Long-Term Survivor Mixture Models

In the third situation, the aim is to assess the proportion of patients who recover after treatment for some particular disease. This is the case in applications where failure is defined to be the occurrence of a certain event such as serious infection or relapse of a particular disease, a certain fraction of the population may never experience this event, which is characterized by the overall survival curve being levelled at non-zero probability; see, for example, Yau and Ng (2001). In some circumstances, the surviving fractions are said to be "cured", as described in Gordon (1990a). For example, in a clinical trial for assessing the efficacy of a treatment therapy for breast cancer, a patient may be considered as cured if she lives an apparently normal life span and dies of causes other than breast cancer and shows no evidence of the latter disease at death. This is the definition of "personal cure" in Haybittle (1983). A mixture model-based approach provides a framework to estimate the cured proportion and identify the risk factors that influence the cured proportion.

Define a binary indicator random variable Λ, where $\Lambda = 1$ indicates that an individual will eventually experience a failure event (i.e., uncured) and $\Lambda = 0$ indicates that the individual will never experience such event (i.e., cured). We assume that an individual has probability π of experiencing the failure event. For an individual with covariate vector x_j (as risk factors), the probability π can be specified to be a logistic function as

$$\pi(x_j) = \text{pr}(\Lambda = 1|x_j) = \frac{1}{1+\exp(v^T x_j)}, \tag{5.41}$$

where the first element of x_j is assumed to be 1 to account for an intercept term and v is a vector of regression parameters in the logistic model (5.41). Let T be the random variable denoting the time to failure from the cause of interest. We take $T = \infty$ where $\Lambda = 0$ so that the corresponding conditional survival function given $\Lambda = 0$ is equal to one for all finite values of T. The unconditional survival function of T can then be expressed as

$$S(t;x) = 1 - \pi(x) + \pi(x)S_2(t;x), \tag{5.42}$$

where $S_2(t;x)$ is the conditional survival function for the group of uncured patients. Model (5.42) is sometimes referred to as the long-term survivor (LTS) mixture model for the obvious reason that individuals who never experience the failure event can be viewed as being LTSs; see Maller and Zhou (1996) and Yau and Ng (2001). There are a number of variations on the LTS mixture model (5.42), including different forms of the lifetime distribution function for $S_2(t;x)$ of the uncured patients group considered in Peng, Dear, and Deham (1998), Sy and Taylor (2000), and Lambert et al. (2010), and extensions to accelerated hazards model settings (Zhang and Peng, 2009). Mixture models incorporating LTS are encountered in many biomedical and health scientific studies, especially in cancer research; see, for example, Perperoglou, Keramopoullos, and Houwelingen (2007), Dal Maso et al. (2014), Jácome et al. (2014), and Sanchez et al. (2014). The monograph by Maller and Zhou (1996) provides a wide range of applications of the LTS mixture model.

In the typical situation in which the LTS mixture model is applied, the observed data are of the form $(t_j, x_j^T, D_j)^T$ for the jth individual (see Section 5.1), where $D_j = 1$ implies that the jth individual was observed to fail from the cause of interest at time t_j during the follow-up period (that is, $\Lambda = 1$), and $D_j = 0$ implies that the failure time is censored at time t_j. That is, the jth individual had not experienced failure by time t_j. In this context, model (5.42) can be fitted by maximum likelihood after the specification of a parametric model for $S_2(t;x)$. An individual who fails from the cause of interest at time t_j $(j = 1, \dots, n)$ contributes a likelihood factor,

$$\pi(x_j)f_2(t_j;x_j), \tag{5.43}$$

where $f_2(t_j;x_j)$ is the p.d.f. of T given that the individual will experience failure from the cause of interest. An individual who has been followed to time t_j without failure contributes a likelihood factor

$$1-\pi(x_j)+\pi(x_j)S_2(t_j;x_j), \tag{5.44}$$

which is the probability that an individual either never experiences failure or, if so, after time t_j.

Denote the unknown parameter vector Ψ as

$$\Psi=(v^T,\beta^T,\psi^T)^T,$$

where β and ψ are, respectively, the vectors of regression coefficients and unknown parameters in the baseline hazard function for the conditional survival function of the uncured group $S_2(t;x)$. From (5.43) and (5.44), the log likelihood for Ψ is given by

$$\begin{aligned}\log L(\Psi)=\sum_{j=1}^{n}\big[&D_j\log\pi(x_j)f_2(t_j;x_j)+\\&(1-D_j)\log\{1-\pi(x_j)+\pi(x_j)S_2(t_j;x_j)\}\big].\end{aligned} \tag{5.45}$$

Assuming a proportional hazards model in (5.45), the Q-function on the $(k+1)$th iteration can be expressed as

$$\begin{aligned}Q(\Psi,\Psi^{(k)})=\sum_{j=1}^{n}\big[&D_j\log\pi(x_j;v)h_{20}(t_j;\psi)\exp(x_j^T\beta)S_{20}(t_j;\psi)^{\exp(x_j\beta)}+\\&(1-D_j)\{\tau(t_j;x_j,\Psi^{(k)})\log(1-\pi(x_j;v))+\\&(1-\tau(t_j;x_j,\Psi^{(k)}))\log\{\pi(x_j;v)S_{20}(t_j;\psi)^{\exp(x_j\beta)}\}\big],\end{aligned} \tag{5.46}$$

where

$$\tau(t_j;x_j,\Psi^{(k)})=\frac{1-\pi(x_j;v^{(k)})}{1-\pi(x_j;v^{(k)})+\pi(x_j;v^{(k)})S_2(t_j;x_j,\beta^{(k)},\psi^{(k)})} \tag{5.47}$$

is the posterior probability that the jth individual with censored survival time t_j is a LTS. The M-step follows closely the procedure described in Section 5.3.1, with the Q-function replaced by (5.46).

In the above application of the model (5.42), the only information available on individuals with $\Lambda=0$ is in the censored data, where a relatively large censored survival time will be suggestive of a LTS; see Maller and Zhou (1994) for more details. Also, it is assumed that an individual will not fail from other causes, as described in Farewell (1982). Hence $\pi(x)$ represents the probability of failing from the cause of interest in the absence of any competing risk. Ghitany and Maller (1992) considered the large-sample properties of the ML estimators for exponential mixture models with LTS under random censorship. They showed that the consistency and asymptotic normality of the ML estimators of the parameters are anticipated if the data are not too heavily censored, relative to the underlying failure rate and the proportion of LTS. However, any inferences from an LTS mixture model should be considered with caution because there may be a degree of non-identifiability between the location parameter of the logistic model and the shape parameters of the survival function; see Farewell (1986) and Laska and Meisner (1992). As a long-tailed survival curve could mimic the effect of non-zero probability of cure, it is not possible to distinguish whether an individual, who does not experience a failure event, belongs to the cured component or just a censored observation in the uncured component. Problems with convergence for the LTS exponential mixture model can be readily found when the Kaplan-Meier curve of the survival data does not have a clear level plateau (Cantor and Shuster, 1992). The likelihood function should be examined carefully to identify potential non-identifiability.

5.5.1 Example 5.3: Long-Term Survivors Mixture Model

We consider the melanoma data from the Eastern Cooperative Oncology Group (ECOG) phase III clinical trial E1690 (Kirkwood et al., 2000). The data set is available at `http://merlot.stat.uconn.edu/~mhchen/survbook`, consisting of 427 patients having surgery for deep primary or regionally metastatic melanoma. The E1690 clinical trial was a subsequent phase III trial involving identical treatments and patient populations as an earlier trial E1684. The E1690 trial was intended as a confirmatory trial to E1684 for confirming a relapse-free survival benefit of high dose interferon (IFN) alpha-2b as an adjuvant therapy post-surgery; see Ibrahim, Chen, and Sinha (2001) and Ibrahim, Chen, and Chu (2012). In these trials, the relapse-free survival was defined as the time from randomization until progression of tumour or death, whichever comes first. Completed information was available for $n = 408$ patients (96% of 427 patients) on risk variables: Treatment (IFN alpha-2b versus control), age in years (mean centered), gender, performance status, nodule category (Levels 1 to 4 corresponding to increased risk of recurrence), and Breslow depth in mm. The Kaplan-Meier relapse-free survival curves by nodule category presented in Figure 5.4 were all levelled at non-zero probability, indicating the presence of LTS. We fitted the LTS Weibull mixture model (5.42) to the data. The estimated regression coefficients for a single Weibull model and a LTS Weibull mixture model were provided in Table 5.5 for comparison.

With the LTS Weibull mixture model (Table 5.5), the treatment effect increased the cured proportion (adjusted OR of 1.43, marginally significant), but did not significantly reduce the mortality risk for those uncured patients (adjusted HR of 1.05). The estimated survival curves by treatment groups for male and female patients with nodule category 4 versus category 1 are displayed in Figure 5.5. The cured probabilities for male patients treated with IFN alpha-2b were estimated to be 0.599 (node 1) and 0.168 (node 4), respectively, compared to those in the control group (0.511 for node 1 and 0.124 for node 4). For female patients, the estimated cured probabilities were 0.636 (node 1) and 0.191 (node 4) for those treated with IFN alpha-2b, whereas for those in the control group, the estimates were 0.549 (node 1) and 0.141 (node 4). There were no significant differences between males and females. Besides

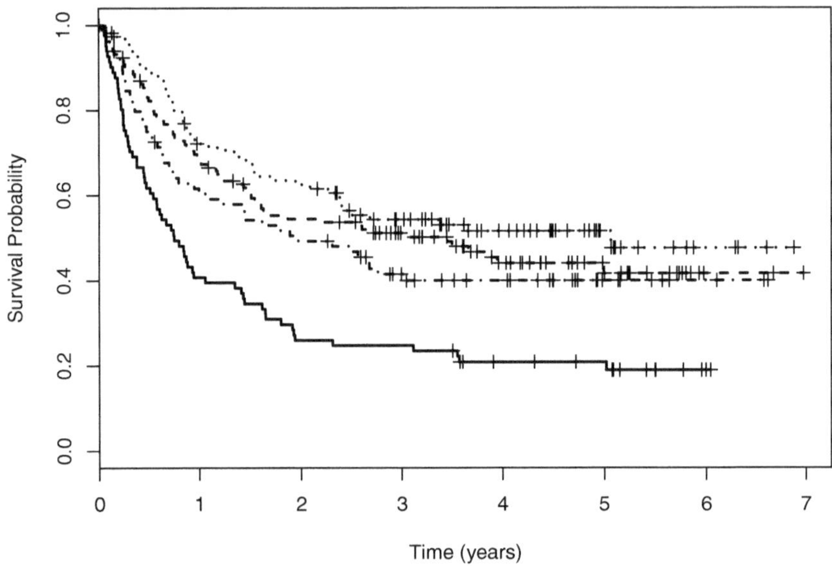

FIGURE 5.4
Kaplan-Meier relapse-free survival curves by nodule category (Increasing risk of recurrence: Level 1 (Dotted line), Level 2 (Dashed line), Level 3 (Dashed-dotted line), and Level 4 (Solid line).

treatment, significant factors for an increased cured proportion were age (adjusted OR of 1.02 per year increase of age), Breslow depth (adjusted OR of 1.16), nodule category (adjusted ORs for Levels 2 and 4 versus Level 1: 2.53 and 7.40, respectively). The Cox-Snell residual plot is given in Figure 5.6. The cumulative hazard function formed from the residuals is quite close to the 45° line, except in the tail where the variability in the estimate of the cumulative hazard function is large. This suggests that the LTS Weibull mixture model is adequate. With a single Weibull model, it was found that the baseline hazard is also increasing. However, as the presence of cured patients was not assumed, the estimate of λ ($\hat{\lambda} = 0.070$) in the baseline hazard function is much smaller than that obtained using the LTS Weibull mixture model ($\hat{\lambda} = 0.758 = \exp(-0.277)$).

It is noted that the E1684 trial data set has been analysed using a SAS macro called "PSPMCM" (Corbière and Joly, 2007) with a Weibull mixture cure model and a Cox's proportional hazard mixture model. The E1684 data set has also been used in Cai et al. (2012) to illustrate the R-package "smcure" for estimating semi-parametric mixture cure models with the assumption of proportional hazards.

TABLE 5.5
Estimates and standard errors (in parentheses) for fitting a single Weibull and a LTS Weibull mixture model (Example 5.3).

Variable	**A single Weibull**	**LTS Weibull mixture model**	
		Logistic (uncured)	**Survival (uncured)**
Constant	-2.658* (0.230)	-0.628 (0.501)	-0.277 (0.331)
Age (years)	0.011 (0.007)	0.024* (0.012)	-0.008 (0.010)
Female	-0.242 (0.150)	-0.154 (0.287)	-0.240 (0.291)
Performance	0.059 (0.217)	-0.094 (0.399)	0.090 (0.363)
Breslow depth (mm)	0.045 (0.024)	0.147* (0.075)	-0.029 (0.036)
Nodule category			
Node1	Reference	Reference	
Node2	0.470* (0.208)	0.929* (0.463)	0.034 (0.335)
Node3	0.580* (0.242)	0.882 (0.453)	0.471 (0.287)
Node4	1.069* (0.242)	2.001* (0.520)	0.498 (0.340)
Treatment	-0.171* (0.085)	-0.360* (0.178)	0.052 (0.245)
Shape parameter α_h	1.284* (0.043)	—	1.031* (0.053)
Log likelihood	-615.999	-482.458	
BIC	1292	1079	

* p-value < 0.05.

Table 5.6 presents the result obtained by an LTS proportional hazards mixture model using the PSPMCM macro. The treatment effects were not significant in both parts of the cured proportion or the mortality risk for uncured patients.

5.6 Diagnostic Procedures

As illustrated in the previous examples, the overall fit of the final parametric mixture model can be checked by graphical examination of the Cox-Snell residuals (Yau and Ng, 2001), where the residual for the jth individual is defined as

$$r_j = H(t_j; \hat{\Psi}, x_j), \tag{5.48}$$

where $H(t_j; \hat{\Psi}, x_j)$ is the cumulative hazard function calculated at the estimated parameters $\hat{\Psi}$ for $j = 1, \ldots, n$. If the mixture regression model is correct and the estimated parameters $\hat{\Psi}$ are close to Ψ, then the residuals $\{r_1, \ldots, r_n\}$ will behave like a censored sample from a unit exponential distribution. Thus the Nelson-Aalen estimate of the cumulative hazard function obtained from the Cox-Snell residuals r_j $(j = 1, \ldots, n)$ can be plotted against the Cox-Snell residuals for diagnostic check of the final model. A computer graphical user interface coded using Microsoft visual C++ has been developed by Lee et al. (2009) to display the Cox-Snell

TABLE 5.6
Estimates and accelerated bias corrected bootstrap 95% CIs (in parentheses) for fitting a LTS proportional hazards mixture model using SAS macro PSPMCM (Example 5.3).

Variable	**LTS proportional hazards mixture model**	
	Logistic (uncured)	**Survival (uncured)**
Constant	-0.375 (-1.36, 0.72)	—
Age (years)	0.023* (0.002, 0.044)	-0.007 (-0.024, 0.007)
Female	-0.231 (-0.744, 0.299)	-0.118 (-0.551, 0.285)
Performance	-0.025 (-0.766, 0.810)	0.002 (-0.625, 0.494)
Breslow depth (mm)	0.111* (0.001, 0.238)	-0.004 (-0.055, 0.048)
Nodule category		
Node1	Reference	
Node2	0.760 (-0.113, 1.591)	0.162 (-0.377, 0.758)
Node3	0.753 (-0.174, 1.583)	0.495 (-0.023, 1.030)
Node4	1.737* (0.662, 2.749)	0.645* (0.105, 1.206)
Treatment	-0.295 (-0.839, 0.207)	-0.038 (-0.389, 0.371)

* p-value < 0.05.

residual plot, where $\log(-\log KM(r_j))$ are plotted against $\log(r_j)$ $(j = 1,\ldots,n)$, and where $KM(r_j)$ denotes the Kaplan-Meier estimate of the survival function of r_j $(j = 1,\ldots,n)$. This interface also provides a diagnostic measure of the overall goodness-of-fit of the final model in terms of the Euclidean norm of the difference between the vectors of $\log(-\log KM(r_j))$ and $\log(r_j)$ for comparing various models. A drawback of the Cox-Snell residual plot is that the residuals r_j $(j = 1,\ldots,n)$ do not indicate the type of departure from the final model when the residual plot does not show a straight line with slope 1. Moreover, it relies on the validity of the proportional hazards assumption. Checking the proportional hazards assumption for component survival distributions is not straightforward. An informative way is to plot $\log(-\log(1-\hat{F}_h(t)/\hat{\pi}_h))$ versus time for each level of the covariate, where $\hat{F}_h(t)$ is the estimated cumulative incidence function, such as the Aalen-Johansen estimator (Aalen and Johansen, 1978), for the hth component and $\hat{\pi}_h$ is the estimated final levelled cumulative incidence function; see Ng and McLachlan (2003a) for more details. The lines with respect to the levels of the covariates should be approximately parallel to support the proportional hazards assumption for the component survival distributions. In practice, this method requires large samples and long-term follow-up.

For LTS mixture models, the SAS PSPMCM macro developed by Corbière and Joly (2007) provides goodness-of-fit measures in terms of comparison between the fitted survival model and the Kaplan-Meier estimate of the (empirical) marginal survival curve (both graphically and via the correlation coefficient between the two sets of estimates); see Maller and Zhou (1996).

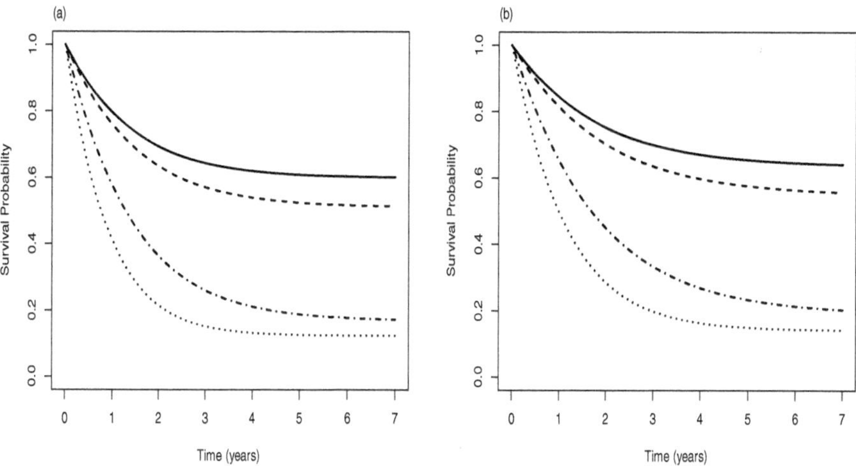

FIGURE 5.5
Estimated survival curves by treatment groups for (a) male patients and (b) female patients (the IFN treatment group: Node 1 (Solid line), Node 4 (Dashed-dotted line); the control group: Node 1 (Dashed line), Node 4 (Dotted line)).

5.7 Fortran Programs for Fitting Mixtures of Survival Models

The Fortran programs "survweicov.f" (for mixture models of Weibull distributions), "semiparamix3ecmnew_final.f" (for semi-parametric mixture survival models), and "lts_mixweiregrsort.f" (for long-term survivor mixture models) as well as external Fortran programs "random.f" and "hybrid.f" can be downloaded from the book's online webpage. The survweicov.f is compiled with random.f and hybrid.f in Cygwin to obtain the executable program MIXWEIBULL, which fits a mixture of parametric survival model (5.14) via the EM algorithm, assuming a Weibull distribution (5.6) with a Cox's proportional hazards form (5.15). The semiparamix3ecmnew_final.f is compiled with random.f and hybrid.f to obtain the executable program SEMIMIXSURV, which fits an ECM-based semi-parametric mixture survival model (5.26) on competing-risk data in the form given by (5.25). The component hazard functions are unspecified, but assumed to follow a proportional hazards model (5.27). Finally, the lts_mixweiregrsort.f is compiled with random.f and hybrid.f to obtain the executable program LTSMIXSURV, which fits a long-term survivor mixture model (5.42) via the EM algorithm, with the proportional hazards assumption.

The program code for fitting the mortality data of colon cancer (Example 5.1) using MIXWEIBULL is provided in Figure 5.7. In Example 5.2, the program code

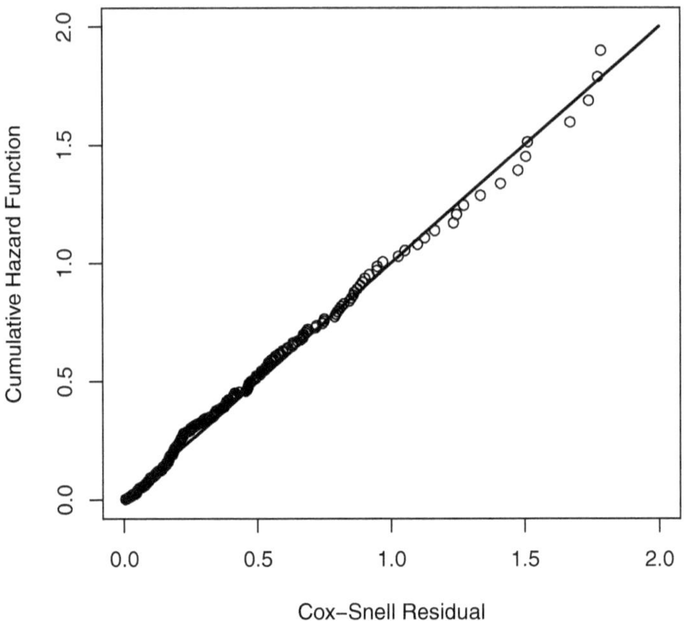

FIGURE 5.6
Cox-Snell residual plot – estimated cumulative hazard function of residuals versus residuals (Imposed is the 45° line).

for fitting the prostate cancer data of competing risks using SEMIMIXSURV is given in Figure 5.8. The program code for fitting the ECOG E1690 clinical trial data using LTSMIXSURV is provided in Figure 5.9.

Main Analysis for Example 5.1

```
$ cd c:/Chapter5              # change to the work directory

$ gfortran -c random.f        # compile a Fortran file to obtain an .o file

$ gfortran -c hybrid.f

$ gfortran -c survweicov.f    # compile the survweicov.f to obtain survweicov.o

# to obtain the executable program MIXWEIBULL from the .o files
$ gfortran -o MIXWEIBULL survweicov.o random.o hybrid.o

$ ./MIXWEIBULL                 # call the MIXWEIBULL program

# output file is "output"


# calculate standard errors using B=100 bootstrap replications

$ ./MIXWEIBULL                 # call the MIXWEIBULL program
 Input: maximum number of iterations
100                                        # consider a maximum of 100 iterations

 Input the seeds for random number generation: input 3 integers between 1 and 30,000,
 e.g. 23 120 3411.
23 120 3411                                # enter the seeds

 Do you want to print out the estimates
   for each bootstrap?     INPUT: 1(Yes); 0(No)
1                                          # to print out the estimates

# output file is "output"
```

FIGURE 5.7
The program codes for Example 5.1.

Main Analysis for Example 5.2

```
$ cd c:/Chapter5             # change to the work directory

$ gfortran -c random.f       # compile a Fortran file to obtain an .o file

$ gfortran -c hybrid.f

# compile the semiparamix3ecmnew_final.f to obtain semiparamix3ecmnew_final.o
$ gfortran -c semiparamix3ecmnew_final.f

# to obtain the executable program SEMIMIXSURV from the .o files
$ gfortran -o SEMIMIXSURV semiparamix3ecmnew_final.o random.o hybrid.o

$ ./SEMIMIXSURV               # call the SEMIMIXSURV program

# output file is "output"


# calculate standard errors using B=100 bootstrap replications

$ ./SEMIMIXSURV               # call the SEMIMIXSURV program
 Input the seeds for random number generation: input 3 integers between 1 and 30,000,
 e.g. 23 120 3411.
23 120 34                                # enter the seeds

 Do you want to print out the estimates
   for each bootstrap?     INPUT: 1(Yes); 0(No)
1                                        # to print out the estimates

# output file is "output"


# for three-component Weibull mixture model
$ gfortran -c random.f       # compile a Fortran file to obtain an .o file

$ gfortran -c hybrid.f

$ gfortran -c mixweiregr.f    # compile the mixweiregr.f to obtain mixweiregr.o

# to obtain the executable program MIXWEIREGR from the .o files
$ gfortran -o MIXWEIREGR mixweiregr.o random.o hybrid.o

$ ./MIXWEIREGR                # call the MIXWEIREGR program

# output file is "output"


# calculate standard errors using B=100 bootstrap replications

$ ./MIXWEIREGR                # call the MIXWEIREGR program
 Input the seeds for random number generation: input 3 integers between 1 and 30,000,
 e.g. 23 120 3411.
23 120 3411                              # enter the seeds

 Do you want to print out the estimates
   for each bootstrap?     INPUT: 1(Yes); 0(No)
1                                        # to print out the estimates

# output file is "output"
```

FIGURE 5.8
The program codes for Example 5.2.

Main Analysis for Example 5.3

```
$ cd c:/Chapter5              # change to the work directory

$ gfortran -c random.f        # compile a Fortran file to obtain an .o file

$ gfortran -c hybrid.f

$ gfortran -c lts_weiregr.f    # compile the lts_weiregr.f to obtain lts_weiregr.o

# to obtain the executable program LTSMIXSURV from the .o files
$ gfortran -o LTSMIXSURV lts_weiregr.o random.o hybrid.o

$ ./LTSMIXSURV                 # call the LTSMIXSURV program

# output file is "output"


# calculate standard errors using B=100 bootstrap replications

$ ./LTSMIXSURV                 # call the LTSMIXSURV program
 Do you need stratified bootstrap? (0/1)
1                                            # consider stratified bootstrap
 Input: which covariate?
9                                            # consider the 9th covariate
 How many strata?
4                              # consider 4 strata (treatment by censoring status)
 Input: number in each of the      4 strata
77 121 97 113
 Sort the data file on Covariate:       9
 Input: maximum number of iterations in bootstrap
100                                          # consider a maximum of 100 iterations

 Input the seeds for random number generation: input 3 integers between 1 and 30,000,
 e.g. 23 120 3411.
23 120 34                                    # enter the seeds

 Do you want to print out the estimates
   for each bootstrap?     INPUT: 1(Yes); 0(No)
1                                            # to print out the estimates

# output file is "output"
```

FIGURE 5.9
The program codes for Example 5.3.

6

Advanced Mixture Modelling with Random-Effects Components

6.1 Why is Random Effects Modelling Needed?

Extending the generalized linear model (GLM) (Nelder and Wedderburn, 1972; McCullagh and Nelder, 1989), in the past three decades, there were vast literature exploring and developing statistical methods in a direction to relax the independence of the response variable assumption. Why is relaxing such an assumption important? In practice, when collecting data to conduct statistical data analysis, it is rather common that the data at hand have some kind of dependence. For example, panel data in a longitudinal study, repeated measurements from the same subject, observation in clusters which share the same/similar environmental cluster-specific effects, etc. Sometimes, such data dependence may even exhibit a hierarchical (or multilevel) clustering structure. Research efforts to tackle this important problem in statistical modelling may broadly be categorized into two main approaches: (i) Generalized estimation equations (GEE) and (ii) Generalized linear mixed models (GLMM).

The GEE method (Liang and Zeger, 1986; Zeger and Liang, 1986; Zeger, Liang, and Albert, 1988) extends the GLM by directly incorporating within-subject associations among repeated measurements. This method is also termed as marginal models or population-average models (in contrast to subject-specific models). The term marginal is employed to illustrate that the model for the mean response at each occasion depends only upon the covariates of interest, and does not incorporate dependence on random effects or previous responses. In other words, this type of model is characterized by the marginal expectation as a function of covariates. The method can estimate the parameters of GLM with an unknown correlation structure between responses, and is viewed as a type of semi-parametric regression technique, relying on the first and second moment assumptions of marginal distributions. Under mild regularity conditions together with the correct mean model specification assumption, estimators obtained from GEE are consistent even when the covariance structure is misspecified. Parameter estimates are typically obtained by using the Newton-Raphson iterative algorithm. Efficiency of parameter estimates is established for a choice of variance structure, including independent, unstructured, exchangeable, and autoregressive. The GEE method is commonly used due to its ease of application and its neatly established asymptotic properties under mild regularity conditions.

On the other hand, GLMM is a popular alternative when particular subject-specific predictions are of interest. GLM has provided a unified framework for a variety of response data types. A natural extension is the generalization of GLM by incorporating random effects such that the dependence (correlation) structure can explicitly be modelled. In essence, GLMM built on the developments in GLM and linear mixed models (LMM), and merged into an advanced modelling technique which allowed non-Gaussian response and relaxation of the independence assumption on the response variable. The premise is that random effects are introduced to the linear predictor so that a specific dependence structure among the responses can be modelled explicitly. Parameter estimation may be achieved by Newton-Raphson, EM or other numerical iterative algorithms. Regression and variance component parameter estimates, as well as random effects prediction, can be obtained. Prediction of random effects (subject-specific predictions) are especially important in some research studies, for example, in identifying high or low risk subjects/clusters.

6.2 Fundamentals for GLMM Formulation and Derivation

For linear mixed models (LMMs) with normally distributed random components, best linear unbiased predictors (BLUPs) were developed by Henderson (1975) for simultaneously estimating the fixed components of a mixed model as well as the realized values of random components of the model. Such estimators or predictors were shown to be best, linear, and unbiased. The estimation procedure has been reviewed by Robinson (1991). When the BLUP estimators of the random components are used to estimate the variances of those random components, the estimated variance components are severely biased towards zero. Harville (1977) has noted that the BLUP computational procedure may be used as a stepping stone to find both maximum likelihood (ML) and residual maximum likelihood (REML) estimators as described in Patterson and Thompson (1971). The relationship is further developed in Fellner (1986, 1987) and reviewed in Speed (1991). Such theory has then been extended to generalized linear mixed models (GLMM) based on the quadratic approximation of the likelihood in the region of the maximum, where the response variable is not necessarily normally distributed, but the model may be fitted using a penalized quasi-likelihood approach which mirrors the development in normal theory models. Modelling starts with the BLUP estimators as the initial step and extends to finding ML and REML estimators. Developments of the GLMM are given by Schall (1991), Breslow and Clayton (1993), McGilchrist (1994), Kuk (1995), Lee and Nelder (1996), and Yau and Kuk (2002), among others. The following subsections show the key derivations drawn from McGilchrist and Yau (1995).

6.2.1 Normally Distributed Random Components and BLUP Estimation

Let y be a response vector which is generated by the linear mixed model

$$y = \eta + \varepsilon, \qquad \eta = X\beta + Zu, \tag{6.1}$$

where ε is distributed as $N(0, \sigma^2 D)$, D is a known symmetric matrix of a dimension equal to the number of observations n in the response vector y and X, Z are matrices of values of regression variables. The unknown parameter β has dimension v while the random component u may be partitioned

$$u^T = [u_1^T, u_2^T, \ldots, u_k^T], \qquad Z = [Z_1, Z_2, \ldots, Z_k],$$

where Z and u are partitioned conformally and u_j are independent random components distributed as $N[0, \sigma_j^2 A_j(\phi)]$. For convenience we write $\sigma_j^2 = \sigma^2\theta_j$ and the σ^2, θ_j and ϕ are unknown parameters. The parameter ϕ may be a vector parameter of dimension p, which describes the covariance structure of the vectors u_j.

Estimation of the parameters in this model is accomplished by firstly developing best linear unbiased prediction (BLUP) estimators and then using those estimators as an initial computation in finding maximum likelihood (ML) and residual maximum likelihood (REML) estimators.

The log-likelihood of y taking u as conditionally fixed, is denoted by l_1 and the logarithm of the probability density function of u is denoted by l_2. Expression for these quantities are

$$\begin{aligned}
l_1 &= -(1/2)[n \log 2\pi\sigma^2 + \log|D| \\
&\quad + \sigma^{-2}(y - X\beta - Zu)^T D^{-1}(y - X\beta - Zu)] \\
l_2 &= -(1/2)\sum_{j=1}^{k}[v_j \log 2\pi\sigma_j^2 + \log|A_j(\phi)| + \sigma_j^{-2} u_j^T A_j^{-1}(\phi) u_j],
\end{aligned} \tag{6.2}$$

where v_j is the dimension of u_j. BLUP estimates of the parameters and realizations of the random components u are obtained by maximizing $l = l_1 + l_2$. For convenience let A be the block diagonal matrix

$$A = \begin{bmatrix} \theta_1 A_1 & & & \\ & \theta_2 A_2 & & \\ & & \ddots & \\ & & & \theta_k A_k \end{bmatrix}.$$

The derivatives with respect to the parameters are

$$\begin{aligned}
\partial l/\partial\beta &= \partial l_1/\partial\beta = \sigma^{-2} X^T D^{-1}(y - X\beta - Zu) \\
\partial l/\partial u &= \sigma^{-2}[Z^T D^{-1}(y - X\beta - Zu) - A^{-1}u] \\
\partial l/\partial\sigma^2 &= -(1/2)[n\sigma^{-2} - \sigma^{-4}(y - X\beta - Zu)^T D^{-1}(y - X\beta - Zu)] \\
\partial l/\partial\sigma_j^2 &= -(1/2)[v_j\sigma_j^{-2} - \sigma_j^{-4} u_j^T A_j^{-1} u_j] \\
\partial l/\partial\phi_s &= -(1/2)\sum_{j=1}^{k}[v_j^{(s)} - \sigma_j^{-2} u_j^T A_j^{-1}(\partial A_j/\partial\phi_s) A_j^{-1} u_j]
\end{aligned}$$

where $v_j^{(s)} = \mathrm{tr}(A_j^{-1}\partial A_j/\partial \phi_s)$, and where $\mathrm{tr}(\cdot)$ is the trace of a matrix. Equating the above derivatives to zero gives the BLUP matrix equation for estimating β and u which is analogous to that given by Henderson et al. (1959), viz.

$$\begin{bmatrix} X^TD^{-1}X & X^TD^{-1}Z \\ Z^TD^{-1}X & Z^TD^{-1}Z+A^{-1} \end{bmatrix} \begin{bmatrix} \widetilde{\beta} \\ \widetilde{u} \end{bmatrix} = \begin{bmatrix} X^TD^{-1}y \\ Z^TD^{-1}y \end{bmatrix};$$

$$\begin{aligned} \widetilde{\sigma}^2 &= n^{-1}(y - X\widetilde{\beta} - Z\widetilde{u})^TD^{-1}(y - X\widetilde{\beta} - Z\widetilde{u}) \\ \widetilde{\sigma}_j^2 &= v_j^{-1}\widetilde{u}_j^TA_j^{-1}\widetilde{u}_j, \quad j = 1,2,\ldots,k; \end{aligned}$$

and

$$\sum_{j=1}^{k}[v_j^{(s)} - \widetilde{\sigma}_j^{-2}\widetilde{u}_j^TA_j^{-1}(\partial A_j/\partial \phi_s)A_j^{-1}\widetilde{u}_j]|_{\phi=\widetilde{\phi}} = 0, \quad s = 1,2,\ldots$$

The BLUP equation for ϕ_s may not be solvable explicitly. Using the first matrix equation and letting

$$\Sigma = D + ZAZ^T, \qquad K = D^{-1} - D^{-1}X(X^TD^{-1}X)^-X^TD^{-1}$$

The matrix equation can be solved to give

$$\widetilde{\beta} = (X^T\Sigma^{-1}X)^-X^T\Sigma^{-1}y, \qquad \widetilde{u} = (Z^TKZ + A^{-1})^-Z^TKy.$$

Note that K is symmetric and $X^TKX = 0$ implies $KX = 0$.

6.2.2 Maximum Likelihood Estimation

The likelihood function for y formed by integration over the distribution of u is

$$l_{\mathrm{ML}} = -(1/2)[n\log 2\pi\sigma^2 + \log|\Sigma| + \sigma^{-2}(y - X\beta)^T\Sigma^{-1}(y - X\beta)] \tag{6.3}$$

The derivatives with respect to the parameters β, σ^2, θ and ϕ are

$$\begin{aligned} \partial l_{\mathrm{ML}}/\partial \beta &= \sigma^{-2}X^T\Sigma^{-1}(y - X\beta) \\ \partial l_{\mathrm{ML}}/\partial \sigma^2 &= -(1/2)[n\sigma^{-2} - \sigma^{-4}(y - X\beta)^T\Sigma^{-1}(y - X\beta)] \\ \partial l_{\mathrm{ML}}/\partial \theta_j &= -(1/2)[\mathrm{tr}\,\Sigma^{-1}\partial\Sigma/\partial\theta_j + \sigma^{-2}(y - X\beta)^T(\partial\Sigma^{-1}/\partial\theta_j)(y - X\beta)] \\ \partial l_{\mathrm{ML}}/\partial \phi_s &= -(1/2)[\mathrm{tr}\,\Sigma^{-1}\partial\Sigma/\partial\phi_s + \sigma^{-2}(y - X\beta)^T(\partial\Sigma^{-1}/\partial\phi_s)(y - X\beta)], \end{aligned}$$

giving

$$\begin{aligned} \hat{\beta}_{\mathrm{ML}} &= \widetilde{\beta} = H^-X^T\Sigma^{-1}y, \quad \text{where} \quad H = X^T\Sigma^{-1}X \\ \hat{\sigma}^2_{\mathrm{ML}} &= n^{-1}(y - X\hat{\beta}_{\mathrm{ML}})^T\Sigma^{-1}(y - X\hat{\beta}_{\mathrm{ML}}) \end{aligned}$$

In general, the equations for ϕ are not explicitly solvable although they may be for particular Σ. The information matrix l_{ML} is

$$\begin{bmatrix} \sigma^{-2}H & 0 & 0 & 0 \\ . & n/2\sigma^4 & (1/2\sigma^2)[\mathrm{tr}\,\Sigma^{-1}\partial\Sigma/\partial\theta_j] & (1/2\sigma^2)[\mathrm{tr}\,\Sigma^{-1}\partial\Sigma/\partial\phi_t] \\ . & . & (1/2)\mathrm{tr}[\Sigma^{-1}\partial\Sigma/\partial\theta_i\Sigma^{-1}\partial\Sigma/\partial\theta_j] & (1/2)\mathrm{tr}[\Sigma^{-1}\partial\Sigma/\partial\theta_i\Sigma^{-1}\partial\Sigma/\partial\phi_t] \\ . & . & . & (1/2)\mathrm{tr}[\Sigma^{-1}\partial\Sigma/\partial\phi_s\Sigma^{-1}\partial\Sigma/\partial\phi_t]. \end{bmatrix}$$

In the current model, the variance matrix has the form

$$\Sigma = D + ZAZ^T = D + \sum_{j=1}^{k} \theta_j Z_j A_j(\phi) Z_j^T$$

so that

$$\partial\Sigma/\partial\theta_j = Z_j A_j(\phi) Z_j^T, \quad \partial\Sigma/\partial\phi_s = \sum_{j=1}^{k} \theta_j Z_j (\partial A_j/\partial\phi_s) Z_j^T$$

Note that if the ϕ parameter is not present in A_j, then that derivative is zero. Using some matrix results to simplify, we have

$$\hat{\sigma}^2_{\text{ML}} = n^{-1} y^T \Sigma^{-1}(y - X\hat{\beta}_{\text{ML}}) = n^{-1} y^T D^{-1}(y - X\tilde{\beta} - Z\tilde{u}) = n^{-1} y^T Q y$$

$$\begin{aligned}
\partial l_{\text{ML}}/\partial\theta_j|_{\beta=\tilde{\beta}} &= -(1/2\theta_j)(v_j - r_j^* - \sigma_j^{-2}\tilde{u}_j^T A_j^{-1}\tilde{u}_j) = 0 \\
&\quad \text{giving } \hat{\sigma}^2_{\text{j(ML)}} = \tilde{u}_j^T A_j^{-1}\tilde{u}_j/(v_j - r_j^*) \\
\partial l_{\text{ML}}/\partial\phi_s|_{\beta=\beta} &= -(1/2)\sum_{j=1}^{k} [v_j^{(s)} + r_j^{*(s)} + \hat{\sigma}^{-2}_{\text{j(ML)}}\tilde{u}_j^T(\partial A_j^{-1}/\partial\phi_s)\tilde{u}_j]|_{\phi=\hat{\phi}_{\text{(ML)}}} = 0
\end{aligned}$$

giving equations which may be solved iteratively for ϕ. The information matrix, multiplied by 2, becomes

$$\begin{bmatrix}
2\sigma^{-2}H & 0 & 0 & 0 \\
. & n/\sigma^4 & \sigma_j^{-2}(v_j - r_j^*) & \sigma^{-2}\sum_{j=1}^{k}(v_j^{(t)} + r_j^{*(t)}) \\
. & . & [\theta_i^{-2}(v_i - 2r_i^*)\delta_{ij} & \theta_i^{-1}(v_i^{(t)} + 2r_i^{*(t)}) - \Sigma_{j=1}^{k}\theta_i^{-1}\theta_j^{-1}r_{ij}^{*(t)} \\
. & . & \quad +\theta_i^{-2}\theta_j^{-2}r_{ij}^*] & \\
. & . & . & \Sigma_{j=1}^{k}[-v_j^{(st)} + 2r_j^{*(st)} + \Sigma_{m=1}^{k}\theta_j^{-1}\theta_m^{-1}r_{jm}^{*(st)}]
\end{bmatrix}.$$

6.2.3 Residual Maximum Likelihood Estimation

Let l_{REML} be the log-likelihood function for residual maximum likelihood techniques. Since the matrix K defined in Section 6.2.1 satisfies $KX = 0$, an expression for l_{REML} given by Patterson and Thompson (1971) is

$$l_{\text{REML}} = -(1/2)[(n - v)\log 2\pi\sigma^2 + \log|K\Sigma K| + \sigma^{-2}y^T K(K\Sigma K)^- Ky] \tag{6.4}$$

where $|K\Sigma K|$ must be interpreted as the determinant of linearly independent rows and columns of $K\Sigma K$. The following development parallels Thompson (1980). The first-order derivatives of l_{REML} are

$$\begin{aligned}
\partial l_{\text{REML}}/\partial\sigma^2 &= -(1/2)[(n-v)\sigma^{-2} - \sigma^{-4}y^T K(K\Sigma K)^{-}Ky] \\
&= -(1/2)[(n-v)\sigma^{-2} - \sigma^{-4}y^T Qy] \\
\partial l_{\text{REML}}/\partial\theta_j &= -(1/2)\{\text{tr }[(K\Sigma K)^{-}K\partial\Sigma/\partial\theta_j K] \\
&\quad -\sigma^{-2}y^T K(K\Sigma K)^{-}K\partial\Sigma/\partial\theta_j K(K\Sigma K)^{-}Ky\} \\
&= -(1/2)[\text{tr }Q\partial\Sigma/\partial\theta_j - \sigma^{-2}y^T Q\partial\Sigma/\partial\theta_j Qy] \\
\partial l_{\text{REML}}/\partial\phi_s &= -(1/2)\{\text{tr }[(K\Sigma K)^{-}K\partial\Sigma/\partial\phi_s K] \\
&\quad -\sigma^{-2}y^T K(K\Sigma K)^{-}K\partial\Sigma/\partial\phi_s K(K\Sigma K)^{-}Ky\} \\
&= -(1/2)[\text{tr }Q\partial\Sigma/\partial\phi_s - \sigma^{-2}y^T Q\partial\Sigma/\partial\phi_s Qy]
\end{aligned}$$

Thus $\hat{\sigma}^2_{\text{REML}} = (n-v)^{-1}y^T Qy$ and the REML information matrix I_{REML} is

$$\begin{bmatrix}
(n-v)/2\sigma^4 & (1/2\sigma^2)\text{tr }Q\partial\Sigma/\partial\theta_j & (1/2\sigma^2)\text{tr }Q\partial\Sigma/\partial\phi_t \\
. & (1/2)\text{tr }Q\partial\Sigma/\partial\theta_i Q\partial\Sigma/\partial\theta_j & (1/2)\text{tr }Q\partial\Sigma/\partial\theta_i Q\partial\Sigma/\partial\phi_t \\
. & . & (1/2)\text{tr }Q\partial\Sigma/\partial\phi_s Q\partial\Sigma/\partial\phi_t
\end{bmatrix}.$$

Using $\partial\Sigma/\partial\theta_j = Z_j A_j(\phi)Z_j^T$, $\partial\Sigma/\partial\phi_s = \sum_{j=1}^k \theta_j Z_j(\partial A_j/\partial\phi_s)Z_j^T$ and some matrix results, we have

$$\begin{aligned}
&\partial l_{\text{REML}}/\partial\theta_j = -(1/2\theta_j)[v_j - r_j - \sigma_j^{-2}\tilde{u}_j^T A_j^{-1}\tilde{u}_j] = 0 \\
&\qquad \text{giving} \quad \hat{\sigma}^2_{j(\text{REML})} = \tilde{u}_j^T A_j^{-1}\tilde{u}_j/(v_j - r_j) \\
&\partial l_{\text{REML}}/\partial\phi_s = -(1/2)\sum_{j=1}^k [v_j^{(s)} + r_j^{(s)} + \hat{\sigma}^{-2}_{j(\text{REML})}\tilde{u}_j^T(\partial A_j^{-1}/\partial\phi_s)\tilde{u}_j]|_{\phi=\hat{\phi}(\text{REML})} = 0
\end{aligned}$$

Again this last equation may have to be solved iteratively for ϕ. The REML information matrix, multiplied by 2, is

$$\begin{bmatrix}
(n-v)/\sigma^4 & \sigma_j^{-2}(v_j - r_j) & \sigma^{-2}\sum_{j=1}^k(v_j^{(t)} + r_j^{(t)}) \\
. & [\theta_i^{-2}(v_i - 2r_i)\delta_{ij} + \theta_i^{-2}\theta_j^{-2}r_{ij}] & \theta_i^{-1}(v_i^{(t)} + 2r_i^{(t)} - \sum_{j=1}^k \theta_i^{-1}\theta_j^{-1}r_{ij}^{(t)}) \\
. & . & \sum_{j=1}^k[-v_j^{(st)} + 2r_j^{(st)} + \sum_{m=1}^k \theta_j^{-1}\theta_m^{-1}r_{jm}^{(st)}]
\end{bmatrix}.$$

6.2.4 Generalized Linear Mixed Models

The extension of the theory to generalised linear mixed models has now been accomplished for a response vector y, which is not necessarily normally distributed, but has a distribution dependent on a vector quantity η which is related to vector of regression variables through the equation

$$\eta = X\beta + Zu.$$

If $\text{f}(y;\beta|u)$ is the probability (density) function of y conditional on fixed u, then $l_1 = \ln \text{f}(y;\beta|u)$ is the log-likelihood of y conditional on fixed u. Taking u to be normally distributed as in the previous section, the logarithm of its probability density is l_2 as given in (6.2) in Section 6.2.1. The sum $l = l_1 + l_2$ is then termed a penalised likelihood function and carries over the spirit of BLUP into a non-normal framework. In this sense l_2 is a penalty function for the conditional log-likelihood l_1. The penalised likelihood estimators of β, u, equivalent to BLUP, are obtained by finding the likelihood derivatives

$$\partial l/\partial\beta = X^T \mathrm{d}l_1/\mathrm{d}\eta,\ \partial l/\partial u_j = Z_j^T \mathrm{d}l_1/\mathrm{d}\eta - \sigma_j^{-2}A_j^{-1}u_j,\ j = 1,2,\ldots,k.$$

$$\partial^2 l/\partial u_j \partial u_j^T = \partial^2 l_1/\partial u_j \partial u_j^T - \sigma_j^{-2}A_j^{-1} = -Z_j^T B Z_j - \sigma_j^{-2}A_j^{-1},$$

where $B = -\mathrm{d}^2 l_1/\mathrm{d}\eta \mathrm{d}\eta^T$. The Newton-Raphson iterative procedure for estimating β, u is

$$\begin{bmatrix} \widetilde{\beta} \\ \widetilde{u} \end{bmatrix} = \begin{bmatrix} \beta_0 \\ u_0 \end{bmatrix} + V^- \begin{bmatrix} X^T \\ Z^T \end{bmatrix} \mathrm{d}l_1/\mathrm{d}\eta - V^- \begin{bmatrix} 0 \\ \sigma^{-2}A^{-1}u_0 \end{bmatrix}, \tag{6.5}$$

where

$$V = \begin{bmatrix} X^T \\ Z^T \end{bmatrix} B[X, Z] + \begin{bmatrix} 0 & 0 \\ 0 & \sigma^{-2}A^{-1} \end{bmatrix}.$$

If V is replaced by $E(V)$, then the iterative procedure becomes the method of scoring.

The estimation procedure is as follows. For any given application, write down the log-likelihood l_1 as a function of $\eta = X\beta + Zu$, taking u to be conditionally fixed. Using initial estimates of θ, ϕ and letting $\beta = \beta_0, u = u_0$ be initial estimates of β, u solve Equation (6.5) for $\widetilde{\beta}, \widetilde{u}$. Initial values are replaced by estimates in a new iteration and so on until convergence. Estimates of σ_j^2, ϕ are obtained from

$$\hat{\theta}_{j(\mathrm{ML})} = \widetilde{u}_j^T A_j^{-1}\widetilde{u}_j/(v_j - r_j^*) \quad \text{or} \quad \hat{\theta}_{j(\mathrm{REML})} = \widetilde{u}_j^T A_j^{-1}\widetilde{u}_j/(v_j - r_j)$$

and

$$\Sigma_{j=1}^k [v_j^{(s)} + r_j^{*(s)} + \hat{\sigma}_j^{-2}\widetilde{u}_j^T \partial A_j^{-1}/\partial\phi_s \widetilde{u}_j] = 0 \quad \text{for } \hat{\phi}_{s(\mathrm{ML})}, s = 1,2,\ldots,p$$
$$\Sigma_{j=1}^k [v_j^{(s)} + r_j^{(s)} + \hat{\sigma}_j^{-2}\widetilde{u}_j^T \partial A_j^{-1}/\partial\phi_s \widetilde{u}_j] = 0 \quad \text{for } \hat{\phi}_{s(\mathrm{REML})}, s = 1,2,\ldots,p.$$

Information matrices are given for ML and REML in Sections 6.2.2 and 6.2.3.

In addition to the conventional GLM with random effects, the above GLMM formulation has been developed in order to address a broader class of problems. Essentially, the method only requires that the distribution of the response variable conditional on fixed random components is expressible in terms of the linear predictor. To change the method from one problem to another requires only the reprogramming of the first and second derivatives of l_1. In contrast to the GLM with random effects, an explicit adjusted dependent variable, which depends on the specification of the link function, is not required. The current GLMM formulation provides flexibility in various applications as exemplified in the next section.

6.3 Application of GLMM to Mixture Models with Random Effects

In this section, we highlight some applications of GLMM to mixture models, including Poisson mixture and survival mixture. In fact, its applicability has a broad scope so long as the distribution of the response variable conditional on fixed random components is expressible in terms of the linear predictor. Other applications include normal mixture (Yau, Lee, and Ng, 2003) and Gamma mixture (Lee et al., 2007).

6.3.1 Poisson Mixture Models

Poisson mixture regression is typically used to model heterogeneous count outcomes that arise from two or more underlying sub-populations.

Let Y denote the count random variable. Without loss of generality, the probability distribution of Y is assumed to be a two-component Poisson mixture

$$\Pr(Y=y)=\pi\frac{e^{-\mu_1}(\mu_1)^y}{y!}+(1-\pi)\frac{e^{-\mu_2}(\mu_2)^y}{y!} \qquad y=0,1,2,\ldots$$

where $0<\pi<1$ gives the proportion of subjects belonging to the first component or sub-population, and parameters μ_1 and μ_2 are the means of the respective Poisson components. The mean and variance of Y are, respectively,

$$\begin{aligned} E(Y) &= \pi\mu_1+(1-\pi)\mu_2 \\ var(Y) &= E(Y)+\{\pi\mu_1^2+(1-\pi)\mu_2^2-E(Y)^2\} \end{aligned}$$

Poisson mixture distributions with more than two components can be defined analogously.

When the observed counts are nested in clusters, let Y_{ij} be the jth observed count in the ith cluster, where m is the number of clusters and $\sum_{i=1}^{m} n_i = n$ is the sample size. In the regression setting, both $\log(\mu_{1ij})$ and $\log(\mu_{2ij})$ are assumed to be a linear function of covariates. The covariates appearing in these two parts are not necessarily the same. With random effects, the linear predictors η_{1ij} and η_{2ij} are defined as

$$\begin{aligned} \log(\mu_{1ij}) &= \eta_{1ij}=x_{1ij}^T\beta_1+v_{1,i} \\ \log(\mu_{2ij}) &= \eta_{2ij}=x_{2ij}^T\beta_2+v_{2,i} \end{aligned}$$

where x_{1ij} and x_{2ij} are vectors of covariates with corresponding regression coefficients β_1 and β_2. Let $v_1=(v_{1,1},\ldots,v_{1,m})^T$ and $v_2=(v_{2,1},\ldots,v_{2,m})^T$. For simplicity, v_1 and v_2 are assumed to be independent and distributed as $\mathrm{N}(0,\sigma_{v_1}^2 I_m)$ and $\mathrm{N}(0,\sigma_{v_2}^2 I_m)$, respectively, where I_m denotes an $m\times m$ identity matrix.

Following the GLMM formulation of McGilchrist (1994), the best linear unbiased prediction (BLUP) type log-likelihood is given by $l = l_1 + l_2$, where

$$\begin{aligned} l_1 &= \sum_{i,j} \log\left(\pi \frac{e^{-\mu_{1ij}}(\mu_{1ij})^{y_{ij}}}{y_{ij}!} + (1-\pi)\frac{e^{-\mu_{2ij}}(\mu_{2ij})^{y_{ij}}}{y_{ij}!}\right) \\ l_2 &= -\frac{1}{2}\left[m\log(2\pi\sigma_{v_1}^2) + \sigma_{v_1}^{-2} v_1^T v_1 + m\log(2\pi\sigma_{v_2}^2) + \sigma_{v_2}^{-2} v_2^T v_2\right]. \end{aligned}$$

Estimation may be performed iteratively. In the initial step, coefficients in the linear predictors are estimated for fixed variance components by maximizing the above BLUP log-likelihood. Estimation of variance component parameters is then achieved using approximate residual maximum likelihood (REML) estimating equations.

Detailed estimation procedures and applications can be found in Section 7.2 and in Wang, Yau, and Lee (2002a) and Wang et al. (2007b).

6.3.2 Zero-Inflated Poisson Mixture Models

In general, zero-inflated Poisson (ZIP) regression is a model for count data with excess zeros. Consider a discrete random variable Y with ZIP distribution:

$$\begin{aligned} \Pr(Y=0) &= \pi + (1-\pi)e^{-\mu} \\ \Pr(Y=y) &= (1-\pi)\frac{e^{-\mu}(\mu)^y}{y!} \qquad y = 1, 2, \ldots, \end{aligned}$$

where $0 < \pi < 1$, so that it incorporates more zeros than those allowed by the Poisson distribution.

The ZIP distribution may also be regarded as a mixture of a Poisson (μ) and a degenerate component putting all its mass at zero. A plausible interpretation is in relation to its two-point heterogeneity: a sub-population who is subject to the structural zeros, and another sub-population with members having an outcome following the Poisson distribution. It can be shown that:

$$\begin{aligned} E(Y) &= (1-\pi)\mu \\ var(Y) &= E(Y) + E(Y)\{\mu - E(Y)\}, \end{aligned}$$

so that the ZIP model incorporates the extra variation unaccounted for by the Poisson distribution.

For independent counts Y_j, $j = 1, \ldots, n$, a ZIP regression model (Lambert, 1992) was proposed to study the effects of risk factors or confounders by assuming both log (μ_j) and logit (π_j) =log ($\pi_j/(1-\pi_j)$) to be linear functions of some covariates.

To extend the ZIP regression by incorporating random effects in the model, let Y_{ij} ($i = 1, 2, \ldots, m; j = 1, 2, \ldots, n_i$) and $\sum_{i=1}^m n_i = n$ be the count random variable of the jth observation in the ith cluster. In the regression setting, both logit(π_{ij}) and log(μ_{ij}) are assumed to depend on a linear function of covariates. The covariates appearing in these two parts are not necessarily the same. With random effects, the linear predictors ξ_{ij} and η_{ij} are defined as follows:

$$\begin{aligned} \text{logit}(\pi_{ij}) &= \xi_{ij} = w_{ij}^T\alpha + u_i, \\ \log(\mu_{ij}) &= \eta_{ij} = x_{ij}^T\beta + v_i, \end{aligned}$$

where w_{ij} and x_{ij} are respectively vectors of covariates for the logistic and the Poisson components, and α and β are the corresponding vectors of regression coefficients. Letting $u = (u_1, \ldots, u_m)^T$ and $v = (v_1, \ldots, v_m)^T$, we assume that u, v are independent and are distributed as $N(0, \sigma_u^2 I_m)$ and $N(0, \sigma_v^2 I_m)$, respectively, where I_m denotes an $m \times m$ identity matrix.

Following the GLMM method, the best linear unbiased prediction (BLUP) type log-likelihood is given by $l = l_1 + l_2$, where:

$$\begin{aligned} l_1 &= \sum_{y_{ij}=0} \log\left(\frac{\exp\xi_{ij} + \exp(-\exp\eta_{ij})}{1+\exp\xi_{ij}}\right) \\ &\quad + \sum_{y_{ij}>0} \left(y_{ij}\eta_{ij} - \exp\eta_{ij} - \log(y_{ij}!) - \log(1+\exp\xi_{ij})\right) \\ l_2 &= -\frac{1}{2}\left[m \log(2\pi\sigma_u^2) + \sigma_u^{-2}u^T u + m \log(2\pi\sigma_v^2) + \sigma_v^{-2}v^T v\right], \end{aligned}$$

so that l can be viewed as a penalized log-likelihood function with l_2 being the penalty for the conditional log-likelihood l_1 when the random effects are conditionally fixed.

Estimation commences with the maximization of the above BLUP log-likelihood. Approximate residual maximum likelihood (REML) estimates are then obtained by adjusting the variance components using REML estimating equations.

The above ZIP formulation is categorized as the unconditional modelling approach. An equivalent way to build the ZIP regression model with random effects is via the conditional modelling approach in which the two components are viewed as a mixture of zero point mass and truncated Poisson distribution. The conditional approach has an advantage in that the two components can be completely separated and thus numerical estimation is computationally more efficient. Of course, interpretation of results, especially the effect of covariates, will be different.

Detailed estimation procedures and applications can be found in Section 7.3 and in Yau and Lee (2001), Wang, Yau, and Lee (2002b), and Lee et al. (2006).

6.3.3 Frailty Models in Survival Analysis

One of the important applications of GLMM is on frailty models in survival analysis. In general, the failure time response variable in survival data analysis does not fall into the exponential family of distribution due to primarily its censoring feature. The proportional hazards models for fixed-effect regression models were popularized by Cox (1972, 1975) in his partial likelihood estimation method. The most appealing aspect of the partial likelihood approach is the cancellation of the unknown baseline hazard function and thus it will be eliminated from the estimation process for the regression parameters. However, when random effects (frailties) are included, the failure time distribution obtained by integration over the distribution of frailties loses the cancellation property. Adopting the GLMM formulation, random effects (frailties) are added to the linear predictor and the corresponding partial log-likelihood l_1 of failure times with the random effects conditionally fixed can be constructed. Such an approach retains the cancellation of baseline hazard property which makes the estimation procedures relatively straightforward and efficient.

A total of m subjects are followed over time and T_{ij} is the jth time to failure or censoring for subject i ($j = 1,2,\ldots,n_i$). Associated risk variables for the jth failure/censoring time interval of subject i are collected into a vector x_{ij}. The proportional hazards model is

$$h(t_{ij};x_{ij}) = h_0(t_{ij})g(\eta_{ij}), \qquad \eta_{ij} = x_{ij}^T\beta + U_i,$$

where $h_0(\cdot)$ is the baseline hazard function, $g(\cdot) = \exp(\cdot)$ for the Cox hazard function, and U_i is the random frailty effect for subject i. The random frailty terms are taken to be independent $N(0,\theta)$ and are constant over time.

The BLUP estimation of the hazard model consists of choosing estimates to maximize the sum of two components:

$$\begin{aligned} l_1 &= \text{Partial log-likelihood of failure times taking } U \text{ fixed}, \\ l_2 &= -(\frac{1}{2})\left[m \ln 2\pi\theta + \sum_{i=1}^{m} \frac{U_i^2}{\theta}\right]. \end{aligned}$$

To write down the partial log-likelihood, arrange the failure/censoring times in increasing order; such time is denoted by T_j ($j = 1,2,\ldots,n$) and a corresponding variable D_j takes the value of 1 if a failure occurs at T_j and 0 if censoring occurs there. The value of η for the subject failing or being censored at T_j is denoted by η_j. Using this notation, the partial log-likelihood for the Cox proportional hazards model, conditional on U fixed, is

$$l_1 = \sum_{j=1}^{n} D_j \left[\eta_j - \ln \sum_{s=1}^{n} \exp(\eta_s)\right].$$

Detailed estimation procedures and applications can be found in McGilchrist (1993), Yau and McGilchrist (1998, 1999), and Yau (2001).

6.3.4 Survival Mixture Models

A natural extension of the (one-component) frailty model is the g (> 1) component mixture of failure time distributions.

Without loss of generality, consider a two-component survival mixture model with random effects.

Let T_{ij} denote the observable failure/censoring time of the jth individual within the ith hospital; let m denote the number of hospitals and n_i the number of observations in the ith hospital; the total number of observations is therefore $n = \sum_{i=1}^{m} n_i$. Let x_{ij} be a vector of covariates associated with T_{ij}. The survival function of T is modelled by a two-component mixture model as

$$S(t_{ij};x_{ij}) = \pi S_1(t_{ij};x_{ij}) + (1-\pi)S_2(t_{ij};x_{ij}) \quad (i=1,\ldots,m; j=1,\ldots,n_i) \tag{6.6}$$

where π denotes the proportion of patients belonging to the first component, the sub-population of patients in an acute condition, and $S_h(t_{ij};x_{ij})$ is the conditional survival function of the hth component $(h=1,2)$. Under the Cox proportional hazards model (Cox, 1972), the conditional hazard function for the hth component $(h=1,2)$ is given by,

$$h_h(t_{ij};x_{ij}) = h_{h0}(t_{ij})\exp(\eta_h(x_{ij})) \quad (h=1,2) \tag{6.7}$$

where $h_{h0}(t_{ij})$ is the baseline hazard function and $\eta_h(x_{ij})$ is the linear predictor relating to the covariate x_{ij}. The commonly used Weibull distribution is assumed for $h_{h0}(t_{ij})$ because it is flexible as either a monotonic increasing, constant, or monotonic decreasing baseline hazard. That is,

$$h_{h0}(t_{ij}) = \lambda_h \alpha_h t_{ij}^{\alpha_h - 1} \quad (i=1,\ldots,m; j=1,\ldots,n_i; h=1,2) \tag{6.8}$$

where λ_h, $\alpha_h > 0$ are unknown parameters.

Observations collected from the same cluster are often correlated. Accordingly, an unobserved random term is introduced multiplicatively in each conditional hazard function to explain the variability shared by patients within a hospital. With reference to Equation (6.7) the random effect U_{hi} of the ith hospital on the hth component hazard function can be accommodated through the linear predictor, via

$$\eta_h(x_{ij}) = x_{ij}^T\beta_h + U_{hi} \quad (i=1,\ldots,m; j=1,\ldots,n_i; h=1,2) \tag{6.9}$$

where β_h is the vector of regression coefficients. The unobservable random hospital effects $U_{hi}(i=1,\ldots,m)$ are taken to be i.i.d. $N(0,\theta_h)$. A positive value of U_{hi} indicates that patients in the ith hospital will experience a higher risk of failure if they belong to the hth sub-population $(h=1,2)$. Thus if θ_h differs from zero, it implies a significant difference in the survival for patients of the hth sub-population between the participating hospitals. Under the formulation based on Equations (6.7), (6.8), and (6.9), the vector of unknown parameters becomes

$$\Psi = (\pi, \beta_1^T, \beta_2^T, u_1^T, u_2^T, \lambda_1, \lambda_2, \alpha_1, \alpha_2)^T$$

where $u_1^T = [U_{11}, U_{12}, \ldots, U_{1m}]$ and $u_2^T = [U_{21}, U_{22}, \ldots, U_{2m}]$. The GLMM method commences with the BLUP at the initial step and proceeds to obtain approximate REML estimators of the parameters θ_h in the variance component (McGilchrist, 1994; Breslow and Clayton, 1993; Schall, 1991). For a given initial value of

$\theta_h (h = 1, 2)$, the BLUP estimator of Ψ maximizes $l = l_1 + l_2$, where

$$\begin{aligned} l_1 &= \sum_{i=1}^{m} \sum_{j=1}^{n_i} [D_{ij} \log f(t_{ij}; x_{ij}) + (1 - D_{ij}) \log S(t_{ij}; x_{ij})] \\ l_2 &= -(1/2)[m \log(2\pi\theta_1) + (1/\theta_1) u_1^T u_1] \\ &\quad + (-1/2)[m \log(2\pi\theta_2) + (1/\theta_2) u_2^T u_2] \end{aligned} \tag{6.10}$$

Here $D_{ij} = 1$ and $D_{ij} = 0$ indicate a failure and a censored observation, respectively, and

$$f(t_{ij}; x_{ij}) = \pi f_1(t_{ij}; x_{ij}) + (1 - \pi) f_2(t_{ij}; x_{ij})$$

is the probability density function of T based on Equation (6.6), where $f_h(t_{ij}; x_{ij})$ is the conditional probability density function given that the patient belongs to the hth component ($h = 1, 2$). In Equation (6.10), l_1 is the log-likelihood based on the failure and censored times conditional on u_1 and u_2, and l_2 is the logarithm of the joint probability density function of u_1 and u_2, with u_1 and u_2 taken to be independent.

Detailed estimation procedures and applications can be found in Section 7.4.

6.3.5 Long-Term Survivor Models with Random Effects

The long-term survivor model (or cured model) can be viewed as a limiting case of the two-component survival model, which is characterized by the existence of a subpopulation being cured and thus will not experience the failure event with reference to the duration of the research study framework.

Define a binary variable Λ, where $\Lambda = 1$ indicates that an individual will eventually experience a failure event (uncured) and $\Lambda = 0$ indicates that the individual will never experience such event (cured). For an individual with covariate vector x_j, the proportion π can be specified to be a logistic function of x such that the conditional distribution of Λ is given by

$$\pi(x_j) = \Pr(\Lambda = 1 | x_j) = \frac{1}{1 + \exp \xi_j},$$

where $\xi_j = w_j^T \gamma$, $w_j = [1 \ x_j^T]^T$, and γ is a vector of logistic parameters. Let T be the time of occurrence of the failure event. We take $T = \infty$ where $\Lambda = 0$ so that the corresponding conditional survival function given $\Lambda = 0$ is equal to one for all finite values of T. The unconditional survival function of T can then be expressed as

$$S(t; x) = 1 - \pi(x) + \pi(x) S_2(t; x), \tag{6.11}$$

where $S_2(t; x)$ is the conditional survival function for the uncured group. Model (6.11) is also known as the long-term survivor mixture model; see Section 5.5.

If we further assume that the conditional hazard function of T for the uncured group, $h_2(t; x)$, follows Cox proportional hazard model (Cox, 1972), then we have

$$h_2(t; x) = h_{20}(t) \exp \eta,$$

where $\eta = x^T\beta$ is the linear predictor, β is a vector of regression parameter and $h_{20}(t)$ is the baseline hazard function for the uncured group. The unconditional hazard function of T is then given by

$$h(t;x) = \pi(x)h_{20}(t)\exp\eta \frac{S_{20}(t)^{\exp\eta}}{S(t;x)}.$$

Denote D_j as the failure indicator for the jth observation. The log-likelihood is given by

$$l_1 = \sum_{j=1}^{n}\{(1-D_j)\log S(t_j;x_j) + D_j \log \pi(x_j)h_{20}(t_j)\exp\eta_j S_{20}(t_j)^{\exp\eta_j}\}$$

For clustered multivariate failure time data, the most intuitive random effect modelling is to assume that the above effects are determined by the unobservable random effects present in each clinic. Therefore, for a total of m clinics, T_{ij} is the observable failure/censoring time for the jth individual in the ith clinic. Let n_i be the number of observations in the ith clinic, the total number of observations is $n = \sum_{i=1}^{m} n_i$. Denote x_{ij} a vector of risk variables corresponding to T_{ij}. The logistic model becomes

$$\pi(x_{ij}) = \frac{1}{1+\exp\xi_{ij}}, \quad \xi_{ij} = w_{ij}^T\gamma + U_i$$
$$(i = 1,2,\ldots,m; j = 1,2,\ldots,n_i)$$

where U_i is the unobservable random effect of the ith clinic determining the proportion cured, and is taken to be i.i.d. $N(0,\theta_u)$. A larger value of U_i indicates that those patients having treatments in the ith clinic will experience a higher cured probability. Similarly, the Cox proportional hazards model becomes

$$h_2(t_{ij};x_{ij}) = h_{20}(t_{ij})\exp\eta_{ij}, \quad \eta_{ij} = x_{ij}^T\beta + V_i$$
$$(i = 1,2,\ldots,m; j = 1,2,\ldots,n_i)$$

where V_i is the unobservable random effect of the ith clinic determining the risk of event (such as, tumour formation) for the uncured individuals and is taken to be i.i.d. $N(0,\theta_v)$.

Let Ω be a vector of the parameters γ, β, v and u. Assuming a known $h_{20}(t)$, estimator of Ω is found by maximizing the BLUP likelihood in the initial step and then is extended to obtain REML estimators of Ω, θ_u and θ_v. Let $u = [U_1, U_2, \ldots, U_m]$ and $v = [V_1, V_2, \ldots, V_m]$, the BLUP estimators for given initial values of θ_u and θ_v, maximize $l = l_1 + l_2$, where

$$\begin{aligned} l_1 &= \text{log-likelihood of failure times with } u \text{ and } v \text{ conditionally fixed} \\ l_2 &= -(1/2)[m\log 2\pi\theta_u + (1/\theta_u)u^T u] + (-1/2)[m\log 2\pi\theta_v + (1/\theta_v)v^T v] \end{aligned}$$

Detailed estimation procedures and applications can be found in Section 7.5; see also Yau and Ng (2001), Lai and Yau (2008, 2009), and Xiang, Ma, and Yau (2011).

7

Advanced Mixture Models for Multilevel or Repeated-Measured Data

7.1 Introduction

This chapter describes the applications of GLMM to mixture models with random effects. These mixture models are useful to analyze multilevel data (data exhibit a hierarchical structure, such as patients nested in hospitals) or recurrent event data, where the response outcome can be measured by a count variable or a failure-time variable with censored observations. For analyzing multilevel count data arising from two or more latent sub-populations, Poisson mixture regression models can be adopted to handle heterogeneity in different higher-level units (such as hospitals) of such data; see, for example, Wang, Yau, and Lee (2002a) and Xiang et al. (2008). On the other hand, survival mixture regression models under Cox proportional hazards setting have been proposed to analyze multilevel failure-time data or recurrent event data; see, for example, Ng et al. (2004), Lai and Yau (2009), and Wang et al. (2007a). As outlined in Chapter 6, the GLMM approach provides an advanced modelling technique to handle response variables that are dependent with complex correlation structures, as arisen from the aforementioned study designs. Assuming normally distributed random components for the random effects, the GLMM approach commences with developing the BLUP estimators in the initial step and proceeds to obtain REML estimators of regression and variance component parameters. By introducing random effects in mixture modelling to conceptualize the heterogeneity in higher-level units, the GLMM approach can explicitly model a specific correlation structure among the outcomes and predict subject-specific random effects. In many research studies involving real-world problems, prediction of random effects is often of particular interest to identify high-risk individuals or higher-level units such as hospitals.

7.2 Poisson Mixture Regression Model with Random Effects

Poisson mixture regression models (Wang, Yau, and Lee, 2002a) can be adopted to analyze multilevel count response variables, where random effects are introduced to account for the inherent correlation among observations within a hierarchical unit. Let Y_{ij} be the jth response of the ith hierarchical unit. Denote m as the number of hierarchical units and n_i as the number of subjects in the ith unit, the total number of subjects is $n=\sum_{i=1}^{m} n_i$. Let $\{(y_{ij},x_{ij});\ i=1,\ldots,m;\ j=1,\ldots,n_i\}$ be a set of observed data, where y_{ij} is the realization of Y_{ij} and x_{ij} is the $p\times 1$ covariate vector associated with the discrete counts Y_{ij}. A Poisson mixture regression model with random effects can be defined as

$$f(y_{ij};x_{ij}) = \sum_{h=1}^{g} \pi_h f_h(y_{ij};\eta_{hij}), \tag{7.1}$$

where $f_h(y_{ij};\eta_{hij}) = \exp\{y_{ij}\eta_{hij} - \exp(\eta_{hij})\}/y_{ij}!$, $y_{ij} = 0,1,2,\ldots$, is the Poisson distribution with mean $\mu = \exp(\eta_{hij})$. Covariate x_{ij} is related to the mean via a linear predictor η_{hij} and

$$\eta_{hij} = \eta_h(x_{ij}) = x_{ij}^T\beta_h + u_{ih} \tag{7.2}$$

for $i=1,\ldots,m; j=1,\ldots,n_i$; and $h=1,\ldots,g$, where β_h is the vector of regression coefficients. In (7.2), an unobserved random effect u_{ih} is incorporated for accommodating the random effect of the ith hierarchical unit on the hth component, where u_{ih} is commonly assumed to be i.i.d. $N(0,\theta_h)$; see, for example, Xiang et al. (2008).

The BLUP estimates of the mixing proportions π_h, fixed-effect parameters β_h, and realizations of the random components u_{ih} can be obtained by maximizing $l = l_1 + l_2$, via the EM algorithm with fixed variance components θ_h $(i=1,\ldots,m;\ h=1,\ldots,g)$. For a Poisson component density, l_1 and l_2 are given by

$$\begin{aligned} l_1 &= \sum_{i,j}\log\Big(\sum_{h=1}^{g}\pi_h\exp\{y_{ij}\eta_{hij}-\exp(\eta_{hij})\}/y_{ij}!\Big) \\ l_2 &= -\tfrac{1}{2}[m\textstyle\sum_{h=1}^{g}\log(2\pi\theta_h)+\sum_{h=1}^{g}\theta_h^{-1}u_h^T u_h], \end{aligned} \tag{7.3}$$

where u_h denotes the vector of random effects $(u_{1h},\ldots,u_{mh})^T$ for the hth component $(h=1,\ldots,g)$. Estimation of θ_h is then achieved using approximate REML estimating equations, following the GLMM formulation of McGilchrist (1994).

Within the EM framework, the complete-data log-likelihood can be decomposed as $l_c = l_\pi + \sum_{h=1}^{g} l_{\eta_h}$, where

$$l_\pi = \sum_{i,j}\sum_{h=1}^{g} z_{hij}\log\pi_h \tag{7.4}$$

and

$$l_{\eta_h} = \sum_{i,j} z_{hij}(y_{ij}\eta_{hij} - \exp(\eta_{hij}) - \log(y_{ij}!)) - \tfrac{1}{2}[m\log(2\pi\theta_h)+\theta_h^{-1}u_h^T u_h] \tag{7.5}$$

for $h=1,\ldots,g$, where z_{hij} is one or zero accordingly as y_{ij} does or does not belong to the hth component $(h=1,\ldots,g)$. The E-step involves the calculation of the partitioned Q-functions corresponding to Q_π and Q_{η_h} $(h=1,\ldots,g)$, by replacing z_{hij} with the current posterior probabilities of component membership τ_{hij}, where

$$\tau_{hij}^{(k)} = \frac{\pi_h^{(k)}\exp\{y_{ij}\eta_{hij}^{(k)} - \exp(\eta_{hij}^{(k)})\}/y_{ij}!}{\sum_{l=1}^{g}\pi_l^{(k)}\exp\{y_{ij}\eta_{lij}^{(k)} - \exp(\eta_{lij}^{(k)})\}/y_{ij}!}. \tag{7.6}$$

The M-step updates the estimates for π_h, β_h, and the random effects u_h that maximize the corresponding Q-functions. In particular, closed-form equation is available for π_h $(h = 1, \ldots, g)$, as

$$\pi_h^{(k+1)} = \sum_{i,j} \tau_{hij}^{(k)}/n \tag{7.7}$$

for $h = 1, \ldots, g-1$ and $\pi_g^{(k+1)} = 1 - \sum_{h=1}^{g-1} \pi_h^{(k+1)}$. For β_h and u_h $(h = 1, \ldots, g)$, the maximization involves solving the following set of non-linear equations:

$$\begin{aligned} &\sum_{i,j} \tau_{hij}^{(k)} \{y_{ij} - \exp(\eta_{hij})\} x_{ij} = 0 \\ &\sum_{j=1}^{n_i} \tau_{hij}^{(k)} \{y_{ij} - \exp(\eta_{hij})\} - \theta_h^{-1} u_{ih} = 0 \qquad (i = 1, \ldots, m). \end{aligned} \tag{7.8}$$

The MINPACK routine HYBRD1 of Moré, Garbow, and Hillstrom (1980) can be used to find a solution to (7.8). The E- and M-steps are alternated repeatedly until convergence in this "inner" loop corresponding to the development of BLUP estimators. The approximate REML estimates of the variance components θ_h and the asymptotic variances of estimates $\hat{\pi}_h$ and $\hat{\beta}_h$ $(h = 1, \ldots, g)$ are obtained based on the GLMM formulation as described in Sections 6.2.3 and 6.2.4. With reference to (7.3) and assuming there are $g = 2$ Poisson components without loss of generality, we denote Ω the negative second derivative of $l = l_1 + l_2$ with respect to the conformal partition of $\pi|\beta_1|\beta_2|u_1|u_2$ in the BLUP procedure (here, $\pi_1 = \pi$ and $\pi_2 = 1 - \pi$). Let the inverse matrix Ω^{-1} be

$$\Omega^{-1} = \begin{pmatrix} A_{11} & A_{12} & A_{13} & A_{14} & A_{15} \\ A_{21} & A_{22} & A_{23} & A_{24} & A_{25} \\ A_{31} & A_{32} & A_{33} & A_{34} & A_{35} \\ A_{41} & A_{42} & A_{43} & A_{44} & A_{45} \\ A_{51} & A_{52} & A_{53} & A_{54} & A_{55} \end{pmatrix}, \tag{7.9}$$

where the partitioned sub-matrices in Ω for Poisson mixture regression models with random effects can be obtained according to the derivation in Wang, Yau, and Lee (2002a). We then have

$$\begin{aligned} \theta_1 &= m^{-1}[\text{tr}\, A_{44} + u_1^T u_1] \\ \theta_2 &= m^{-1}[\text{tr}\, A_{55} + u_2^T u_2], \end{aligned} \tag{7.10}$$

where tr A_{44} is the trace of matrix A_{44}. Using (7.10), the estimates of the variance components θ_1 and θ_2 are updated in an "outer" loop. The inner and outer loops are then repeated to update the coefficient parameters and the variance component parameters, respectively, until convergence. The asymptotic variance of the parameter estimates are obtained from the corresponding partition of Ω^{-1}, as

$$\text{var}\begin{pmatrix} \hat{\pi} \\ \hat{\beta}_1 \\ \hat{\beta}_2 \end{pmatrix} = \begin{pmatrix} A_{11} & A_{12} & A_{13} \\ A_{21} & A_{22} & A_{23} \\ A_{31} & A_{32} & A_{33} \end{pmatrix} \tag{7.11}$$

and

$$\text{var}\begin{pmatrix}\hat{\theta}_1\\ \hat{\theta}_2\end{pmatrix} =$$

$$2\begin{pmatrix} \frac{(m-2\hat{\theta}_1^{-1}\text{tr}\,A_{44})}{\hat{\theta}_1^2}+\frac{\text{tr}(A_{44}A_{44})}{\hat{\theta}_1^4} & \frac{\text{tr}(A_{45}A_{54})}{\hat{\theta}_1^2\hat{\theta}_2^2} \\ \frac{\text{tr}(A_{45}A_{54})}{\hat{\theta}_1^2\hat{\theta}_2^2} & \frac{(m-2\hat{\theta}_2^{-1}\text{tr}\,A_{55})}{\hat{\theta}_2^2}+\frac{\text{tr}(A_{55}A_{55})}{\hat{\theta}_2^4} \end{pmatrix}^{-1}. \tag{7.12}$$

7.2.1 Robust Estimation Using Minimum Hellinger Distance

In some applications of Poisson mixture models with random effects, the REML estimation of regression coefficients and variance component parameters is sensitive to potential outliers or data contamination. For uncontaminated data, this problem may be overcome by specifying good initial values in the EM algorithm and regarding the solution as a local likelihood estimate; see Section 1.3. For contaminated data, the problem may persist even though good starting values are adopted. The likelihood-based estimation method has a tendency to fit an extra component to the contamination; see Section 1.2. Alternative to the REML method, robust estimation for Poisson mixture models with random effects using the minimum Hellinger distance (MHD) estimation method has been developed to downweight the contribution of atypical counts; see Xiang et al. (2008) and the references therein. In contrast to the M-estimation-type approach, the MHD method has various robustness and second-order efficiency properties, providing estimators with a balance between efficiency and robustness, as discussed in Simpson (1987), Lindsay (1994), and Xiang et al. (2008).

Let Ψ denote the vector of unknown parameters. Lu, Hui, and Lee (2003) considered MHD estimation for Poisson mixture models without random effects, where robust MHD estimates of Ψ were obtained by minimizing the Hellinger distance between a consistent estimator of the probability function $f(y|x_{ij};\Psi)$ and the empirical probability function $f_n(y)$ with respect to Ψ. With unobservable random effects u_{ih} in the mixture model, $f(y|x_{ij};\Psi)$ becomes a joint probability function of y and u_{ih} for given x_{ij} and Ψ. The MHD estimation developed by Xiang et al. (2008) is implemented through an approximation of the marginal probability function by summing a set of s Poisson mixtures. Specifically, the marginal probability function is approximated using the Gaussian quadrature technique for integrating out the random effects u_{ih} in $f(y|x_{ij};\Psi)$, as

$$\begin{aligned} g(y|x_{ij};\Psi) &= \int_{-\infty}^{\infty} f(y|\tilde{u}_{ih},x_{ij};\Psi)\phi(\tilde{u}_{ih})d\tilde{u}_{ih}, \\ &\approx \sum_{q=1}^{s} w_q f(y|\tilde{u}_q,x_{ij};\Psi) \\ &= \sum_{q=1}^{s}\sum_{h=1}^{g} w_q\pi_h f_h(y;\eta_h(x_{ij})|\tilde{u}_q), \end{aligned} \tag{7.13}$$

where $\tilde{u}_{ih} = u_{ih}/\sqrt{\theta_h}$ is i.i.d. $N(0,1)$, $\phi(\tilde{u}_{ih})$ is the standard normal density function, and $f(y|\tilde{u}_{ih}, x_{ij}; \Psi)$ is the so-called Poisson log-normal distribution. The values of Gaussian quadratic mass points $\tilde{u}_q$ and their masses w_q are available in standard references, such as Abramowitz and Stegun (1964); see also Xiang et al. (2008) and Xiang, Yau, and Lee (2012). With (7.13), the MHD approach previously developed for Poisson mixture models by Lu, Hui, and Lee (2003) can be applied analogously. Specifically, a consistent estimator of the marginal probability function of y based on the sample data is given by

$$g(y;\Psi) = \frac{1}{n}\sum_{i,j} g(y|x_{ij};\Psi) = \sum_{i,j}\sum_{q=1}^{s}\sum_{h=1}^{g} \frac{w_q \pi_h}{n} f_h(y;\eta_h(x_{ij})|\tilde{u}_q). \tag{7.14}$$

Denote the empirical distribution $f_n(y) = N_y/n$, where N_y is the frequency of y among $Y_{11}, \ldots, Y_{mn_m}$ for $y = 0, 1, 2, \ldots$. Assuming the sample size n is sufficiently large, the Hellinger distance between $f_n(y)$ and $g(y;\Psi)$ is

$$d(f_n(y), g(y;\Psi)) = \sum_{y=0}^{\infty} \{f_n^{1/2}(y) - g^{1/2}(y;\Psi)\}^2. \tag{7.15}$$

From (7.15), the MHD estimator of Ψ can be obtained as a solution of the equation $\partial \rho_n(\Psi)/\partial\Psi = 0$, where $\rho_n(\Psi) = \sum_{y=0}^{M} f_n^{1/2}(y) g^{1/2}(y;\Psi)$, with $M = \max_{ij}\{y_{ij}\}$; see Xiang et al. (2008) for details. In some situations, there is little information regarding the joint distributional form of the random effects u_{ih}. To this end, a nonparametric maximum likelihood (NPML) approach within the MHD estimation procedure has been developed for robust estimation of Poisson mixture models with random effects (Xiang, Yau, and Lee, 2012). Without the need of specifying a parametric form for the random effects distribution, this NPML approach avoids the problem of numerical integration in deriving the MHD estimating equation and improves robustness against misspecification of the parametric distribution for the random effects.

7.2.2 Assessment of Model Adequacy and Influence Diagnostics

Goodness of fit statistics, such as Pearson and deviance residual diagnostics, can be used to assess adequacy of a Poisson mixture regression model with random effects. From Lee and Nelder (1996) and Wang, Yau, and Lee (2002a), the Pearson residuals for a two-component Poisson mixture model with random effects are defined as

$$r_{ij}^P = \frac{y_{ij} - \hat{\mu}_{ij}}{\sqrt{\widehat{\text{var}}(y_{ij})}}, \tag{7.16}$$

where

$$\begin{aligned} \hat{\mu}_{ij} &= \hat{\pi} \exp \hat{\eta}_{1ij} + (1-\hat{\pi}) \exp \hat{\eta}_{2ij} \\ \widehat{\text{var}}(y_{ij}) &= \hat{\mu}_{ij} + \hat{\pi}\exp(2\hat{\eta}_{1ij}) + (1+\hat{\pi})\exp(2\hat{\eta}_{2ij}) - \hat{\mu}_{ij}^2 \end{aligned} \tag{7.17}$$

for $i = 1, \ldots, m;\ j = 1, \ldots, n_i$. Accordingly, the deviance residuals are defined as

$$r_{ij}^D = \text{sign}(y_{ij} - \hat{\mu}_{ij})\sqrt{d_{ij}}, \tag{7.18}$$

where sign$(\cdot)$ is a sign function, $d_{ij} = 2[l_1(y_{ij}, y_{ij}) - l_1(\hat{\mu}_{ij}, y_{ij})]$, with $l_1(y_{ij}, y_{ij}) = -y_{ij} + \log(y_{ij})^{y_{ij}} - \log(y_{ij}!)$ and $l_1(\hat{\mu}_{ij}, y_{ij})$ being the first part of the BLUP log likelihood in (7.3). The Pearson (Deviance) goodness-of-fit statistic is then given by the sum of squared Pearson (Deviance) residuals for all n subjects, where the degrees of freedom are given by

$$\text{d.f.} = n - \text{tr}\left[I - \Omega^{-1} \begin{pmatrix} 0 & 0 & 0 & 0 & 0 \\ 0 & 0 & 0 & 0 & 0 \\ 0 & 0 & 0 & 0 & 0 \\ 0 & 0 & 0 & I_m/\theta_1 & 0 \\ 0 & 0 & 0 & 0 & I_m/\theta_2 \end{pmatrix} \right]. \tag{7.19}$$

Based on the local influence approach of Cook (1986), Xiang et al. (2005) have developed influence diagnostics to assess the sensitivity of the fitted two-component Poisson mixture models with random effects. Identifying those clusters and/or individual observations that have an impact on the estimation of model parameters can provide insight into the source of overdispersion as well as model justification. With a two-component Poisson mixture mixed model (g=2) in (7.1) and (7.2), denote $\phi = \log(\pi/(1-\pi))$ and let $\alpha = (\phi, \beta_1^T, \beta_2^T)^T$ be the parameter vector of interest with random effects u_{i1} and u_{i2} as nuisance parameters. Following Cook (1986), the effect of a small perturbation ω on estimation of parameter vector α can be assessed via the likelihood displacement, $F(\omega) = 2[L(\hat{\alpha}|\omega_0) - L(\hat{\alpha}|\omega)]$, where $\hat{\alpha}$ is REML estimator and ω_0 denote the null perturbation point such that $L(\hat{\alpha}|\omega_0) = L(\hat{\alpha})$. Thus $F(\omega)$ compares REML estimator of $\hat{\alpha}$ under ω of the perturbed data to $\hat{\alpha}$ with respect to the contours of the unperturbed joint log likelihood. The approach of Xiang et al. (2005) investigates the direction of the greatest change in the displacement surface at point $F(\omega_0)$, through the computation of the normal curvature of the displacement surface, $C(d) = 2|d^T \Delta^T \ddot{L} \Delta d|$, where d is a unit vector, $\Delta = \frac{\partial^2 L}{\partial \alpha \partial \omega^T}|_{\alpha=\hat{\alpha}, \omega=\omega_0}$, and $\ddot{L} = \frac{\partial^2 L}{\partial \alpha \partial \alpha^T}|_{\alpha=\hat{\alpha}}$. Let $d_{\max} = \max_d C(d)$ denote the direction of the maximum normal curvature, then $d_{\max}$ is the main diagnostic indicating the direction that can produce the greatest local change of the likelihood displacement at $F(\omega_0)$, which is obtained as the eigenvector corresponding to the largest eigenvalue of the matrix $\Delta^T \ddot{L} \Delta$. The normal curvature function under various perturbations can be obtained specifically at both hierarchical unit- and observation-levels. For influence diagnostics at the unit-level, let the observation vector y, component-specific covariate matrices x_h, and the corresponding linear predictor vector η_h ($h = 1, 2$) be arranged according to the m hierarchical units, such that each subvector y_i and η_{hi}, and submatrix x_{hi} has n_i rows corresponding to the ith unit. The normal curvature of the ith unit for unit-weight perturbation is then given by

$$C_i(\phi) = \frac{2(\sum_{j=1}^{n_i} \hat{\xi}_{ij} - (n_i e^{\hat{\phi}}/(1+e^{\hat{\phi}}))}{\sum_{i,j} \hat{\xi}_{ij}(1-\hat{\xi}_{ij}) + (ne^{\hat{\phi}}/(1+e^{\hat{\phi}})^2)}$$

$$C_i(\beta_1) = 2[x_{1i}^T \hat{\xi}_i (y_i - e^{\eta_{1i}})]^T \left[\frac{\partial^2 l_1}{\partial \beta_1 \partial \beta_1^T} \bigg|_{\alpha=\hat{\alpha}} \right]^{-1} x_{1i}^T \hat{\xi}_i (y_i - e^{\eta_{1i}})$$

$$C_i(\beta_2) = 2[x_{2i}^T (1-\hat{\xi}_i)(y_i - e^{\eta_{2i}})]^T \left[\frac{\partial^2 l_1}{\partial \beta_2 \partial \beta_2^T} \bigg|_{\alpha=\hat{\alpha}} \right]^{-1} x_{2i}^T (1-\hat{\xi}_i)(y_i - e^{\eta_{2i}}), \quad (7.20)$$

where $\xi_{ij} = \frac{e^{\phi} \exp\{y_{ij}\eta_{1ij} - \exp(\eta_{1ij})\}}{e^{\phi} \exp\{y_{ij}\eta_{1ij} - \exp(\eta_{1ij})\} + \exp\{y_{ij}\eta_{2ij} - \exp(\eta_{2ij})\}}$ and ξ_i is the vector of (ξ_{ij}) arranged accordingly for the ith unit $(i = 1, \ldots, m)$. Index plots of the above normal curvature functions can be used to identify hierarchical units that are potentially influential on the parameter estimates due to unit-weight perturbation; see Xiang et al. (2005) for details of unit specific influence diagnostics involving individual covariate perturbation.

For observation-level influence diagnostics, the assessment method based on the local influence approach is appropriate when the study involves moderate-sized samples with a few hierarchical units. As described in Xiang et al. (2005) for case-weight perturbation, the normal curvature of the (i, j)th observation is given by

$$\begin{aligned} C_{ij}(\phi) &= \frac{2(\hat{\xi}_{ij} - e^{\hat{\phi}}/(1+e^{\hat{\phi}}))^2}{\sum_{i,j} \hat{\xi}_{ij}(1-\hat{\xi}_{ij}) + ne^{\hat{\phi}}/(1+e^{\hat{\phi}})^2} \\ C_{ij}(\beta_1) &= 2\hat{\xi}_{ij}^2 (y_{ij} - e^{\eta_{1ij}})^2 x_{1ij}^T (X_1^T \ddot{L}_{\eta_1\eta_1} X_1)^{-1} x_{1ij} \\ C_{ij}(\beta_2) &= 2(1-\hat{\xi}_{ij})^2 (y_{ij} - e^{\eta_{2ij}})^2 x_{2ij}^T (X_2^T \ddot{L}_{\eta_2\eta_2} X_2)^{-1} x_{2ij}, \end{aligned} \quad (7.21)$$

where $X_h = (x_{h11}, \ldots, x_{h1n_1}, \ldots, x_{hm1}, \ldots, x_{hmn_m})^T$. Subject index plots of the above normal curvature functions can be obtained to identify individuals that are potentially influential on the parameter estimates due to case-weight perturbation; see Xiang et al. (2005) for details of observation-specific influence diagnostics involving individual covariate perturbation.

7.2.3 Example 7.1: Recurrent Urinary Tract Infection Data

We consider the recurrent urinary tract infections (UTI) data of Xiang et al. (2008) collected from a retrospective cohort study conducted in 2003 to determine the risk factors associated with recurrent UTI among elderly women in residential aged-care facilities. As described in Nicolle (2002), UTI is one of the most common bacterial infections affecting between 10% and 20% of women aged 60 years and over. In this cohort study, a total of 201 female residents aged 60 years or above with an institutionalization period of at least 6 months were recruited from six residential aged-care facilities (four nursing homes and two hostels with average cluster size

TABLE 7.1
Estimates and standard errors (in parentheses) for a two-component Poisson mixture regression model with random effects (Example 7.1). Adapted from Xiang et al. (2008).

Parameter	REML method		MHD method	
	Cluster 1	Cluster 2	Cluster 1	Cluster 2
Intercept	-3.072* (0.92)	-0.861* (0.22)	-4.386* (0.56)	-0.041 (0.39)
History of UTI	0.355 (0.39)	0.581* (0.18)	1.106 (2.75)	0.509* (0.16)
Fecal incontinence	2.207* (0.59)	-0.901 (0.46)	1.314 (1.78)	0.358 (0.43)
Courses of antibiotics	0.303* (0.06)	0.261* (0.03)	0.249 (0.14)	0.147* (0.06)
Mixing proportion π_h	0.359 (0.09)	0.641	0.467 (0.08)	0.533
Variance component θ_h	0.910	0.044	0.339	0.082
Chi-square statistic	7.468		4.548	
Pearson statistic	175.262		134.020	

* p-value < 0.05.

of 33.5) in the Perth metropolitan area of Western Australia. The outcome variable is the number of UTI episodes (ranged from 0 to 17) during the 2-year follow-up period. We consider the following risk factors of patient's clinical characteristics: history of prior UTI (yes versus no); presence versus absence of fecal incontinence; and number of courses of antibiotics taken.

We fitted the data using a two-component mixture of Poisson regression model with random effects using the REML estimation method (m=6). A two-component (g=2) mixture model was found to provide a better fit than a single component Poisson mixed model according to the chi-square and Pearson statistics; see Xiang et al. (2008) for more information. Table 7.1 presents the estimated regression coefficients and their standard errors for the two-component Poisson mixture regression model with random effects. With the REML method, the first component corresponded to a group of women aged 60 or over (35.9%) who had a very low rate of UTIs (estimated mean = 0.046). This group represents a low-risk (normative) sub-population in which subjects are generally healthy and at low risk of UTI. About 64.1% of women aged 60 or over had a higher rate of UTIs (estimated mean = 0.423; the second component). For this (majority) group of women representing a high-risk (susceptible) sub-population of having UTIs, the rate of UTIs increased significantly for those women with a history of prior UTIs (adjusted rate ratio of 1.788; 95% CI = 1.2 to 2.6) and with increased number of courses of antibiotics (adjusted rate ratio of 1.298; 95% CI = 1.2 to 1.4). The dispersion of the random institution effect was estimated to be θ_1=0.910 for the normative sub-population, which was larger than the corresponding estimate of θ_2=0.044 for the susceptible sub-population, after adjusting for covariate effects. Goodness-of-fit chi-square and Pearson statistics showed that the REML estimation method fitted the UTI data reasonably well. However, with observation-level influence diagnostic assessment described in Section 7.2.2, it was found that Patient 13 and Patient 53 were the two exceptions who appeared to

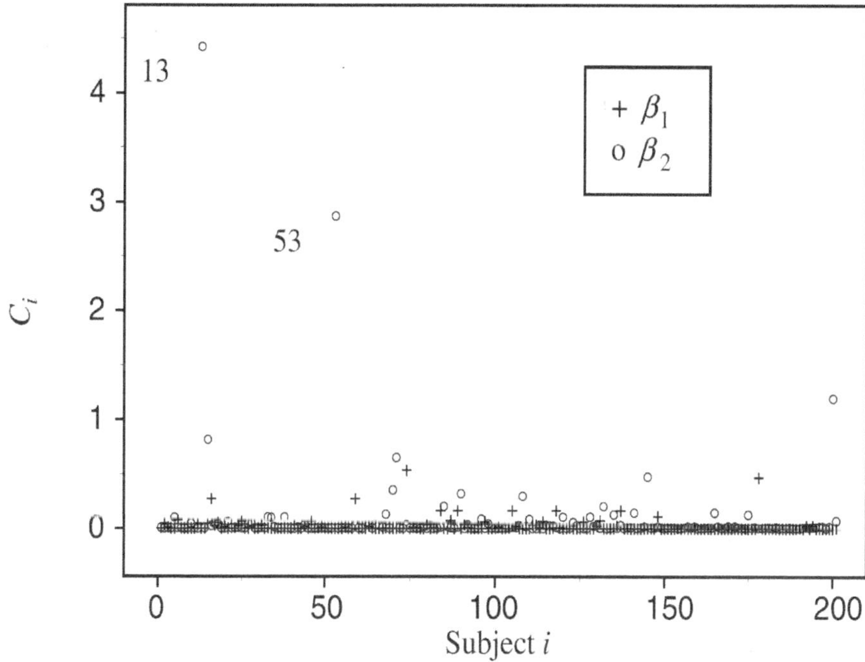

FIGURE 7.1
Subject index plot of curvatures C_{ij} based on case-weight perturbations (Example 7.1). Adapted from Xiang et al. (2005).

be influential on the regression coefficients. An examination of the data revealed that Patient 13 had the second highest UTI count of 10 episodes in the sample, whereas Patient 53 incurred the highest UTI count of 17 and had the greatest number of 21 antibiotic treatments during the follow up period. Figure 7.1 displays a subject index plot of curvatures C_{ij} based on a fitted two-component multilevel Poisson mixture regression model with an extra unit-specific covariate of hostel versus nursing home facilities; see Xiang et al. (2005) and Xiang, Yau, and Lee (2012) for more information. To this end, Xiang et al. (2008) considered the robust MHD estimation approach for two-component mixture of Poisson regression model with random effects, as presented in Section 7.2.1. The results were given in Table 7.1 for comparison. The robust MHD estimation method provided different estimates of regression coefficients and their standard errors, compared to the REML method. In general, MHD estimation results were more conservative in terms of statistical significance of the risk factors included in the mixture model, due to downweighting the contribution of atypical observed counts. In particular, fecal incontinence was no longer a significant risk factor. The number of courses of antibiotics was also not significant in the first component corresponding to the formative sub-population of elderly women with low-risk UTI. In addition, the MHD method led to a smaller estimated mixing proportion for the second component ($\hat{\pi}_2 = 0.533$) corresponding to the high-risk group of women and different estimated variance components, compared to the REML method. Values of the chi-square and Pearson statistics were both smaller, suggesting that the MHD estimation method provided a better fit to the UTI data for the assumed two-component mixture of Poisson regression model with random institution effects; see also the comparative results by Xiang, Yau, and Lee (2012) who applied the NPML approach within the MHD procedure described in Section 7.2.1 to the UTI data.

7.3 Zero-Inflated Poisson Mixture Models with Random Effects

Count data with excess zeros relative to a Poisson distribution are common in many multilevel analytical applications in medical and health sciences. Yau and Lee (2001) and Wang, Yau, and Lee (2002b) illustrated the usefulness of zero-inflated Poisson (ZIP) mixture regression models with random effects for the analysis of multilevel count data with excess zeros. As described in Section 6.3.2, a multilevel ZIP regression model is a mixture of a Poisson distribution and a degenerate component of point mass at zero, where random effects are incorporated in the linear predictors ξ_{ij} and η_{ij} to account for data dependency within hierarchical units in the logistic and the Poisson components, respectively. In particular,

$$
\begin{aligned}
\xi_{ij} &= w_{ij}^T\alpha + u_i \\
\eta_{ij} &= x_{ij}^T\beta + v_i
\end{aligned}
\tag{7.22}
$$

for $i = 1, \ldots, m; j = 1, \ldots, n_i$, where $w_{ij} = (1\ x_{ij}^T)^T$ contain a constant of one for the intercept term in the logistic transform of π as $\log[\pi(x_{ij})/(1-\pi(x_{ij}))] = \xi_{ij}$. It is noted that the mixing proportion of excess zeros is formulated to relate π with covariate x_{ij} $(i = 1, \ldots, m; j = 1, \ldots, n_i)$, where the association is of particular interest in many applications of ZIP models.

Following the GLMM method, the BLUP type log-likelihood $l = l_1 + l_2$ becomes

$$
\begin{aligned}
l_1 &= \sum_{y_{ij}=0} \log\left(\frac{\exp\xi_{ij} + \exp(-\exp\eta_{ij})}{1+\exp\xi_{ij}}\right) + \\
&\quad \sum_{y_{ij}>0} \left(y_{ij}\eta_{ij} - \exp\eta_{ij} - \log(y_{ij}!) - \log(1+\exp\xi_{ij})\right) \\
l_2 &= -\tfrac{1}{2}[m\log(2\pi\theta_u) + \theta_u^{-1}u^Tu + m\log(2\pi\theta_v) + \theta_v^{-1}v^Tv];
\end{aligned}
\tag{7.23}
$$

see Wang, Yau, and Lee (2002b) for more information. In (7.23), u and v denote the vectors of random effects $(u_1, \ldots, u_m)^T$ and $(v_1, \ldots, v_m)^T$ in the logistic and the Poisson components, with the assumption of i.i.d. $N(0, \theta_u I_m)$ and $N(0, \theta_v I_m)$, respectively. The EM algorithm can be used to obtain the BLUP estimates. The E-step calculates the partitioned Q-functions corresponding to

$$\begin{aligned}
Q_\xi &= \sum_{i,j}[\tau_{ij}\xi_{ij} - \log(1+\exp\xi_{ij})] - \tfrac{1}{2}[m\log(2\pi\theta_u) + \theta_u^{-1}u^T u] \\
Q_\eta &= \sum_{i,j}(1-\tau_{ij})(y_{ij}\eta_{ij} - \exp(\eta_{ij}) - \log(y_{ij}!)) - \tfrac{1}{2}[m\log(2\pi\theta_v) + \\
&\qquad \theta_v^{-1}v^T v],
\end{aligned} \tag{7.24}$$

where

$$\tau_{ij}^{(k)} = \begin{cases} \frac{1}{1+\exp\{-w_{ij}^T\alpha^{(k)} - u_i^{(k)} - \exp[x_{ij}^T\beta^{(k)} + v_i^{(k)}]\}} & \text{if } y_{ij} = 0 \\ 0 & \text{if } y_{ij} \geq 1 \end{cases} \tag{7.25}$$

is the current estimated posterior probability of being an excess zero. The M-step updates the estimates for α, β, and the random effects u_i and v_i $(i = 1,\ldots,m)$ that maximize the corresponding Q-functions. From (7.24), it follows that the M-step involves solving the following system of non-linear equations:

$$\begin{aligned}
&\sum_{i,j}\{\tau_{ij}^{(k)}(1-\pi(x_{ij})) - (1-\tau_{ij}^{(k)})\pi(x_{ij})\}w_{ij} = 0 \\
&\sum_{i,j}(1-\tau_{ij}^{(k)})\{y_{ij} - \exp(\eta_{ij})\}x_{ij} = 0 \\
&\sum_{j=1}^{n_i}\{\tau_{ij}^{(k)}(1-\pi(x_{ij})) - (1-\tau_{ij}^{(k)})\pi(x_{ij})\} - \theta_u^{-1}u_i = 0 \qquad (i = 1,\ldots,m) \\
&\sum_{j=1}^{n_i}(1-\tau_{ij}^{(k)})\{y_{ij} - \exp(\eta_{ij})\} - \theta_v^{-1}v_i = 0 \qquad (i = 1,\ldots,m),
\end{aligned} \tag{7.26}$$

which can be solved using the MINPACK routine HYBRD1 of Moré, Garbow, and Hillstrom (1980). After the BLUP estimators are developed via the EM algorithm, the approximate REML estimates of the variance components θ_u and θ_v, and the asymptotic variances of estimates $\hat{\alpha}$ and $\hat{\beta}$ are then obtained based on the GLMM method. Based on the formulation presented in Wang, Yau, and Lee (2002b), we denote $\Omega_{\alpha,u}$ and $\Omega_{\beta,v}$ the negative second derivative of $l = l_1 + l_2$ with respect to the conformal partitions of $\alpha|u$ and $\beta|v$ in the BLUP procedure, respectively, such that the inverse matrices are given by

$$\begin{aligned}
\Omega_{\alpha,u}^{-1} &= \begin{pmatrix} A_{11} & A_{12} \\ A_{21} & A_{22} \end{pmatrix}, \\
\Omega_{\beta,v}^{-1} &= \begin{pmatrix} A_{33} & A_{34} \\ A_{43} & A_{44} \end{pmatrix}.
\end{aligned} \tag{7.27}$$

We then have the update for the variance components in the outer loop as

$$\begin{aligned}
\theta_u &= m^{-1}[\text{tr}\,A_{22} + u^T u] \\
\theta_v &= m^{-1}[\text{tr}\,A_{44} + v^T v].
\end{aligned} \tag{7.28}$$

The asymptotic variance of the parameter estimates are given by

$$\begin{aligned}
\text{var}\,\hat{\alpha} &= A_{11} \\
\text{var}\,\hat{\beta} &= A_{33} \\
\text{var}\,\hat{\theta}_u &= 2\left[\frac{(m-2\hat{\theta}_u^{-1}\text{tr}\,A_{22})}{\hat{\theta}_u^2}+\frac{\text{tr}(A_{22}A_{22})}{\hat{\theta}_u^4}\right]^{-1} \\
\text{var}\,\hat{\theta}_v &= 2\left[\frac{(m-2\hat{\theta}_v^{-1}\text{tr}\,A_{44})}{\hat{\theta}_v^2}+\frac{\text{tr}(A_{44}A_{44})}{\hat{\theta}_v^4}\right]^{-1}
\end{aligned} \tag{7.29}$$

7.3.1 Score Test for Zero-Inflation in Mixed Poisson Models

With the applications of ZIP mixture models with random effects, it is important to assess whether an extra component of point mass at zero is appropriate, relative to a Poisson mixed model. Score tests for zero-inflation have been developed for correlated count data, where a fit of the Poisson mixed model under only the null hypothesis is required; see Xiang et al. (2006), Lee et al. (2006), Xiang and Teo (2011), and the references therein. Let $\gamma = \pi/(1-\pi)$, testing the null hypothesis $H_0 : \pi = 0$ against $H_1 : \pi \neq 0$ is equivalent to testing $H_0^* : \gamma = 0$ against $H_1^* : \gamma \neq 0$. Under this null hypothesis and assuming a multilevel setting with REML estimates $\tilde{\beta}$, $\tilde{v}$, and $\tilde{\theta}_v$, the score function $U(\beta, v, \theta_v, \gamma)$ is then given by

$$U(\tilde{\beta},\tilde{v},\tilde{\theta}_v,0) = \left(0,\ldots,0,\sum_{i,j}\left[\frac{I_{(y_{ij}=0)}}{\exp(-\exp\tilde{\eta}_{ij})}-1\right]\right)^T, \tag{7.30}$$

corresponding to the null Poisson mixed model by setting $\pi = 0$ in l_1, where

$$\begin{aligned}
l_1 = \sum_{y_{ij}=0} &\log[\pi+(1-\pi)\exp(-\exp\eta_{ij})]+ \\
&\sum_{y_{ij}>0}\log[(1-\pi)\exp(y_{ij}\eta_{ij}-\exp\eta_{ij})/(y_{ij}!)]
\end{aligned} \tag{7.31}$$

and in (7.30), $I_{(y_{ij}=0)}$ is an indicator function and $\tilde{\eta}_{ij} = x_{ij}^T\tilde{\beta}+\tilde{v}_i$. The expected Fisher information matrix $\Im(\beta, v, \theta_v, \gamma)$ can be partitioned as

$$\Im(\beta,v,\theta_v,\gamma) = \begin{pmatrix} \Im_{11} & \Im_{12} \\ & \Im_{22} \end{pmatrix}, \tag{7.32}$$

where

$$\Im_{11} = \begin{pmatrix} \Im_{\beta\beta} & \Im_{\beta v} & \Im_{\beta\theta_v} \\ & \Im_{vv} & \Im_{v\theta_v} \\ & & \Im_{\theta_v\theta_v} \end{pmatrix}, \quad \Im_{12} = \begin{pmatrix} \Im_{\beta\gamma} \\ \Im_{v\gamma} \\ \Im_{\theta_v\gamma} \end{pmatrix}, \quad \Im_{22} = \Im_{\gamma\gamma}. \tag{7.33}$$

All entities in the sub-matrices in (7.33) can be obtained from the first and second derivatives of l with respect to β, v, and θ_v, as given in Xiang et al. (2006).

Consequently, the score statistic for testing zero-inflation ($H_0^* : \gamma = 0$) is given by

$$\begin{aligned} S(\tilde{\beta},\tilde{v},\tilde{\theta}_v,0) &= U(\tilde{\beta},\tilde{v},\tilde{\theta}_v,0)^T \tilde{\mathfrak{I}}^{-1} U(\tilde{\beta},\tilde{v},\tilde{\theta}_v,0) \\ &= \frac{\{\sum_{i,j}[I_{(y_{ij}=0)}/\exp(-\exp\tilde{\eta}_{ij})-1]\}^2}{\sum_{i,j}[1/\exp(-\exp\tilde{\eta}_{ij})-1]-\tilde{\mathfrak{I}}_{12}^T\tilde{\mathfrak{I}}_{11}^{-1}\tilde{\mathfrak{I}}_{12}}, \end{aligned} \tag{7.34}$$

where the $\tilde{\mathfrak{I}}$ matrices are evaluated at the REML estimates ($\tilde{\beta}$, $\tilde{v}$, $\tilde{\theta}_v$) and $\gamma = 0$; see Xiang and Teo (2011) for more information. Under the null hypothesis, the score statistic S has an asymptotic χ_1^2 distribution. For the simple case when $v = 0$ for testing zero-inflation of ZIP mixture models without random effects, S reduces to the ordinary score test statistic of Van den Broek (1995).

In some applications involving multilevel count data with excess zeros, the non-zero counts may be overdispersed in relation to a Poisson distribution. A zero-inflated negative binomial (ZINB) mixed regression model may be more appropriate than the ZIP mixed model to analyze overdispersed correlated count data with excess zeros; see, for example, Yau, Wang, and Lee (2003). In this context, a score test to assess overdispersion is proposed by Xiang et al. (2007) for testing the ZIP mixed regression model against the ZINB alternative. The formulation of this score test parallels that of Ridout, Hinde, and Demétrio (2001) for models without random effects.

Goodness-of-fit Pearson and deviance residual statistics to assess adequacy of a ZIP mixture regression model with random effects can be derived in a similar manner as in Section 7.2.2; see Yau and Lee (2001) and Xiang et al. (2007) for more details.

7.3.2 Example 7.2: Revisit of the Recurrent UTI Data

A feature of the UTI data considered in Example 7.1 is the presence of a high proportion (53.7%) of zero UTI counts (Xiang et al., 2008). This preponderance of zero counts suggested that a multilevel ZIP regression model described in Section 7.3 may be used to analyze the UTI data, whereas the score test formulated in Section 7.3.1 can be adopted to assess whether a multilevel ZIP model provides a better fit compared to a multilevel Poisson model. In this regard, Xiang et al. (2006) considered a multilevel ZIP regression model with ξ_{ij} in (7.22) assumed to be an unknown constant such that the mixing proportion for excess zeros is π, independent of any risk factors. They worked on the UTI data, with two additional clinical risk factors (the presence versus absence of any anatomical abnormalities and immuno-compromised treatment of oral corticosteroids or chemotherapeutic agents) and exclusion of the risk factor regarding courses of antibiotics. Besides these four risk factors, the duration of follow-up was included as an offset term to adjust for the individual exposure; see Xiang et al. (2006) for more information. Table 7.2 presents the results, which suggest that the proportion of extra zeros π was estimated to be 47.9%. Elderly women with a history of UTI (adjusted rate ratio of 2.177; 95% CI = 1.7 to 2.8) or presence of anatomical abnormalities (adjusted rate ratio of 1.495; 95% CI = 1.1 to 2.1) were positively associated with the rate of UTIs. The score statistic $S = 6.484$ was significant (p=0.01) with respect to the asymptotic χ_1^2 distribution, providing strong evidence of zero inflation in this UTI data set. Xiang et al. (2006)

TABLE 7.2
Estimates and standard errors (in parentheses) for multilevel ZIP regression models (Example 7.2). Adapted from Xiang et al. (2006).

Parameter	ZIP with constant π	ZIP with logistic transformed π Poisson part	ZIP with logistic transformed π Logistic part
Intercept	-5.839* (0.140)	-5.781* (0.134)	0.411 (0.351)
History of UTI	0.778* (0.134)	0.734* (0.132)	-0.947* (0.409)
Fecal incontinence	0.247 (0.127)	0.216 (0.128)	-0.290 (0.359)
Anatomical abnormalities	0.402* (0.166)	0.376* (0.169)	-1.163 (0.721)
Immuno-compromised	0.179 (0.251)	0.166 (0.259)	-0.467 (1.040)
Mixing proportion π	0.479 (0.087)		
Variance component θ_u or θ_v	0.095 (0.057)	0.091 (0.056)	0.587 (0.339)

* p-value < 0.05.

also considered a multilevel ZIP regression model with π modelled by a logistic function of risk factors and random institution effects; see (7.22). The results are given in Table 7.2 for comparison. Specifically, as shown in the logistic part, the risk of having a recurrent UTI was significantly higher for elderly women with a history of UTI (adjusted OR = 2.578; 95% CI = 1.2 to 5.7).

7.4 Survival Mixture Models with Random Effects

As outlined in Section 6.3.4, survival mixture models described in Section 5.3 can be extended to incorporate random effects in the linear predictor with reference to the component hazard functions. These survival mixture models with random effects are useful to tackle problems with censored failure time data exhibiting multilevel correlation structures (for example, patients nested within hospitals). Let T_{ij} denote the observable failure/censoring time of the jth individual within the ith hospital, where $i = 1,\ldots,m$ and $j = 1,\ldots,n_i$ such that the total number of observations is $n = \sum_{i=1}^{m} n_i$. Let x_{ij} be a vector of covariates associated with T_{ij}. Without loss of generality, we consider a two-component survival mixture model

$$S(t_{ij};x_{ij}) = \pi S_1(t_{ij};x_{ij}) + (1-\pi)S_2(t_{ij};x_{ij}) \tag{7.35}$$

for $i = 1,\ldots,m; j = 1,\ldots,n_i$, under Cox proportional hazards model (Cox, 1972) where the conditional hazard function for the hth component ($h = 1,2$) is given by

$$h_h(t_{ij};x_{ij}) = h_{h0}(t_{ij})\exp(\eta_h(x_{ij})) \quad (h = 1,2). \tag{7.36}$$

In (7.36), $h_{h0}(t_{ij})$ is the baseline hazard function and $\eta_h(x_{ij})$ is the linear predictor relating to the covariates x_{ij} and the hospital-specific random effects as

$$\eta_h(x_{ij}) = x_{ij}^T\beta_h + u_{hi} \quad (i = 1,\ldots,m; j = 1,\ldots,n_i; h = 1,2). \tag{7.37}$$

The unobservable random effects u_{hi} $(i = 1,\ldots,m)$ are taken to be i.i.d. $N(0,\theta_h)$, introduced multiplicatively in each component hazard function to explain the variability shared by patients within a hospital. Following the GLMM method described in Section 6.3.4, the BLUP type log-likelihood is $l = l_1 + l_2$, where

$$\begin{aligned} l_1 &= \sum_{i,j}[D_{ij}\log f(t_{ij};x_{ij}) + (1-D_{ij})\log S(t_{ij};x_{ij})] \\ l_2 &= -\tfrac{1}{2}[m\log(2\pi\theta_1) + \theta_1^{-1}u_1^T u_1 + m\log(2\pi\theta_2) + \theta_2^{-1}u_2^T u_2], \end{aligned} \tag{7.38}$$

where $D_{ij} = 1$ and $D_{ij} = 0$ indicate a failure and a censored observation, respectively, and $f_h(t_{ij};x_{ij})$ are the component density functions conditioned that the patient belongs to the hth component $(h = 1,2)$.

The EM algorithm is adopted to obtain the BLUP estimates, where the E-step calculates the partitioned Q-functions corresponding to

$$\begin{aligned} Q_\pi &= \sum_{i,j}[\tau_{ij}\log(\pi/(1-\pi)) + \log(1-\pi)] \\ Q_{\eta_1} &= \sum_{i,j}\{\tau_{ij}[D_{ij}\log f_1(t_{ij};x_{ij}) + (1-D_{ij})\log S_1(t_{ij};x_{ij})] - \tfrac{1}{2}[m\log(2\pi\theta_1) + \\ &\qquad \theta_1^{-1}u_1^T u_1]\} \\ Q_{\eta_2} &= \sum_{i,j}\{(1-\tau_{ij})[D_{ij}\log f_2(t_{ij};x_{ij}) + (1-D_{ij})\log S_2(t_{ij};x_{ij})] - \\ &\qquad \tfrac{1}{2}[m\log(2\pi\theta_2) + \theta_2^{-1}u_2^T u_2]\} \end{aligned} \tag{7.39}$$

where

$$\tau_{ij}^{(k)} = \frac{\pi^{(k)}(f_1^{(k)})^{D_{ij}}(S_1^{(k)})^{(1-D_{ij})}}{\pi^{(k)}(f_1^{(k)})^{D_{ij}}(S_1^{(k)})^{(1-D_{ij})} + (1-\pi^{(k)})(f_2^{(k)})^{D_{ij}}(S_2^{(k)})^{(1-D_{ij})}} \tag{7.40}$$

is the current estimated posterior probability that t_{ij} belongs to the first component, and where $f_h^{(k)} = f_h(t_{ij};x_{ij})$; $S_h^{(k)} = S_h(t_{ij};x_{ij})$ $(h = 1,2)$ are evaluated based on the current estimates; see Ng et al. (2004) for details. The M-step updates the estimates for π, β_h, and the random effects u_h $(h = 1,2)$ that maximize the corresponding Q-functions.

Assuming a Weibull distribution baseline hazard $h_{h0}(t_{ij}) = \lambda_h \alpha_h t_{ij}^{\alpha_h - 1}$ ($\lambda_h, \alpha_h > 0$), it follows that the M-step involves solving the following nonlinear equations:

$$\sum_{i,j} (\tau_{ij}^{(k)})^{(2-h)} (1 - \tau_{ij}^{(k)})^{(h-1)} [D_{ij} + \log S_h(t_{ij}; x_{ij})] x_{ij} = 0$$

$$\sum_{j=1}^{n_i} [(\tau_{ij}^{(k)})^{(2-h)} (1 - \tau_{ij}^{(k)})^{(h-1)} (D_{ij} + \log S_h(t_{ij}; x_{ij})] - u_{hi}/\theta_h = 0 \quad (i = 1, \ldots, m)$$

$$\sum_{i,j} (\tau_{ij}^{(k)})^{(2-h)} (1 - \tau_{ij}^{(k)})^{(h-1)} [D_{ij}/\lambda_h - \exp(\eta_h(x_{ij})) t_{ij}^{\alpha_h}] = 0$$

$$\sum_{i,j} (\tau_{ij}^{(k)})^{(2-h)} (1 - \tau_{ij}^{(k)})^{(h-1)} \frac{D_{ij} + (D_{ij}\alpha_h - h_h(t_{ij}; x_{ij}) t_{ij}) \log t_{ij}}{\alpha_h} = 0 \qquad (7.41)$$

for β_h, u_h, λ_h, and α_h, respectively, ($h = 1, 2$). The MINPACK routine HYBRD1 of Moré, Garbow, and Hillstrom (1980) can be adopted to solve (7.41). A closed form equation is available for π as

$$\pi^{(k+1)} = \sum_{i,j} \tau_{ij}^{(k)} / n; \qquad (7.42)$$

see Ng et al. (2004) for details. Based on the BLUP estimators obtained via the EM algorithm, the approximate REML estimates of the variance components θ_h ($h = 1, 2$) and the asymptotic variances of estimates $\hat{\pi}$, $\hat{\beta}_1$, and $\hat{\beta}_2$ are obtained based on the GLMM method, via the negative second derivative of $l = l_1 + l_2$ with respect to the conformal partition of $\pi | \beta_1 | \beta_2 | u_1 | u_2$, as outlined in (7.9) to (7.12) in Section 7.2.

7.4.1 Example 7.3: rhDNase Clinical Trial Data

We consider the data set from the randomized trial of rhDNase (a recombinant deoxyribonuclease I enzyme) for the treatment of cystic fibrosis (Therneau and Hamilton, 1997). In this trial, patients with cystic fibrosis were recruited at $m = 51$ institutions and randomized to either treatment rhDNase or placebo. The data set is available at `https://vincentarelbundock.github.io/Rdatasets/datasets.html`, consisting of 647 patients (322 and 325 patients were randomized to rhDNase and placebo groups, respectively). Patients with cystic fibrosis often experience exacerbations of respiratory symptoms and progressive deterioration of lung function due to accumulation of extracellular DNA in the airways. As indicated by Therneau and Hamilton (1997), more than 90% of cystic fibrosis patients eventually die of lung disease. This randomized clinical trial was conducted to evaluate the treatment of rhDNase in reducing exacerbations of respiratory symptoms, where an exacerbation was defined as an infection that required the use of intravenous (IV) antibiotics. Patients were monitored for pulmonary exacerbations and during the follow-up period, there were a total of 767 observations of recurrent events. However, in this example, we consider only the primary endpoint regarding the time to first pulmonary exacerbation (possibly censored) for each of $n = 647$ patients. Two covariates were

TABLE 7.3
Estimates and standard errors (in parentheses) for fitting a two-component Weibull mixture model with random institution effects (Example 7.3).

Variable	**first component**	**second component**
Constant ($\log \lambda_h$)	-5.8* (0.29)	-5.7* (0.90)
rhDNase	-0.450* (0.187)	0.122 (0.781)
Baseline FEV1	-0.026* (0.005)	-0.137* (0.014)
Mixing proportion π_h	0.889	0.111
Shape parameter α_h	1.272	3.554
Variance component θ_h	0.377* (0.163)	1.390* (0.405)

* p-value < 0.05.

considered: treatment (1: rhDNase versus 0: placebo) and baseline level of Forced Expiratory Volume in 1 second (FEV1), which is a measure of lung capacity.

With a two-component Weibull mixture model described in Chapter 5, it was found that the larger sub-population consists of 90.6% of patients with an increasing hazard ($\alpha_1 = 1.187$), whereas the smaller sub-population consists of 9.4% of patients with a bigger increasing hazard of $\alpha_2 = 2.943$. This latter group may be considered as patients who suffer an acute pulmonary exacerbation. The results of this preliminary analysis were then treated as initial values for fitting the rhDNase trial data by the two-component Weibull mixture model with random institution effects adjustment presented in Equations (7.35) to (7.37). Table 7.3 displays the results. It can be seen that 11.1% of patients were identified as a sub-population suffering an acute pulmonary exacerbation with a large increasing hazard of α_2=3.554. This sub-population did not benefit from rhDNase. Indeed, benefit from rhDNase appeared only for those patients having a smaller hazard of α_1=1.272 (a majority of 88.9% of patients), where rhDNase significantly reduced the incidence of pulmonary exacerbations (adjusted HR of 0.638; 95% CI = 0.529 to 0.769). To illustrate such a difference, fitted stratified Cox proportional hazards models with adjustment of FEV1 are presented in Figure 7.2 separately for patients in the two components. The effects of baseline FEV1 were also different between the two components. While both sub-populations showed a reduced hazard of exacerbations with a larger FEV1, the effect was smaller in the majority of patients (the first component; adjusted HR of 0.974 per unit increase in FEV1; 95% CI = 0.970 to 0.979) compared to those patients with an acute pulmonary exacerbation (the second component; adjusted HR of 0.872 per unit increase in FEV1; 95% CI = 0.860 to 0.884).

Significant institution variation was detected in both components; see Figure 7.3 for predicted random effects from the $m = 51$ institutions. The heterogeneity in time to exacerbation due to the differences among institutions was higher (estimated variance component of 1.390) for patients with an acute pulmonary exacerbation (the second component) than in the first component (estimated variance component of

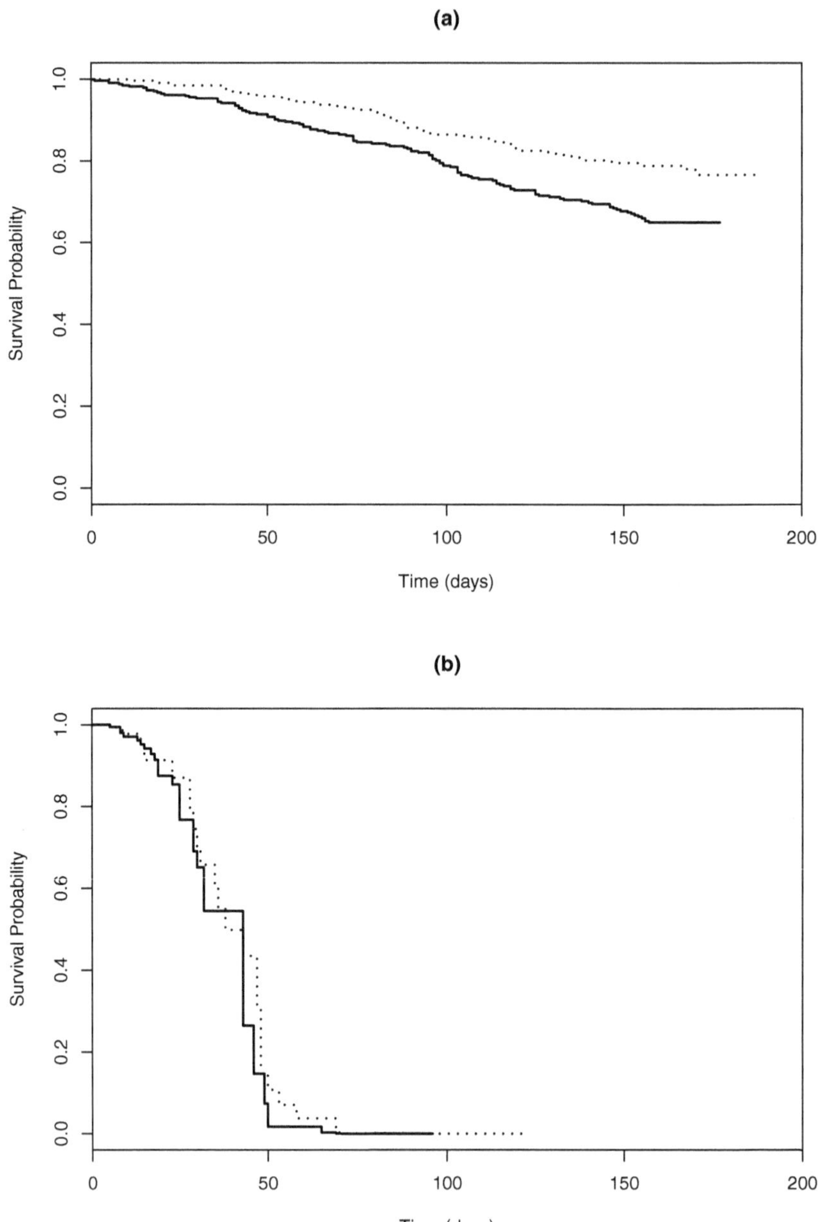

FIGURE 7.2
Survival curves by treatment group (rhDNase: Dotted line; placebo: Solid line) with adjustment for baseline FEV1 for patients in (a) the first component and (b) the second component (acute exacerbations).

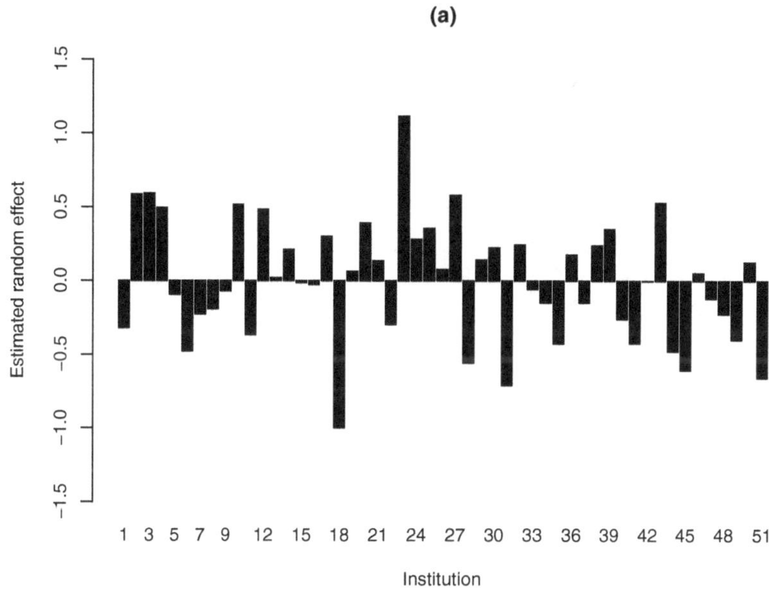

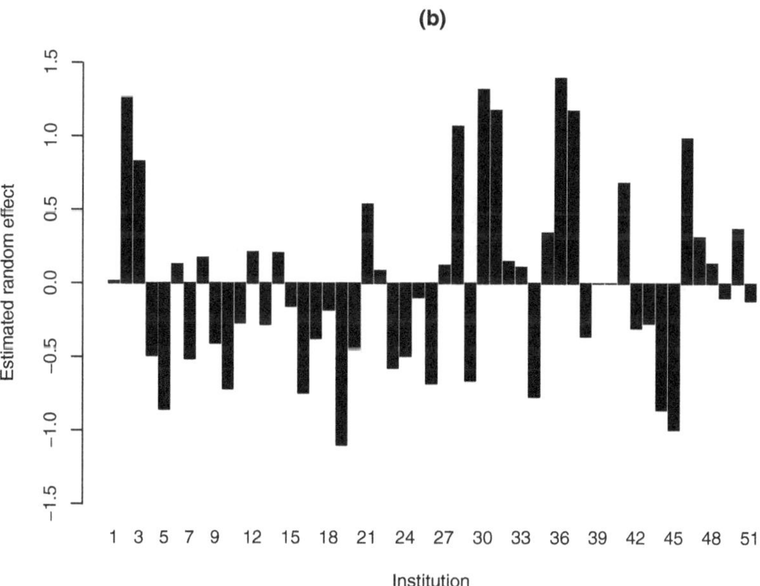

FIGURE 7.3
Prediction of random institution effects for (a) the first component and (b) the second component (acute exacerbations).

0.377). Additional insights to assess the variation among institutions on time to pulmonary exacerbation can be obtained by assessing the random institution effects in Figure 7.3.

In this example, we consider only the time to first pulmonary exacerbation. To analyze the total of 767 recurrent observations from $n = 647$ patients, the multilevel survival mixture model (7.35) can be extended to incorporate multivariate random patient effects (frailties) to account for the serial dependence between recurrent events recorded on the same patient. For example, using the approach of Wang et al. (2007a) the linear predictor in (7.37) can be formulated to relate to the covariate x_{ij}, the institution-specific random effects u_{hi}, and the patient-specific random effects v_{hijk}, as

$$\eta_h(x_{ij}) = x_{ij}^T\beta_h + u_{hi} + v_{hijk} \tag{7.43}$$

for $i = 1,\dots,m;\ j = 1,\dots,n_i;\ h = 1,2$, where $k = 1,2,\dots,n_{ij}$ indicates the number of observed recurrent times of the jth patient nested within the ith institution. The unobservable random institution effects u_{hi} are taken to be i.i.d. $N(0,\theta_h)$, whereas the unobservable random patient effects $(v_{hij1},\dots,v_{hijn_{ij}})^T$ follow a first-order autoregressive AR(1) correlation structure to accommodate a time-varying frailty effect. More recently, Tawiah et al. (2019) have developed a multilevel frailty model with a random treatment-by-institution interaction term for the analysis of the rHDNase recurrent-event data.

7.5 Long-Term Survivor Mixture Models with Random Effects

As presented in Section 5.5, a long-term survivor mixture model is a limiting case of a two-component survival mixture model, where one component corresponds to a group of patients who are cured and thus do not experience the event of interest (e.g., relapse of a disease). Long-term survivor mixture models can be extended in a similar manner described in previous sections to handle multilevel survival data with existence of long-term survivors, characterized by the overall survival curve being levelled at non-zero probability (Yau and Ng, 2001). As outlined in Section 6.3.5, the formulation of multilevel long-term survivor mixture models is achieved by incorporating random effect terms into the linear predictors with reference to the logistic model and the Cox proportional hazards model. Precisely, based on the long-term survivor model

$$S(t;x) = 1 - \pi(x) + \pi(x)S_2(t;x),$$

the linear predictor in the logistic model regarding the probability that an individual will eventually experience a failure event (uncured), $\pi(x_{ij}) = 1/(1+\exp\xi_{ij})$, is

$$\xi_{ij} = w_{ij}^T\gamma + u_i; \tag{7.44}$$

see, for example, Lai and Yau (2009). The linear predictor in the hazard model for the hazard function of those uncured patients, $h_2(t_{ij};x_{ij}) = h_{20}(t_{ij})\exp\eta_{ij}$, is

$$\eta_{ij} = x_{ij}^T\beta + v_i \tag{7.45}$$

for $i = 1,\ldots,m; j = 1,\ldots,n_i$, where $w_{ij} = (1\ x_{ij}^T)^T$, $u = (u_1,\ldots,u_m)^T$ and $v = (v_1,\ldots,v_m)^T$ are unobservable random effects, which are taken to be i.i.d. $N(0,\theta_u I_m)$ and $N(0,\theta_v I_m)$, respectively.

Following the GLMM method, the BLUP type log likelihood $l = l_1 + l_2$ becomes

$$\begin{aligned} l_1 &= \sum_{i,j}[D_{ij}\log\pi(x_{ij})f_2(t_{ij};x_{ij}) + (1-D_{ij})\log S(t_{ij};x_{ij})] \\ l_2 &= -\tfrac{1}{2}[m\log(2\pi\theta_u) + \theta_u^{-1}u^T u + m\log(2\pi\theta_v) + \theta_v^{-1}v^T v], \end{aligned} \tag{7.46}$$

where $D_{ij} = 1$ and $D_{ij} = 0$ indicate a failure and a censored observation, respectively, and $f_2(t_{ij};x_{ij})$ is the component density functions of T for the uncured individuals. The EM algorithm is used to obtain the BLUP estimates, where the E-step calculates the partitioned Q-functions corresponding to

$$\begin{aligned} Q_\xi &= \sum_{i,j}[\tau_{ij}\xi_{ij} - \log(1+\exp\xi_{ij})] - \tfrac{1}{2}[m\log(2\pi\theta_u) + \theta_u^{-1}u^T u] \\ Q_\eta &= \sum_{i,j}\{\tau_{ij}[D_{ij}\log f_2(t_{ij};x_{ij}) + (1-D_{ij})\log S_2(t_{ij};x_{ij})]\} - \tfrac{1}{2}[m\log(2\pi\theta_v) + \\ &\qquad \theta_v^{-1}v^T v] \end{aligned} \tag{7.47}$$

where

$$\tau_{ij}^{(k)} = \begin{cases} \dfrac{\pi^{(k)}S_2^{(k)}}{1-\pi^{(k)}+\pi^{(k)}S_2^{(k)}} & \text{if } D_{ij} = 0 \\ 1 & \text{if } D_{ij} = 1 \end{cases} \tag{7.48}$$

is the current estimated posterior probability of uncured, and where $S_2^{(k)} = S_2(t_{ij};x_{ij})$ is evaluated based on the current estimates. The M-step updates the estimates for γ, β, and the random effects u and v that maximize the corresponding partitioned Q-functions.

Assuming a Weibull distribution baseline hazard $h_{20}(t_{ij}) = \lambda\alpha t_{ij}^{\alpha-1}$ $(\lambda,\alpha > 0)$ for the uncured patient component, it follows that the M-step involves solving the following non-linear equations:

$$\begin{aligned} &\sum_{i,j}\{\tau_{hj}^{(k)}(1-\pi(x_{ij})) - (1-\tau_{hj}^{(k)})\pi(x_{ij})\}w_{ij} = 0 \\ &\sum_{i,j}\tau_{ij}^{(k)}[D_{ij} + \log S_2(t_{ij};x_{ij})]x_{ij} = 0 \\ &\sum_{j=1}^{n_i}\{\tau_{ij}^{(k)}(1-\pi(x_{ij})) - (1-\tau_{ij}^{(k)})\pi(x_{ij})\} - u_i/\theta_u = 0 \qquad (i = 1,\ldots,m) \\ &\sum_{j=1}^{n_i}\{\tau_{ij}^{(k)}(D_{ij} + \log S_2(t_{ij};x_{ij})\} - v_i/\theta_v = 0 \qquad (i = 1,\ldots,m) \\ &\sum_{i,j}\tau_{ij}^{(k)}[D_{ij}/\lambda - \exp(\eta(x_{ij}))t_{ij}^{\alpha}] = 0 \\ &\sum_{i,j}\tau_{ij}^{(k)}\frac{D_{ij} + (D_{ij}\alpha - h_2(t_{ij};x_{ij})t_{ij})\log t_{ij}}{\alpha} = 0 \end{aligned} \tag{7.49}$$

for γ, β, u, v, λ, and α, respectively. The MINPACK routine HYBRD1 of Moré, Garbow, and Hillstrom (1980) can be adopted to solve (7.49). After the BLUP estimators are obtained via the EM algorithm, the approximate REML estimates of the variance components θ_u and θ_v, and the asymptotic variances of estimates $\hat{\gamma}$ and $\hat{\beta}$ are then computed based on the GLMM method, via the negative second derivative of $l = l_1 + l_2$ with respect to the conformal partition of $\gamma|u$ and $\beta|v$, as outlined in (7.27) to (7.29) in Section 7.3.

Alternatively to the specification of a Weibull distribution baseline hazard as above, Lai and Yau (2009) had the form of the baseline hazard $h_{20}(t_{ij})$ unspecified. In particular, the marginal likelihood approach of Kuk and Chen (1992) was adopted so that l_1 in the BLUP-type log-likelihood (7.46) can be rewritten as

$$\sum_{r=1}^{k} \left[\eta_{(r)} - \log \left(\sum_{j \in R_{(r)}} D_j \exp \eta_j \right) \right], \tag{7.50}$$

where the failure/censoring times were arranged in increasing order and $\eta_{(r)}$ and $R_{(r)}$ were denoted, respectively, the linear predictor and risk set corresponding to the distinct event times at $t_{(r)}$ for $r = 1, \ldots, k$. They also considered the extension of the multilevel long-term survivor mixture model in (7.45) for analysis of recurrent-event data. This was done by incorporating an additional random patient effect v_{ij} in the hazard function part to model the correlation among repeated events nested within patients; see Lai and Yau (2009) for more details.

Diagnostic assessment for parametric regression models based on the Cox proportional hazards model can be implemented according to the procedures described in Section 5.6, where the residual becomes

$$r_{ij} = H(t_{ij}; \hat{\Psi}, x_{ij}) \qquad (i = 1, \ldots, m; \; j = 1, \ldots, n_i)$$

to account for the multilevel setting; see Yau and Ng (2001) for more information.

7.5.1 Example 7.4: Chronic Granulomatous Disease Data

The chronic granulomatous disease (CGD) data (Fleming and Harrington, 1991) were from a placebo-controlled randomized trial of gamma interferon (γ-INF) in CGD conducted by the International CGD Co-operative Study Group to assess the efficacy of γ-INF in reducing the incidence of serious infections in CGD patients. The data set consists of 128 patients from 13 hospitals (the number of patients per hospital ranged from 4 to 26) and is available at `https://vincentarelbundock.github.io/Rdatasets/datasets.html`. Each patient might experience more than one CGD infection and the survival times were measured as gap times between recurrent CGD infections for each patient. Patients were followed for about 1 year and during the follow-up period, there were a total of 203 observations. Of 63 patients randomized to the γ-INF treatment group, 20 CGD infections among 14 patients were observed, which suggested the presence of a cured proportion of patients who are long-term survivors regarding CGD infections. Lai and Yau (2009) fitted the multilevel semi-parametric cure model with

TABLE 7.4
Estimates and standard errors (in parentheses) for fitting a multilevel semi-parametric mixture cure model with random effects (Example 7.4). Adapted from Lai and Yau (2009).

Parameter	Logistic part	Hazard part
Intercept	1.140* (0.545)	–
γ-INF	-1.445* (0.676)	-0.522 (0.325)
Variance component θ_u or θ_v	0.027 (0.185)	0.065 (0.109)
Variance component θ_v^*	–	0.134 (0.204)

* p-value < 0.05.

random effects (7.50) to the recurrent CGD data, with $n = \sum_{i=1}^{m} n_i = 128$ patients and $N = \sum_{i,j} n_{ij} = 203$ observations. Here n_{ij} denotes the number of recurrent CGD infections for the jth patient of the ith hospital. Random hospital effects u_i and v_i $(i = 1,\ldots,m)$ were incorporated in the logistic model (7.44) and the hazard function (7.45), respectively, to account for the dependence of observations within hospitals. To model the correlation among repeated infections nested within patients, an additional random patient effect v_{ij} was introduced in the hazard function so that (7.45) becomes

$$\eta_{ijr} = x_{ij}^T\beta + v_i + v_{ij} \qquad (i = 1,\ldots,m;\ j = 1,\ldots,n_i;\ r = 1,\ldots,n_{ij}), \tag{7.51}$$

where $(v_{i1},\ldots,v_{in_i},\ldots,v_{m1},\ldots,v_{mn_m})$ were assumed to be i.i.d. $N(0,\theta_v^* I_n)$ and that $\theta_v^*\phi = \theta_v$; see Lai and Yau (2009) for more details.

The results are presented in Table 7.4. It was found that the γ-INF treatment significantly increased the patient's cured probability of CGD infections (OR of 4.242; 95% CI = 1.128 to 15.96); the corresponding cured probability for the treatment group was 57.6% compared with 24.2% for the placebo group. Although the γ-INF treatment was not significant in the hazard function part, the negative sign in the coefficient indicated an effect in reducing the hazard rate of recurrent CGD infections in the uncured patients. The small values in the variance component of random hospital effects indicated that heterogeneity of hospitals was essentially negligible. However, serial dependence among times to recurrent CGD infections was stronger (with the estimated variance component of $\hat{\theta}_v^* = 0.134$), though it was not statistically significant (possibly due to a relatively small sample of uncured patients and a few numbers of recurrent infections among those uncured patients); see Lai and Yau (2009) for more discussion.

7.6 Computing Programs for Fitting Multilevel Mixture Models

Computing programs for parameter estimation of multilevel Poisson mixture regression models (Section 7.2) and multilevel ZIP mixture regression models (Section 7.3) were developed in S-plus macros and are available on request from the corresponding author of Wang, Yau, and Lee (2002a) and Wang, Yau, and Lee (2002b), respectively. The S-plus program implemented for the MHD estimation (Section 7.2.1) is available from the *Biometrics* website `http://www.biometrics.tibs.org`.

Computing programs for fitting multilevel survival mixture models are available at the book's online webpage. They include the Fortran program "survweirandom.f" and external Fortran program "hybrid.f" (for multilevel mixture models of Weibull distributions described in Section 7.4), as well as the S-plus program implemented for the parameter estimation of multilevel semi-parametric mixture cure models with random effects (Equation (7.50) in Section 7.5) for analyzing recurrent-event survival data.

The survweirandom.f is compiled with hybrid.f in Cygwin to obtain the executable program MIXWEIRANDOM, which fits a multilevel mixture of parametric survival model (7.35) using the GLMM estimation method, assuming a Weibull distribution with a Cox's proportional hazards form (7.36).

The program code for fitting the rhDNase clinical trial data (Example 7.3) using MIXWEIRANDOM is provided in Figure 7.4.

Main Analysis for Example 7.3

```
$ cd c:/Chapter7                  # change to the work directory

$ gfortran -c hybrid.f            # compile a Fortran file to obtain an .c file

$ gfortran -c survweirandom.f     # compile the survweirandom.f to get survweirandom.o

# to obtain the executable program MIXWEIRANDOM from the .o files
$ gfortran -o MIXWEIRANDOM survweirandom.o hybrid.o

$ ./MIXWEIRANDOM                  # call the MIXWEIRANDOM program

# output file is "output"         # output includes standard errors of estimates
```

FIGURE 7.4
The program codes for Example 7.3.

8

Advanced Mixture Models for Correlated Multivariate Continuous Data

8.1 Introduction

The previous two chapters described advanced mixture models with the use of random-effects terms for modelling correlations among observations obtained in a multilevel or repeated measure setting. This chapter introduces mixtures of linear mixed models with two sets of random effects and describes how the models provide a flexible approach for clustering multivariate continuous data collected using various experimental settings, including cross-sectional repeated-measure data, time-course data, and multilevel longitudinal data.

Let $y = (y_1^T, \ldots, y_n^T)^T$ be the observed p-dimensional multivariate continuous data such that the data can be represented by an $n \times p$ matrix. With a mixture model-based clustering approach on the basis of the n rows of the data matrix, it is assumed that y_j has come from a mixture model of component densities (say, a multivariate normal distribution given its computational tractability; see Chapter 2). That is, the mixture probability density function is given by

$$f(y;\Psi) = \sum_{h=1}^{g} \pi_h \phi(y;\mu_h,\Sigma_h), \tag{8.1}$$

where $\phi(y;\mu_h,\Sigma_h)$ denotes the p-dimensional multivariate normal distribution with mean μ_h and covariance matrix Σ_h. Here the vector of unknown parameters Ψ consists of the mixing proportions $\pi_1,\ldots,\pi_{g-1}$, the elements of the component means μ_h, and the distinct elements of the component-covariance matrices Σ_h $(h = 1,\ldots,g)$.

As described in Chapter 2, it is common to assume in Model (8.1) that individual observed data y_j $(j = 1,\ldots,n)$ are independent of one another such that the likelihood function can be expressed as a multiplication of individual density functions. However, due to modern study designs, it is now commonplace to have observations collected in a setting where there are non-negligible correlations among observed data. As a consequence, the aforementioned independence assumption may no longer be valid.

Mixtures of linear mixed models with two sets of random effects provide a flexible way to capture various forms of correlation among observed data. With reference to multivariate normal mixtures (8.1), the mean of y_j conditional on realized random effects b_{hj} and c_h can be expressed as

$$E(y_j|b_{hj}, c_h) = W\beta_h + Ub_{hj} + Vc_h, \tag{8.2}$$

given that y_j belongs to the hth component ($h = 1, \ldots, g$), where W, U, and V denote the known design matrices corresponding to the fixed and the two sets of random effects, respectively.

With (8.2), the elements of β_h (a p-dimensional vector) are fixed effects (unknown constants) modelling the conditional mean of y_j in the hth component, the vectors b_{hj} and c_h contain the unobservable row-specific random effects and column-specific random effects, respectively. The first set of random effects b_{hj} captures within-subject variation in observations that vary around the overall mean (row) vector. On the other hand, the second set of random effects c_h captures variation between subjects, inducing dependency among the observed vectors y_j from the same component. By allowing the observed vectors in the same component to be correlated, there can be greater individual variation exhibited by the individuals in the same component than would otherwise be possible under a simpler model that specifies the component-means as fixed-effects; see, for example, Ng et al. (2006) and Ng et al. (2015). The formulation of the two sets of random effects can be performed to incorporate specific experimental design information such as disease status of microarray tissue samples in which the genes are measured in cross-sectional studies; covariate information such as the time ordering of the gene measurements in time-course studies; or the structure of the hierarchical longitudinal data as in multilevel studies. These three different design settings will be considered in Sections 8.3 to 8.5. In next section, we describe the ML estimation of the mixture of linear mixed models (8.2) via the EM algorithm.

8.2 Maximum Likelihood Estimation via the EM Algorithm

With the linear mixed models in (8.2), the vectors b_{hj} ($j = 1, \ldots, n$) and c_h of random effects terms are taken to be multivariate normal $N(0, B_h)$ and $N(0, C_h)$, respectively. Here we assume B_h and C_h ($h = 1, \ldots, g$) are all diagonal matrices, where $B_h = \theta_{bh}I$ and $C_h = \theta_{ch}I$, and where I denotes an identity matrix. This assumption implies that the elements in random-effect vectors $b_{h1}, \ldots, b_{hn}$, and c_h are independently distributed. In some applications, the independence assumption of elements in random effects b_{hj} ($j = 1, \ldots, n$) can be relaxed by adopting a non-diagonal matrix for B_h to account for the correlation among the elements of b_{hj}; see, for example, Wang, Ng, and McLachlan (2012) and Ng et al. (2015) for more information. The dimensions of B_h and C_h depend on specific design settings as illustrated in the next three sections. The measurement error vector in (8.2), denoted as ε_{hj}, is also taken to be multivariate normal $N_p(0, A_h)$, where $A_h = \theta_{ah}I$ is a diagonal matrix ($h = 1, \ldots, g$).

We now let the vector of unknown parameters $\Psi = (\psi_1^T, \ldots, \psi_g^T, \pi_1, \ldots, \pi_{g-1})^T$, where ψ_h is the vector containing the unknown parameters in β_h, A_h, B_h, and C_h

$(h=1,\ldots,g)$. Maximum likelihood estimation of Ψ can be undertaken by proceeding conditionally on the column-specific random effects c_h within the framework of the EM algorithm, as described in Ng et al. (2006) and Ng (2013). By treating the unobservable component indicator variables z and the random effects b and c as missing data, the complete-data log likelihood is given by

$$\log L_c(\Psi) = \sum_{h=1}^{g}\left[\sum_{j=1}^{n} z_{hj}\log\pi_h - \tfrac{1}{2}\left\{\sum_{j=1}^{n} z_{hj}\log|\theta_{bh}I| + \log|\theta_{ch}I| + \sum_{j=1}^{n} z_{hj}\log|A_h| + \frac{b_h^T b_h}{\theta_{bh}} + \frac{c_h^T c_h}{\theta_{ch}} + \varepsilon_h^T\Omega_h\varepsilon_h\right\}\right], \tag{8.3}$$

apart from an additive constant, where $\Omega_h = I\otimes A_h^{-1}$, $b_h = (b_{h1}^T,\ldots,b_{hn}^T)^T$, and $\varepsilon_h = (\varepsilon_{h1}^T,\ldots,\varepsilon_{hn}^T)^T$ for $h=1,\ldots,g$; see Ng et al. (2006) for more details. Since $b_h^T b_h$, $c_h^T c_h$, $\varepsilon_h^T\varepsilon_h$, and $(y_j - Ub_{hj} - Vc_h)$ are sufficient statistics (Searle, Casella, and McCulloch, 1992; McCulloch and Searle, 2001) for the complete-data likelihood (8.3), the conditional expectation of the complete-data log likelihood is obtained simply by replacing these sufficient statistics in (8.3) by their conditional expectations, where the expectation is with respect to the joint distribution of the missing data given the observed data and the current estimates of Ψ. On the $(k+1)$th iteration of the EM algorithm, the conditional expectation, $E_{\Psi^{(k)}}(z_{hj}|y)$ is given by

$$E_{\Psi^{(k)}}(z_{hj}|y) = \int \frac{\pi_h^{(k)} f(y_j|z_{hj}=1,c_h;\psi_h^{(k)})}{\sum_{i=1}^{g}\pi_i^{(k)} f(y_j|z_{ij}=1,c_i;\psi_i^{(k)})} f(c|y)dc, \tag{8.4}$$

where the fraction in (8.4) is the current estimated posterior probability that y_j belongs to the hth component given c, and where

$$\log f(y_j|z_{hj}=1,c_h;\psi_h^{(k)}) = -\tfrac{1}{2}\left\{\log|A_h^{(k)} + \theta_{bh}^{(k)}UU^T| + (y_j - W\beta_h^{(k)} - Vc_h)^T[A_h^{(k)} + \theta_{bh}^{(k)}UU^T]^{-1}(y_j - W\beta_h^{(k)} - Vc_h)\right\} \tag{8.5}$$

is the log density of y_j conditioned on c_h and the membership of the hth component, apart from an additive constant. Based on the normal theory described in Searle, Casella, and McCulloch (1992), after integrating over the conditional distribution of random effects c given y, the conditional expectation $E_{\Psi^{(k)}}(z_{hj}|y)$ in (8.4) becomes

$$E_{\Psi^{(k)}}(z_{hj}|y) = \tau_{hj}^{(k)} = \frac{\pi_h^{(k)} f(y_j|z_{hj}=1;\psi_h^{(k)})}{\sum_{i=1}^{g}\pi_i^{(k)} f(y_j|z_{ij}=1;\psi_i^{(k)})}, \tag{8.6}$$

where

$$\log f(y_j|z_{hj}=1;\psi_h^{(k)}) = -\tfrac{1}{2}\left\{\log|A_h^{(k)} + \theta_{bh}^{(k)}UU^T + \theta_{ch}^{(k)}VV^T| + (y_j - W\beta_h^{(k)})^T[A_h^{(k)} + \theta_{bh}^{(k)}UU^T + \theta_{ch}^{(k)}VV^T]^{-1}(y_j - W\beta_h^{(k)})\right\} \tag{8.7}$$

is the marginal log density of y_j given that it belongs to the hth component ($h = 1,\ldots,g$), apart from an additive constant.

For the next conditional expectation, $E_{\Psi^{(k)}}(b_h^T b_h|y)$, it is given by

$$E_{\Psi^{(k)}}(b_h^T b_h|y) = \sum_{j=1}^{n} \tau_{hj}^{(k)} b_{hj}^{(k)^T} b_{hj}^{(k)} + F_h^{(k)}, \tag{8.8}$$

where

$$b_{hj}^{(k)} = E_{\Psi^{(k)}}(b_{hj}|y) = \theta_{bh}^{(k)} U^T [A_h^{(k)} + \theta_{bh}^{(k)} UU^T]^{-1} \Bigg[(y_j - W\beta_h^{(k)}) - \theta_{ch}^{(k)} VV^T M_h^{(k)} \sum_{l=1}^{n} \tau_{hl}^{(k)} (y_l - W\beta_h^{(k)})\Bigg], \tag{8.9}$$

and where

$$M_h^{(k)} = \left[A_h^{(k)} + \theta_{bh}^{(k)} UU^T + \sum_{j=1}^{n} \tau_{hj}^{(k)} \theta_{ch}^{(k)} VV^T\right]^{-1}.$$

In (8.8), $F_h^{(k)}$ is the trace of the current conditional covariance matrix $\text{var}(b_h|y)$ and is given by

$$F_h^{(k)} = \sum_{j=1}^{n} \tau_{hj}^{(k)} \text{trace}(\theta_{bh}^{(k)} UU^T) - -\text{trace}(\theta_{bh}^{(k)^2} U^T M_h^{(k)} U) - (\sum_{j=1}^{n} \tau_{hj}^{(k)} - 1)\text{trace}(\theta_{bh}^{(k)^2} U^T [A_h^{(k)} + \theta_{bh}^{(k)} UU^T]^{-1} U). \tag{8.10}$$

For the conditional expectation $E_{\Psi^{(k)}}(c_h^T c_h|y)$, we have

$$E_{\Psi^{(k)}}(c_h^T c_h|y) = c_h^{(k)^T} c_h^{(k)} + G_h^{(k)}, \tag{8.11}$$

where

$$c_h^{(k)} = E_{\Psi^{(k)}}(c_h|y) = \theta_{ch}^{(k)} V^T M_h^{(k)} \sum_{l=1}^{n} \tau_{hl}^{(k)} (y_l - W\beta_h^{(k)}) \tag{8.12}$$

and

$$G_h^{(k)} = \text{trace}(\theta_{ch}^{(k)} VV^T) - \sum_{j=1}^{n} \tau_{hj}^{(k)} \text{trace}(\theta_{ch}^{(k)^2} V^T M_h^{(k)} V) \tag{8.13}$$

is the trace of the conditional covariance matrix $\text{var}(c_h|y)$.

The last conditional expectation $E_{\Psi^{(k)}}(\varepsilon_h^T \Omega_h \varepsilon_h|y)$ is given by

$$E_{\Psi^{(k)}}(\varepsilon_h^T \Omega_h \varepsilon_h|y) = \sum_{j=1}^{n} \tau_{hj}^{(k)} \varepsilon_{hj}^{(k)^T} A_h^{(k)^{-1}} \varepsilon_{hj}^{(k)} + H_h^{(k)}, \tag{8.14}$$

where

$$\varepsilon_{hj}^{(k)} = y_j - W\beta_h^{(k)} - Ub_{hj}^{(k)} - Vc_h^{(k)}$$

and

$$H_h^{(k)} = \sum_{j=1}^{n} \tau_{hj}^{(k)} \text{trace}(A_h^{(k)}) - \text{trace}(A_h^{(k)^T} M_h^{(k)} A_h^{(k)}) - (\sum_{j=1}^{n} \tau_{hj}^{(k)} - 1)\text{trace}(A_h^{(k)^T} [A_h^{(k)} + \theta_{bh}^{(k)} UU^T]^{-1} A_h^{(k)}). \quad (8.15)$$

The M-step updates the estimates to $\Psi^{(k+1)}$ as given by

$$\pi_h^{(k+1)} = \frac{\sum_{j=1}^{n} \tau_{hj}^{(k)}}{n}, \quad (8.16)$$

$$\beta_h^{(k+1)} = \beta_h^{(k)} + \frac{[W^T W]^{(-1)} W M_h^{(k)} A_h^{(k)} \sum_{j=1}^{n} \tau_{hj}^{(k)} (y_j - W\beta_h^{(k)})}{\sum_{j=1}^{n} \tau_{hj}^{(k)}}, \quad (8.17)$$

$$\theta_{bh}^{(k+1)} = \frac{E_{\Psi^{(k)}}(b_h^T b_h|y)}{\sum_{j=1}^{n} \tau_{hj}^{(k)} \text{trace}(UU^T)}, \quad (8.18)$$

$$\theta_{ch}^{(k+1)} = \frac{E_{\Psi^{(k)}}(c_h^T c_h|y)}{\text{trace}(VV^T)}, \quad (8.19)$$

and

$$A_h^{(k+1)} = \frac{E_{\Psi^{(k)}}(\varepsilon_h^T \Omega_h \varepsilon_h|y)}{\sum_{j=1}^{n} \tau_{hj}^{(k)}}. \quad (8.20)$$

As described in Ng et al. (2006) and Ng (2013), the E- and M-steps are alternated repeatedly until convergence of the EM sequence of iterates. To effect a probabilistic or an outright clustering of multivariate data into g components, we calculate the estimated posterior probabilities of component membership conditional on the column-specific random effect vector $c = (c_1^T, \ldots, c_g^T)^T$ and the observed multivariate data y. However, the random effects c_h $(h = 1, \ldots, g)$ are unobservable, so we use the estimated conditional expectation given the observed data, $\hat{c}_h = E_{\hat{\Psi}}(c_h|y)$. Since the multivariate data within a component are independently distributed given c_h, it meets the needs of performing a clustering of multivariate data with each data vector considered individually in terms of its (estimated) posterior probabilities of component membership given y_j and c, which are given by

$$\tau(y_j, \hat{c}; \hat{\Psi}) = \frac{\hat{\pi}_h f(y_j|z_{hj} = 1, \hat{c}_h; \hat{\psi}_h)}{\sum_{i=1}^{g} \hat{\pi}_i f(y_j|z_{ij} = 1, \hat{c}_i; \hat{\psi}_i)}, \quad (8.21)$$

where $\log f(y_j|z_{hj} = 1, \hat{c}_h; \hat{\psi}_h)$ is given by (8.5) with $\psi_h^{(k)}$ replaced by $\hat{\psi}_h$ and c_h replaced by $\hat{c}_h$. An outright clustering of the multivariate data into g components is therefore achieved by assigning each data y_j to the component in which it has the highest estimated conditional posterior probability $\tau(y_j, \hat{c}; \hat{\Psi})$.

Initialization of the EM algorithm can be undertaken by fitting multivariate normal mixture models without random effects b_{hj} and c_h $(h = 1, \ldots, g;\ j = 1, \ldots, n_i)$;

see, for example, Chapter 2 and McLachlan and Peel (2000, Chapter 3). As described in Section 1.3.3, the initialization scheme used here is to run a mixture model with various random starts and the set of parameters corresponding to the largest likelihood value is then used to obtain the final parameter estimates. The number of components in the mixture model (8.1) is assessed using the BIC. The standard errors of the estimates of Ψ are obtained by the bootstrap resampling method with replacement, where the number of bootstrap replications is taken to be 100; see Section 1.3.4.

8.3 Clustering of Gene-Expression Data (Cross-Sectional with Repeated Measurements)

The clustering of gene expression data across some experimental conditions of interest contributes significantly to the elucidation of unknown gene function, the validation of gene discoveries and the interpretation of biological processes, as illustrated in McLachlan, Do, and Ambroise (2004, Chapter 5) and Ng et al. (2006). For instance in the work of Eisen et al. (1998), cluster analysis was used to identify genes that showed similar expression patterns in yeast over a wide range of experimental conditions. The clustering of genes on the basis of their expression levels has an important role to play in the classification of tissue samples and the discovery of genes that belong to the same molecular pathway; see McLachlan, Bean, and Peel (2002) and McLachlan, Do, and Ambroise (2004, Section 5.11) for more details. The identification of clusters of genes that are potentially coregulated also helps in identifying promoter elements by searching for common motifs in upstream regions of the genes in each cluster (Segal, Yelensky, and Koller, 2003), and can be used to form powerful predictors of disease outcomes, as illustrated in Ben-Tovim Jones et al. (2005).

Mixture model-based approaches have been widely used in the cluster analysis of gene expression data (see Yeung et al. (2001), Ghosh and Chinnaiyan (2002), and McLachlan, Bean, and Peel (2002)), given that mixture models provide a sound mathematical framework for clustering. With this approach to clustering, a common assumption is to take the component densities to be multivariate normal, which is computational tractable and ensures that the implied clustering is invariant under changes in location and scale, as well as rotation, of the data (Ng et al., 2006). However, in unmodified form, these mixture models of multivariate normal distributions do not incorporate experimental design information. For example, microarray experiments are now being carried out with replication for capturing either biological or technical variability in expression levels so as to improve the quality of inferences made from a wide variety of experimental studies, as considered in Lee et al. (2000); Pavlidis, Li, and Noble (2003); and Ng et al. (2006). Replicated measurements from each microarray tissue sample are often interdependent and tend to be more alike in characteristics than data chosen at random from the population as a whole. Ignoring the dependence between microarray gene expression data can result in overlooking the importance of certain sources of variability in microarray experiments and can subsequently lead to spurious or misleading clustering of microarray data; see Luan and Li (2003) and Yeung, Medvedovic, and Bumgarner (2003) for more information.

Here we consider a microarray experimental setting, where there are n genes from m known classes of p tissue samples. These m classes may correspond to tissues that are at different stages in some process or in distinct pathological states (such as normal tissues versus tumours). With reference to the mixture of linear mixed models given in (8.1) and (8.2), the random effects $b_{hj} = (b_{1hj}, \ldots, b_{mhj})^T$ can be used to represent the unobservable gene-specific effects shared among the gene expressions within the m tissue classes for each gene. These gene-specific random effects b_{hj} thus allow for correlation between the gene expressions across the tissues both within a class and between classes for the same gene $(j = 1, \ldots, n)$. Another set of random effects is formulated as $c_h = (c_{1h}, \ldots, c_{ph})^T$, containing the unobservable tissue-specific random effects shared among expressions common to all genes from the hth component for each tissue sample. The random effects c_h induce a correlation between those genes from the same component and is an attempt to allow for the fact that in reality the gene profile vectors are not all independently distributed. The expressions on a given gene should be independent, but this may not hold in practice due to poor experimental conditions resulting in batch-effects (Ng et al., 2015). Under this experimental setting, the dimensions of the design matrix X are $p \times m$, the design matrix U is equal to X, and V is an $p \times p$ identity matrix.

To account for potential correlation between gene-specific random effects b_{lhj} $(l = 1, \ldots, m)$, the random components B_h $(h = 1, \ldots, g)$ are assumed to be non diagonal as

$$B_h = \begin{pmatrix} \theta_{b1h} & \theta_{b12h} & \cdots & \theta_{b1mh} \\ \theta_{b12h}^T & \theta_{b2h} & \cdots & \theta_{b2mh} \\ \vdots & \vdots & \vdots & \vdots \\ \theta_{b1mh}^T & \theta_{b2mh}^T & \cdots & \theta_{bmh} \end{pmatrix}, \tag{8.22}$$

where the non-diagonal parameters accounts for the correlation between gene-specific random effects b_{lhj} $(l = 1, \ldots, m;\ j = 1, \ldots, n)$, which are shared, respectively, among the expressions on the jth gene in the lth tissue class; see Ng et al. (2015) for more details. In (8.22), the non-diagonal parameters are assumed to be different from each other. Alternatively, a simpler correlation structure may be adopted, where the covariance θ_{blrh} between gene-specific random effects b_{lhj} and b_{rhj} is given by

$$\theta_{blrh} = \rho\sqrt{\theta_{blh}\theta_{brh}} \qquad (l \neq r) \tag{8.23}$$

and the same parameter ρ is used for all the non-diagonal elements $(l = 1, \ldots, m;\ r = 1, \ldots, m)$.

The assignment of the n genes into g clusters is implemented using the estimated conditional posterior probabilities of component membership given y_j and $\hat{c}_h$, $\tau(y_j, \hat{c}; \hat{\Psi})$, as given in (8.21). While clustering of genes may help to elucidate unknown gene function (as shown in Yeung, Medvedovic, and Bumgarner (2003) and Ng et al. (2006)), another aim in the analysis of gene expression data is to identify "marker" genes that are differentially expressed between the m classes of tissue samples. In this latter context, the simplest method for identification of differentially expressed genes is to assess the log ratio of expression levels between two classes of tissue samples (known as the "fold change"); see, for example, DeRisi, Iyer, and Brown (1997) and Love, Huber, and Anders (2014). Alternatively, multiple hypothesis test-based approaches have been considered by Smyth (2004), Storey (2007), Zhao (2011), and Ritchie et al. (2015) to assess statistical significance of differential expression for each gene separately; see McLachlan, Do, and Ambroise (2004, Chapter 5) for more information. To adjust for multiple testing, the expected proportion of false positives among the genes declared to be differentially expressed, namely the false discovery rate (FDR) (Benjamini and Hochberg, 1995), is controlled at some desired level; see also the discussion in Pawitan et al. (2005). Clustering-based approaches have also been proposed for identification of differentially expressed genes, but these methods either work on gene-specific summary statistics (Pan, Lin, and Le, 2002; Matsui and Noma, 2011) or reduced forms of gene-expression data considered by Dahl and Newton (2007). Alternatively, clustering methods that can handle full gene-expression data rely on the assumption that pure clusters of null (non-differentially expressed) genes and differentially expressed genes exist (He, Pan, and Lin, 2006; Qiu et al., 2008); see also Qi et al. (2011) and Ng and McLachlan (2016). For example, there would be three clusters corresponding to null genes, up-regulated and down-regulated differentially expressed genes. However, the intent of obtaining pure clusters is not always possible or verifiable. There is also the problem of how to identify the clusters corresponding to null genes and up- and down-regulated differentially expressed genes, assuming the clusters are pure; see Li et al. (2007) and Ng et al. (2015) for more details. One way of approaching this problem has been to adopt a hierarchical approach whereby each cluster is decomposed into three subclusters representing null, up-regulated and down-regulated differentially expressed genes as considered by Yuan and Kendziorski (2006).

Recently, a new method has been developed to draw inference on differences between tissue classes using cluster-specific contrasts of mixed effects (Ng et al., 2015). In addition to the class-specific fixed effects terms, this method forms a cluster-specific contrast by including gene-specific random effects terms, which are obtained by initial clustering of genes into a number g of clusters with a fitting of a mixture of linear mixed models given in (8.1) and (8.2). Moreover, this method does not rely on the clusters being pure as to whether all cluster members are differentially expressed or null genes. The proposed test statistic can be adopted to rank the genes in each cluster in order of evidence against the null hypothesis of no differential expression between tissue classes. Furthermore, the correlation structure among those highly ranked differentially expressed genes in each cluster can be further explored using a non-parametric clustering approach. These two methods will be described in the next two sub-sections.

8.3.1 Inference on Differences Between Classes Using Cluster-Specific Contrasts of Mixed Effects

The relevance of each gene for differentiating the m tissue classes is quantified on the basis of cluster-specific contrasts of mixed effects. Under the assumption that z is known, where each gene expression profile y_j is classified with respect to the g components $G_1, \ldots, G_g$ in its mixture distribution with components specified by (8.2), and that the vector ζ_i containing the distinct elements in the component-covariance matrices A_h, B_h, and C_h is known.

We let

$$r_h = (\beta_h^T, b_{G_h}^T, c_h^T)^T \qquad (h = 1, \ldots, g) \tag{8.24}$$

be the vector containing the fixed and random effects for the n_h genes belonging to the hth component $G_h\,(h = 1, \ldots, g)$, where

$$b_{G_h} = (b_{h_1}^T, \ldots, b_{h_{n_h}}^T)^T \tag{8.25}$$

is the $(2n_h)$-dimensional vector containing the gene-specific random effects terms for the n_h genes belonging to $G_h\,(h = 1, \ldots, g)$; see, for example, Ng and McLachlan (2013) and Ng et al. (2015). For an individual gene j belonging to the hth cluster G_h, a cluster-specific normalized contrast s_{hj} can be formed by quantifying the differential expression of gene j in terms of mixed effects r_h. That is,

$$s_{hj} = d_j^T r_h / \lambda_{hj} \qquad (h = 1, \ldots, g), \tag{8.26}$$

where d_j is a vector whose elements sum to zero and λ_{hj} is the normalizing term. By replacing r_h with the BLUP estimator $\hat{r}_h$ in the right-hand side of (8.26) and taking the normalizing term λ_{hj} to be the standard error of $d_j^T \hat{r}_h$ (conditional on membership of the jth gene to the hth component G_h), an estimate of the test statistic s_{hj} is given by

$$\hat{s}_{hj} = d_j^T \hat{r}_h / \sqrt{d_j^T \hat{\Omega}_h d_j}, \tag{8.27}$$

where the covariance matrix of the BLUP estimator of the mixed effects, $\hat{\Omega}_h$, (and d_j) is partitioned conformally corresponding to $\beta_h | b_{G_h} | c_h$ with dimensions m, mn_h, and p, respectively. That is,

$$\hat{\Omega}_h = \begin{bmatrix} \hat{\Omega}_{h\beta} & \hat{\Omega}_{h\beta b} & \hat{\Omega}_{h\beta c} \\ \hat{\Omega}_{h\beta b}^T & \hat{\Omega}_{hb} & \hat{\Omega}_{hbc} \\ \hat{\Omega}_{h\beta c}^T & \hat{\Omega}_{hbc}^T & \hat{\Omega}_{hc} \end{bmatrix}^{-1}, \tag{8.28}$$

with

$$\begin{aligned}
\hat{\Omega}_{h\beta} &= n_h W^T \hat{A}_h^{-1} W, \\
\hat{\Omega}_{h\beta b} &= 1_{n_h} \otimes W^T \hat{A}_h^{-1} U, \\
\hat{\Omega}_{h\beta c} &= n_h W^T \hat{A}_h^{-1} V, \\
\hat{\Omega}_{hb} &= I_{n_h} \otimes (U^T \hat{A}_h^{-1} U + \hat{B}_h^{-1}), \\
\hat{\Omega}_{hbc} &= 1_{n_h} \otimes U^T \hat{A}_h^{-1} V, \\
\hat{\Omega}_{hc} &= n_h (V^T \hat{A}_h^{-1} V + \hat{C}_h^{-1}),
\end{aligned} \tag{8.29}$$

where 1_{n_h} is a n_h-dimensional vector of ones, I_{n_h} is an identity matrix with dimension n_h, and the sign $\otimes$ denotes the Kronecker product of two matrices; see Ng et al. (2015). A test statistic for the jth gene is therefore obtained by weighting the estimated (normalized) contrast $\hat{s}_{hj}$ over the g components in the mixture model,

$$w_j = \sum_{h=1}^{g} \tau_h(y_j, \hat{c}; \hat{\Psi})\, \hat{s}_{hj} \qquad (j = 1, \ldots, n), \tag{8.30}$$

where $\tau_h(y_j, \hat{c}; \hat{\Psi})$ is the posterior probability that the jth gene belongs to the hth component G_h conditional on y_j and c, as given in (8.21).

In (8.27), the choice of d_j has direct implications for the inference space of the contrast s_{ij}, as described in detail by Ng et al. (2015). For m=2 classes of tissue samples, a typical form of d_j is

$$d_j^T = (1 \;\, \text{-}1 \; \vdots \; 0 \;\, 0, \; \ldots, \; 0 \;\, 0, \; 1 \;\, \text{-}1, \; 0 \;\, 0, \; \ldots \; \vdots \; 0 \; \ldots \; 0), \tag{8.31}$$

where only one pair of (1 -1) exists in the second partition corresponding to the gene-specific random effects b_{hj} for a gene in the hth cluster ($h = 1, \ldots, g$). With reference to McLean, Sanders, and Stroup (1991), the contrast (8.31) represents an "intermediate inference space" in that the inference is "narrow" to gene-specific random effects b_{hj} but "broad" to tissue-specific random effects c_h. With d_j being set as in (8.31), this means a contrast of differential expressions between two classes of tissues is considered and the inference applies to the specific genes studied in the experiment (narrow) and to the entire population from which biological tissue samples were obtained (broad); see Ng et al. (2015) for more details.

Based on the weighted contrast w_j ($j = 1, \ldots, n$) in (8.30), the n genes can be ranked in order of their relevance for differentiating the m classes (with respect to the defined form of d_j for the normalized contrast in (8.27)). In the next subsection, we describe a non-parametric clustering approach in an attempt to explore the correlation structure of top-ranked differentially expressed genes in each identified cluster G_h ($h = 1, \ldots, g$), say, for those genes with contrast w_j more extreme than thresholds w_{0u} or w_{0d} for up- and down-regulated genes, respectively. A guide to plausible values of w_{0u} and w_{0d} can be obtained using the method described in Ng et al. (2015). Firstly, the p column labels of the data matrix $H = (y_1, \ldots, y_n)^T$ are permutated in B times randomly and independently. For each permutation data matrix $H^{(b)} = (y_1^{(b)}, \ldots, y_n^{(b)})^T$, we then compute $W_j^{(b)}$ $(b = 1, \ldots, B)$ in the same way that the original observed value w_j is calculated except that H is replaced by $H^{(b)}$; see Ng et al. (2015) for more details. The next procedure is to fit a three-component mixture of t-distributions (see McLachlan and Peel (2000, Chapter 7)) to the replicated values of $W_j^{(b)}$ for $b = 0, \ldots, B$, where $b = 0$ corresponds to the observed values w_j $(j = 1, \ldots, n)$. The component corresponding to the central portion of $W_j^{(b)}$ can be considered as an approximation to the null distribution of W_j, under the null hypothesis of no differential expression. With reference to this estimated component t-distribution that represents the null distribution, the P-value for each original observed value w_j, denoted as P_j, is then calculated for $j = 1, \ldots, n$. As described in McLachlan, Bean, and Ben-Tovim Jones (2006), these P-values can be converted to z-scores, via the $N(0,1)$ distribution function Φ as

$$z(P_j) = \Phi^{-1}(1 - P_j), \tag{8.32}$$

to which a two-component normal mixture model is fitted with the first component corresponding to the null genes and the second component corresponding to the differentially expressed genes (which are determined by the estimated posterior probability for the first (null) component being less than some threshold; for example, $\hat{\tau}_0(z(P_j)) < c_0$ for $c_0 = 0.1$); see also Ng et al. (2015) for using the P-values to control the FDR at a specified level.

8.3.2 A Non-Parametric Clustering Approach for Identification of Correlated Differentially Expressed Genes

We present a non-parametric method to cluster the t_h top-ranked genes into networks of differentially expressed genes that are highly correlated; see also Ng and McLachlan (2017). These t_h genes have the weighted contrasts w_j more extreme than either w_{0u} or w_{0d} in Cluster G_h $(h = 1, \ldots, g)$, being declared to be differentially expressed based on the procedure described above. The first step of the non-parametric clustering method calculates the pairwise correlation coefficient for each pair of the t_h genes in G_h $(h = 1, \ldots, g)$. Whereas the Pearson correlation coefficient is widely used to measure the strength of the linear relationship between gene expressions, the Spearman (rank) correlation coefficient may be more appropriate to depict monotonic relationships when the gene expressions are not linearly related or they are not normally distributed. In the second step, significance of the pairwise correlation coefficients is assessed with the use of a permutation method proposed by Ng, Holden, and Sun (2012) to determine the null distribution of correlation coefficients. Precisely, the m class labels of tissue samples are randomly permuted separately for each gene. Assuming that the null distributions of correlation coefficients are the same for each pair of the t_h genes, we can pool the permutations for all $N_{t_h} = t_h(t_h - 1)/2$ pairs of genes to determine the null distribution of correlation coefficients. With $B = 100$ repetitions of random permutations, the P-value for each pair of genes is estimated by

$$P_l = \sum_{b=1}^{B} \frac{\#\{r : R_{0r}^{(b)} \geq R_l, r = 1, \ldots, N_{t_h}\}}{N_{t_h} B} \qquad (l = 1, \ldots, N_{t_h}), \tag{8.33}$$

where $R_{0r}^{(b)}$ is the null version of correlation coefficient for the rth pair of differentially expressed genes after the bth repetition of permutations $(r = 1, \ldots, N_{t_h}; b = 1, \ldots, B)$. Let $P_{(1)} \leq \cdots \leq P_{(N_{t_h})}$ be the ordered observed P-values obtained from (8.33), the procedure proposed by Benjamini and Hochberg (1995) is used to determine the cut-off value $\hat{k}$, where

$$\hat{k} = \arg \max_{1 \leq k \leq N_{t_h}} \{k : P_{(k)} \leq \alpha k / N_{t_h}\}, \tag{8.34}$$

with control of the FDR at level α. In practice, α is set such that the expected number of false positives among the pairs of genes identified to be significantly correlated is smaller than one; see Ng, Holden, and Sun (2012) and Ng and McLachlan (2017). With (8.34), pairwise correlation coefficients corresponding to P-values $P_{(1)} \leq \cdots \leq P_{(\hat{k})}$ are identified to be significant, providing that their absolute values are greater than the minimum practically important effect size of 0.2 (an absolute value of correlation coefficients between 0-0.2 is regarded as very weak; see Evans (1996)). Significance of the pairwise correlation coefficients is represented by a $t_h \times t_h$ symmetric binary matrix M with elements of one or zero indicating the corresponding correlation coefficients are significant or not. In the final step, we search in M to identify networks of differentially expressed genes in which all members in a group significantly correlate with one another; see Ng, Holden, and Sun (2012) for more details. This three-step clustering approach obtains overlapping groups (networks) of correlated differentially expressed genes.

8.3.3 Example 8.1: Cluster Analysis of a Pancreatic Cancer Gene-Expression Data Set

We illustrate the methodology for clustering cross-sectional gene-expression data with repeated measurements using the pancreatic cancer gene-expression data set of Subramaniam et al. (2012). The data set comprised full expression values of 37,628 genes (after excluding 4 genes with incomplete expression values) for $m = 3$ classes of tissue samples. The three classes correspond to tissue samples from 5 pancreatic cancer patients treated with gemcitabine alone, 6 patients treated with P276 (a novel CDK inhibitor) alone, and 3 patients treated with combination of P276 and gemcitabine (P276-Gem). In this example, we aim to identify networks of correlated genes that are differentially expressed in tissue samples treated with P276-Gem. With the proposed three-step approach described in Section 8.3, we first fitted a mixture model with random-effects terms (8.2) to the gene-expression data set with $g = 6$ to $g = 15$ clusters, taking W to be a 14×3 zero-one matrix where the first 5 rows are $(1,0,0)$, the next 6 rows are $(0,1,0)$, and the final 3 rows are $(0,0,1)$. The design matrix U is set equal to W, while V is set to I_{14}. Based on the BIC for model selection (see Ng et al. (2006)), we identified there are nine clusters of genes, with the mixing

TABLE 8.1
Estimates of the mixture model with random-effects terms (9 components) for pancreatic cancer data (Example 8.1). Adapted from Ng et al. (2015).

h	π_h	β_h	A_h	B_h	C_h
1	0.245	(0.08, -0.21, -0.67)	0.64, 0.70, 0.44	0.33, 0.26, 0.26, 0.29, 0.20, 0.20	0.22
2	0.223	(-0.54, -0.47, -0.10)	0.58, 0.67, 0.71	0.41, 0.31, 0.15, 0.35, -0.14, -0.10	0.24
3	0.039	(-0.61, -0.50, -0.62)	0.74, 1.11, 1.31	0.61, 0.44, 0.59, 0.51, -0.52, -0.42	0.28
4	0.049	(-1.25, -0.87, 0.42)	0.70, 0.81, 0.79	0.38, 0.26, 0.29, 0.31, 0.00, 0.06	0.67
5	0.012	(-0.11, 0.22, 0.29)	2.24, 1.13, 0.97	0.41, 0.17, 0.53, -0.16, -0.43, 0.24	0.59
6	0.105	(0.56, 0.57, 0.89)	0.85, 0.94, 0.78	0.16, 0.16, 0.24, 0.13, 0.14, 0.17	0.46
7	0.007	(-0.29, -0.29, 0.22)	1.28, 2.16, 1.64	1.24, 0.93, 0.57, 1.07, -0.80, -0.69	0.29
8	0.186	(0.47, 0.69, 0.64)	0.67, 0.78, 0.73	0.17, 0.15, 0.26, 0.13, 0.17, 0.17	0.31
9	0.135	(0.32, 0.25, -0.22)	1.05, 0.88, 0.61	0.13, 0.15, 0.43, -0.05, -0.17, 0.22	0.20

Note: The elements in β_h are β_{1h}, β_{2h}, and β_{3h}.
The elements in A_h are $\sqrt{\theta_{a1h}}$, $\sqrt{\theta_{a2h}}$, and $\sqrt{\theta_{a3h}}$.
The elements in B_h are $\sqrt{\theta_{b1h}}$, $\sqrt{\theta_{b2h}}$, $\sqrt{\theta_{b3h}}$, $\sqrt{\theta_{b12h}}$, $\sqrt{\theta_{b13h}}$, and $\sqrt{\theta_{b23h}}$.
The element in C_h is $\sqrt{\theta_{ch}}$.

proportions all greater than 0.5% (that is, about 188 genes). The ML estimates of the unknown parameters are presented in Table 8.1.

The ranking of differentially expressed genes is then implemented on the basis of the weighted estimates of a contrast in the mixed effects (8.30). With $m = 3$ classes of tissue samples, we adopt

$$d_j^T = (\text{-0.5 -0.5 1} \vdots \text{0 0 0}, \ldots, \text{0 0 0, -0.5 -0.5 1, 0 0 0}, \ldots \vdots \text{0} \ldots \text{0}), \quad (8.35)$$

where both tissue classes 1 and 2 contribute half to the contrast as reference to tissue class 3 (P276-Gem). The top 50 ranked up-regulated genes (corresponding to large positive weighted contrast w_j) and the top 50 down-regulated genes (corresponding to large negative w_j) are presented for functional annotation and enrichment analyses, using web-based DAVID Functional Annotation Tool (Dennis et al., 2003). The expression profiles for these top 100 ranked genes are presented in Figure 8.1(a) and 8.1(b), respectively. In Figure 8.1(c), we present these genes based on the average expression levels in Classes 1 and 2 combined versus Class 3 (P276-Gem). We compare the gene lists obtained by our method and existing approaches including Significance Analysis of Microarray (SAM) of Tusher, Tibshirani, and Chu (2001), Linear Models for Microarray Data (LIMMA) of Smyth (2004), Optimal Discovery Procedure (ODP) of Storey (2007), and the t-test. With SAM, ODP, and the t-test, Classes 1 and 2 were combined and were then compared with Class 3. With LIMMA, a linear contrast of (-0.5 -0.5 1) was adopted in multiple testing for all genes. The comparative results are presented in Table 8.2. It can be seen that different gene lists were obtained by the methods. With the Subramaniam data, the top 50 up- and down-regulated gene lists obtained by SAM and LIMMA are more alike; see the Supplementary file of Ng et al. (2015) for more information about the results of the functional annotation and enrichment analyses.

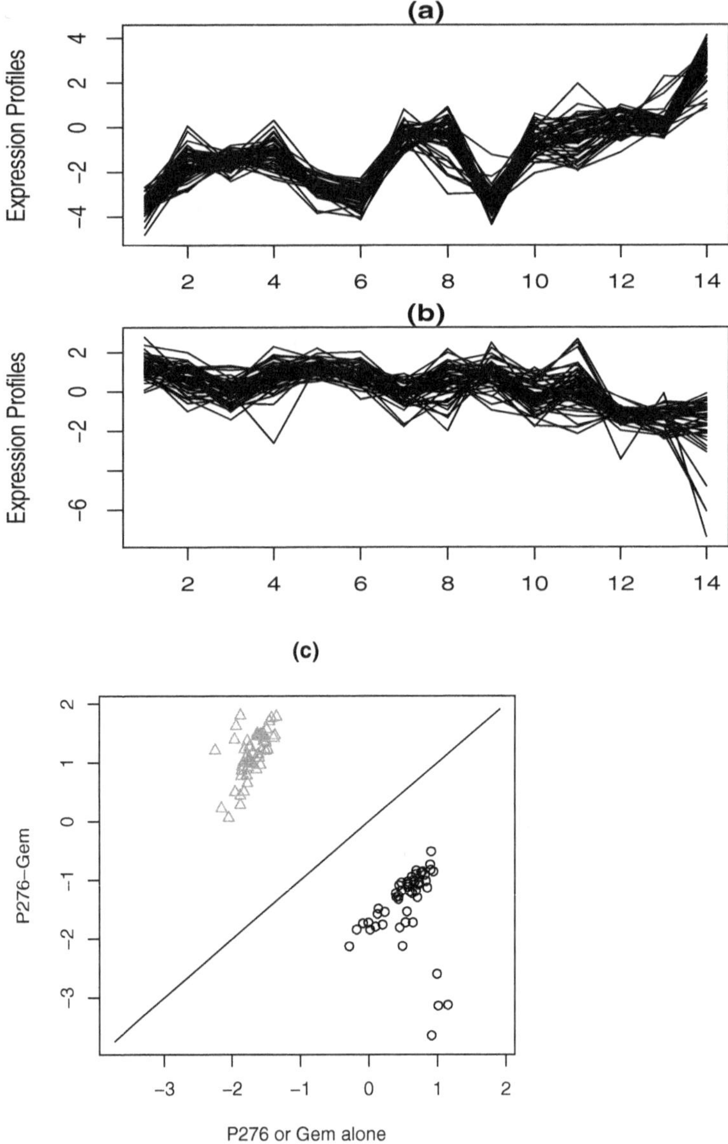

FIGURE 8.1
(a) Expression profiles for top 50 ranked up-regulated genes (Example 8.1: Pancreatic cancer data); (b) Expression profile for top 50 ranked down-regulated genes; (c) The top-ranked genes based on the average expression levels in the Gem or P276 alone and the combined P276-Gem groups ($\triangle$ – up-regulated genes; $\circ$ – down-regulated genes). Note: Expression profiles in (a) and (b) are row-standardized to give a better view on differential expressions. Adapted from Ng et al. (2015).

TABLE 8.2
Number of common genes in top 100 up- or down-regulated genes obtained by different methods (Example 8.1: Pancreatic cancer data). Adapted from Ng et al. (2015).

Number of common genes	*t*-test	SAM	LIMMA	ODP	contrast
Up-regulated genes					
t-test	–	28	24	3	0
SAM	28	–	44	10	7
LIMMA	24	44	–	14	12
ODP	3	10	14	–	18
contrast	0	7	12	18	–
Down-regulated genes					
t-test	–	23	15	4	13
SAM	23	–	42	22	16
LIMMA	15	42	–	26	13
ODP	4	22	26	–	4
contrast	13	16	13	4	–

With reference to Section 8.3.1, we next fitted a three-component heteroscedastic mixture of t-distributions to the replicated values of $W_j^{(b)}$ for $b = 0, \ldots, 19$ and $j = 1, \ldots, 37628$. An approximation of the null distribution of W_j is the estimated (central) component t-distribution with $\hat{\mu} = 0.372$, $\hat{\sigma}^2 = 1.224$, and degrees of freedom $\hat{\nu} = 19.8$. The associated P-values for each gene were then converted to z-scores, to which a two-component normal mixture model was fitted. Here the null density (the first component) was taken to be standard normal (the theoretical null distribution) and the threshold c_0 was set equal to 0.1; see McLachlan, Bean, and Ben-Tovim Jones (2006). There are a total of 11,758 genes declared to be differentially expressed, corresponding to genes with w_j more extreme than $w_{0u} = 3.231$ or $w_{0d} = -2.487$. Among them, 7,787 genes have valid and unique gene symbols (2,557 up-regulated and 5,230 down-regulated). Descriptive statistics of w_j for these 7,787 highly ranked differentially expressed genes are provided in Table 8.3. It can be seen that Clusters 1, 5, and 9 contain down-regulated differentially expressed genes, Cluster 4 contains up-regulated differentially expressed genes, and Clusters 2, 3, and 7 contain both up-regulated and down-regulated differentially expressed genes.

We then identify networks of correlated differentially expressed genes in the final step by applying the non-parametric clustering method separately to the t_h genes in Cluster G_h ($h = 1, \ldots, 9$), where the Spearman correlation coefficient is used to measure the strength of the monotonic relationship between gene expressions. For large clusters including Clusters 1, 4, and 9, only the top 5% of differentially expressed genes in terms of the magnitude of w_j were considered. For Cluster 2, only the top 10% of the 754 up-regulated differentially expressed genes were considered. We set α at different values between 0.0001 to 0.002 so that the expected number of false positives among the pairs of genes identified to be significantly correlated is smaller

TABLE 8.3
Descriptive statistics of w_j for the differentially expressed genes with valid gene identifiers and w_j more extreme than either w_{0u} or w_{0d} (9 clusters)

h	t_h	mean (SD)	median (IQR)	(minimum, maximum)
1	4298	-3.399 (0.560)	-3.349 (0.852)	(-5.546, -2.488)
2	767	3.751 (0.993)	3.725 (0.661)	(-3.222, 6.440)
3	91	-2.648 (2.142)	-3.098 (0.861)	(-5.615, 3.800)
4	1776	4.940 (0.826)	4.949 (1.113)	(3.232, 8.349)
5	1	-2.583 (n.a.)	-2.583 (n.a.)	n.a.
6	0	n.a.	n.a.	n.a.
7	39	0.331 (3.927)	-2.583 (7.306)	(-4.091, 6.772)
8	0	n.a.	n.a.	n.a.
9	815	-3.280 (0.544)	-3.208 (0.787)	(-2.487, -5.122)

Notation: SD - standard deviation; IQR - interquartile range; n.a. - not applicable.

than one; see Section 8.3.2. Networks of correlated differentially expressed genes were displayed using UCINET6 for Windows developed by Borgatti, Everett, and Freeman (2002), where the size of a node (representing a gene) is proportional to the "degree" of the node, which is the number of other genes that are significantly correlated with the gene. Figure 8.2 presents the identified networks of down-regulated differentially expressed genes in Clusters 1 and 9. The network of up-regulated differentially expressed genes (Cluster 4) is provided in Figure 8.3. Clusters 2, 3, and 7 had networks of up- and down-regulated differentially expressed genes (see Figure 8.4).

The identified networks of correlated differentially expressed genes for each cluster are presented in Table 8.4. Eight isolated networks of differentially expressed genes were identified: {DUT, ZFP106} and {K0404_HUMAN, OR51H1P, PPP1R2} down-regulated genes networks from Cluster 1, {ADAM12, FBX06, RBM41} and {PPP2R5C, PPOX, S100Z} up-regulated genes networks and {ELF1, KIR2DL1} down-regulated genes network from Cluster 2, {FBXO41, GSTZ1, NCOA3, SLU7, VSX1} up-regulated and {ASB17, MEOX1, PRNP, Q5NV77_HUMAN} down-regulated genes networks from Cluster 3, and {APOL6, HP, HNRPM, MBD6, PLEKHA8, VILL} up-regulated genes network from Cluster 7. Moreover, for Cluster 7 (Figure 8.4(c)), there is a network that contains a mix of up-regulated and down-regulated differentially expressed genes. The up-regulated genes are ALDOA, ARID2, COASY, and LOC284393, which are significantly correlated with 4, 14, 13, and 0 down-regulated genes, respectively.

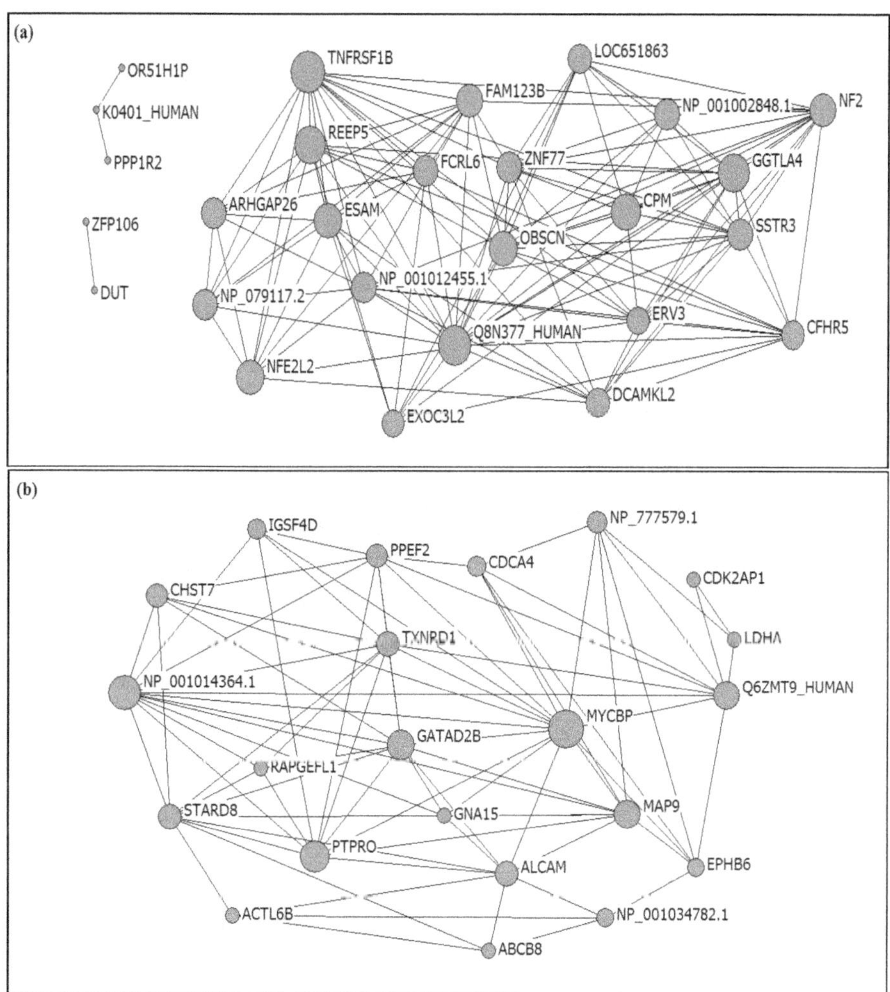

FIGURE 8.2
Network of down-regulated differentially expressed genes in (a) Cluster 1 (only genes with degrees ≥20 were displayed) and (b) Cluster 9 (only genes with degrees ≥5 were displayed). Nodal size is proportional to the degree (the number of genes that are significantly correlated with the gene).

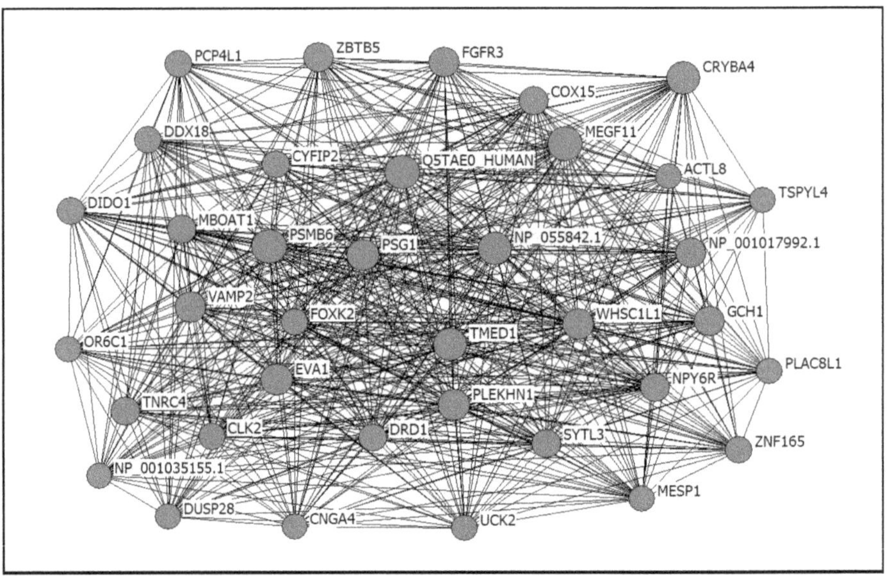

FIGURE 8.3
Network of up-regulated differentially expressed genes in Cluster 4; only genes with degrees ≥ 40 were displayed. Nodal size is proportional to the degree (the number of genes that are significantly correlated with the gene).

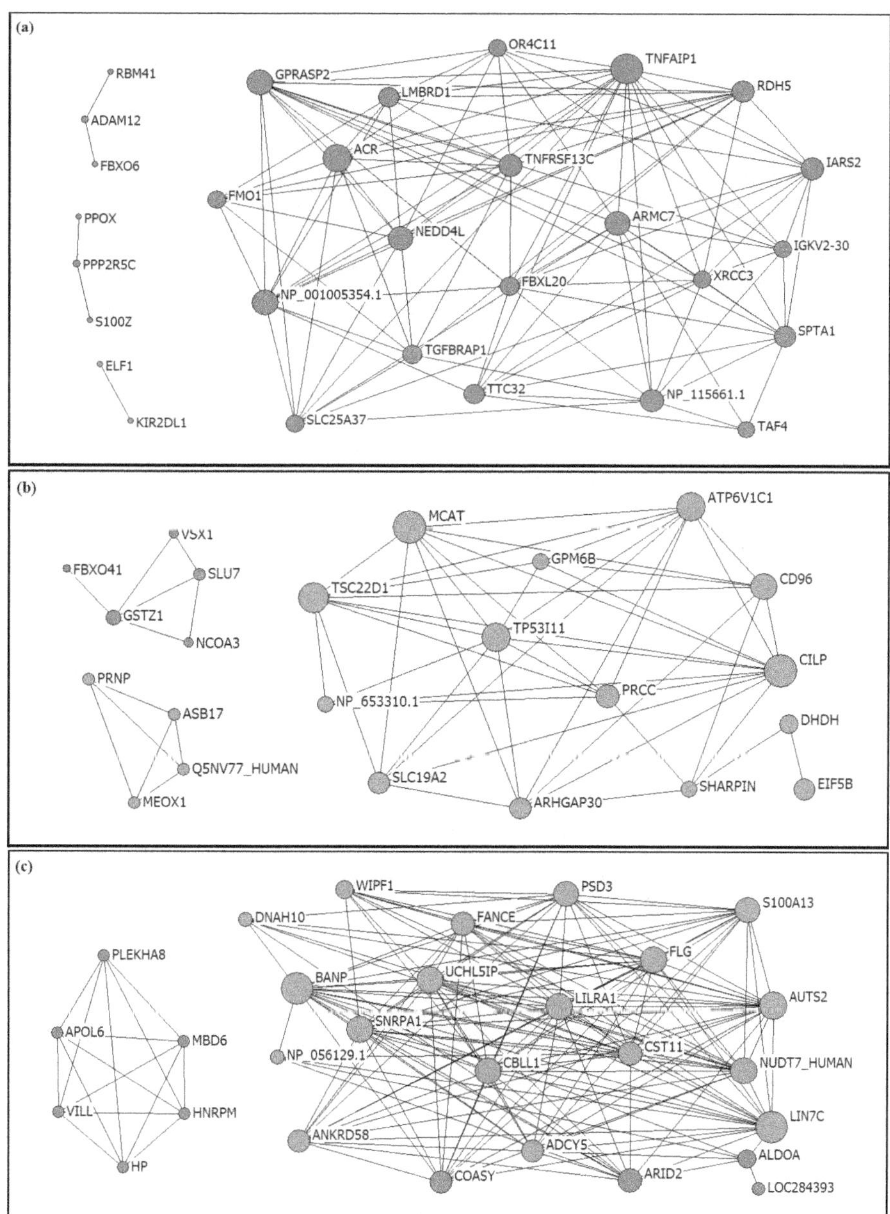

FIGURE 8.4
Networks of up-regulated and down-regulated differentially expressed genes in (a) Cluster 2; (b) Cluster 3; and (c) Cluster 7. Only genes with degrees ≥ 10 were displayed in Cluster 2 and degrees ≥ 5 for Clusters 3 and 7. Nodal size is proportional to the degree (the number of genes that are significantly correlated with the gene).

TABLE 8.4
A summary of networks of highly correlated differentially expressed genes

h	t_h	highly correlated differentially expressed genes
1	215[a,b]	(DUT, ZFP106), (K0401_HUMAN, OR51H1P, PPP1R2), ARHGAP26, CFHR5, CPM, DCAMKL2, ERV3, ESAM, EXOC3L2, FAM123B, FCRL6, GGTLA4, LOC651863, NF2, NFE2L2, NP_001002848.1, NP_001012455.1, NP_079117.2, OBSCN, Q8N377_HUMAN, REEP5, SSTR3, TNFRSF1B, ZNF77
2	89[b,c]	(ELF1, KIR2DL1), (ADAM12, FBX06, RBM41), (PPP2R5C, PPOX, S100Z), ACR, ARMC7, FBXL20, FMO1, GPRASP2, IARS2, IGKV2-30, LMBRD1, NEDD4L, NP_001005354.1, NP_115661.1, OR4C11, RDH5, SLC25A37, SPTA1, TAF4, TGFBRAP1, TNFAIP1, TNFRSF13C, TTC32, XRCC3
3	91[b]	(FBXO41, GSTZ1, NCOA3, SLU7, VSX1), (ASB17, MEOX1, PRNP, Q5NV77_HUMAN), ARHGAP30, ATP6V1C1, CD96, CILP, DHDH, EIF5B, GPM6B, MCAT, NP_653310.1, PRCC, SHARPIN, SLC19A2, TP53I11, TSC22D1
4	89[a]	ACTL8, CLK2, CNGA4, COX15, CRYBA4, CYFIP2, DDX18, DIDO1, DRD1, DUSP28, EVA1, FGFR3, FOXK2, GCH1, MBOAT1, MEGF11, MESP1, NP_001017992.1, NP_001035155.1, NP_055842.1, NPY6R, OR6C1, PCP4L1, PLAC8L1, PLEKHN1, PSG1, PSMB6, Q5TAE0_HUMAN, SYTL3, TMED1, TNRC4, TSPYL4, UCK2, VAMP2, WHSC1L1, ZBTB5, ZNF165
5	1	ALDH3A2
7	39[b]	(APOL6, HP, HNRPM, MBD6, PLEKHA8, VILL), ALDOA, ARID2, COASY, LOC284393, ADCY5, ANKRD58, AUTS2, BANP, CBLL1, CST11, DNAH10, FANCE, FLG, LILRA1, LIN7C, NP_056129.1, NUDT7_HUMAN, PSD3, S100A13, SNRPA1, UCHL5IP, WIPF1
9	41[a]	ABCB8, ACTL6B, ALCAM, CDCA4, CDK2AP1, CHST7, EPHB6, GATAD2B, GNA15, IGSF4D, LDHA, MAP9, MYCBP, NP_001014364.1, NP_001034782.1, NP_777579.1, PPEF2, PTPRO, Q6ZMT9_HUMAN, RAPGEFL1, STARD8, TXNRD1

[a] For large clusters, only the top 5% of differentially expressed genes were considered.

[b] Genes that form an isolated network are grouped within parentheses.

[c] Only the top 10% of up-regulated differentially expressed genes were considered.

8.4 Clustering of Time-Course Gene-Expression Data

Time-course experiments with microarrays are often performed to study dynamic biological systems and genetic regulatory networks (GRNs) that model how genes influence each other in cell-level development of organisms. It is well known that the information encoded in DNA leads to the expression of certain phenotypes (or characteristics) in the organism (Styczynski and Stephanopoulos, 2005). The determination of GRNs provides important insights into the fundamental biological processes such as growth and is useful in disease diagnosis and genomic drug design as well as prediction of response to treatment; see Ng, Wang, and McLachlan (2006) and Hafemeister et al. (2011). Time-course experiments measure gene expressions at several points in time as a serial study. For example, Spellman et al. (1998) published an 18-point time series data set measuring the expression levels of more than 6,000 genes from the budding yeast (Saccharomyces cerevisiae) genome. These temporal profiles of expression levels reflect gene interactions in pathways. The inference from these time-course gene-expression data thus allows researchers to explore gene interactions from a causal perspective. This inference procedure is referred to as "reverse engineering", which attempts to determine the cause (the network of gene interactions) from the outcome (the observed effects on the gene expression profiles); see D'haeseleer, Liang, and Somogyi (2000) and Ng, Wang, and McLachlan (2006) for more information.

As genes sharing the same expression pattern over time are likely to be involved in the same regulatory process, the inference of gene regulation can be accomplished via the clustering of time-course data into groups of co-expressed genes, as illustrated in D'haeseleer, Liang, and Somogyi (2000) and Toh and Horimoto (2002). However, gene expressions obtained at the same time point are often interdependent and existing methods may ignore the dependency of the expression measurements over time and the correlation among gene expression profiles; see Spellman et al. (1998), Cho et al. (1998), Yeung et al. (2001), and Ma et al. (2006). Such assumptions of independence violate regulatory interactions and can result in overlooking certain important subject and temporal effects for GRNs. The mixture of linear mixed models given in (8.1) and (8.2) provides a clustering-based method for the inference of GRNs, where the two sets of random effects are adopted to model the subject and temporal effects in the clustering of time-course data. The method starts with the clustering of genes into groups in which they have similar expression profiles. This process helps to identify sets of genes that are potentially regulated by the same mechanism, as considered by Luan and Li (2003). Other potential applications of co-expressed gene clusters include the inference of functional annotation and the identification of molecular signatures that are potential predictors of disease outcomes; see D'hæseleer, Liang, and Somogyi (2000) and Ben-Tovim Jones et al. (2005) for more information.

Here we consider a time-course gene expression data with measurements of n genes at p time points such that the time-course microarray data are given by $y = (y_1^T, \ldots, y_n^T)^T$, where $y_j = (y_{1j}, \ldots, y_{pj})^T$ is the vector of the time-course gene expression data for the jth gene ($j = 1,\ldots,n$). With reference to (8.1) and (8.2), Ng et al. (2006) considered the random effects b_{hj} to be one-dimensional, representing the unobservable gene-specific effect shared among the gene expressions measured at the p time points for each gene. Another set of random effects $c_h = (c_{1h},\ldots,c_{ph})^T$ containing the unobservable temporal random effects shared among expressions common to all genes from the hth component for each time point. The temporal random effects c_h introduce interdependency among gene expression levels within the same cluster obtained from the same time point. Thus, the design matrices of the two sets of random effects are given by $U = 1_p$ and $V = I_p$, where 1_p is a vector of ones with dimension being specified by the subscript. With this specification, it is assumed that there exists random (one-dimension) gene effect b_{hj} and random temporal effects c_h with p dimensions for each gene $j = 1,\ldots,n$. Here B_h and C_h ($h = 1,\ldots,g$) are diagonal matrices, where $B_h = \theta_{bh}$ (one-dimension) and $C_h = \theta_{ch}I_p$ (p-dimensions). Recently, the gene-specific random effects model for b_{hj} was further extended in Wang, Ng, and McLachlan (2012) to postulate an autoregressive AR(1) variance structure for modelling the correlation of expression values across time points. That is, b_{hj} are now p-dimensional, following a $N(0,B_h)$ distribution with

$$B_h = \theta_{bh}^2 A(\rho) = \frac{\theta_{bh}^2}{1-\rho^2}\begin{pmatrix} 1 & \rho & \ldots & \rho^{p-1} \\ \rho & 1 & \ldots & \rho^{p-2} \\ \vdots & \vdots & \vdots & \vdots \\ \rho^{p-1} & \rho^{p-2} & \ldots & 1 \end{pmatrix}, \tag{8.36}$$

which assumes an autocorrelation covariance structure. Alternatively, Grün, Scharl, and Leisch (2012) considered a mixture of linear additive models (LAMs) for the clustering of time-course gene expression data, where random effects on individual genes are incorporated in the component densities and estimated using regularized likelihood approaches. In contrast to the mixture of LMMs, the mixture of LAMs assume that the repeated observations for the same gene are independent given the component membership.

With time-course data, Fourier basis functions are adopted for W in (8.2) to model a periodic cycle of relationship; see, for example, Ng et al. (2006). That is, the design matrix W is taken to be an $p\times 2$ matrix with the l-th row ($l = 1,\ldots,p$)

$$(\cos(2\pi l/\omega + \Phi)\ \sin(2\pi l/\omega + \Phi)), \tag{8.37}$$

where ω is the period of the cell cycle and Φ is the phase offset; see also Spellman et al. (1998). The least squares estimation approach proposed by Booth et al. (2004) may be applied to obtain the values for ω and/or Φ from the data.

The clustering of time-course data allows us to extract groups of genes that are co-expressed over certain cell-cycle periods and hence infer shared regulatory inputs and functional pathways. The network of regulatory interactions is then determined by searching for regulatory control elements (activators and inhibitors) shared by the clusters of co-expressed genes based on a time-lagged correlation coefficients measurement method, as shown in Li et al. (2002) and Ng, Wang, and McLachlan (2006).

8.4.1 Inference for Gene Regulatory Interactions

Once the co-expressed genes are clustered into groups with similar patterns of expression levels over time, the second step of the method "predicts" the gene expression profile for each gene based on the estimated parameters $\hat{\Psi}$ obtained in the first clustering step. Given that the jth gene belongs to the hth cluster, its gene expression profile at different time points can be expressed as

$$\hat{y}_j = W\hat{\beta}_h + U\hat{b}_{hj} + V\hat{c}_h \tag{8.38}$$

for $j = 1,\ldots,n$. The average gene expression profile for each cluster, $\hat{y}^h = (\hat{y}^h_1,\ldots,\hat{y}^h_p)^T$, is then obtained as

$$\hat{y}^h = \sum_{j=1}^{n} \hat{\tau}_{hj}\hat{y}_j / \sum_{j=1}^{n} \hat{\tau}_{hj} \qquad (h = 1,\ldots,g), \tag{8.39}$$

where the expression profiles for genes from the same cluster may be used to estimate the times to peak expression levels and the cycle periods, using the least square approach of Booth et al. (2004).

The network of gene regulatory interactions is then determined in the final step by searching for candidate regulatory control elements (activators and inhibitors) shared by the clusters of co-expressed genes, as proposed by Ng, Wang, and McLachlan (2006). In the analysis of time-course data, time-lagged correlation coefficients (Li et al., 2002) have been adopted to characterize the time-delayed dependency between genes. Here with the method proposed by Ng, Wang, and McLachlan (2006), the time-lagged correlation coefficients are calculated using the average gene expression profiles for the clusters represented by $\hat{y}^h$. In particular, the correlation coefficient between the hth cluster at time t and the lth cluster at time $t+\kappa$ is defined as

$$R_{hl}(\kappa) = \text{corr}\{(\hat{y}^h_1,\ldots,\hat{y}^h_{(p-\kappa)})^T,\ (\hat{y}^l_{\kappa},\ldots,\hat{y}^l_p)^T\}, \tag{8.40}$$

where $\kappa = 1,2,3,\ldots$ is a constant integer and the function corr(., .) is the standard correlation coefficient between the two $(p-\kappa)$ dimensional vectors; see Li et al. (2002). A large absolute value of $R_{hl}(\kappa)$ thus indicates a pairwise dependence between the genes expression of the hth cluster at time t and that of the lth cluster at time $t+\kappa$. In practice, the value of κ should be so chosen that $p-\kappa$ is not too small, such as $p-\kappa > 10$ suggested by Li et al. (2002). The examination of the significance of these $g*(g-1)$ time-lagged correlation coefficients, after adjusting for multiple testing using the Bonferroni procedure, helps to determine potential regulation between genes that interact with each other directly or via some intermediates; see Toh and Horimoto (2002). For example, upstream DNA sequence patterns specific to each cluster may be identified in the promoter region of gene clusters, through which co-regulation of the genes within the cluster is achieved (Tavazoie et al., 1999). The proposed method identifies the most significant correlations (if present) for a range of values for κ for each pair of gene clusters. The identified pairs of gene clusters are then considered to be potential elements within the network of gene regulatory interactions.

8.4.2 Example 8.2: Cluster Analysis of a Time-Course Gene-Expression Data Set

We consider the yeast Saccharomyces cerevisiae data set of Cho et al. (1998). The data set used Affymetrix oligonucleotide microarrays to query the abundance of 6,220 mRNA species in synchronized Saccharomyces cerevisiae batch cultures, using a temperature-sensitive mutation of cdc28. The yeast cells were sampled at 10 minute intervals for 17 time points across two cell cycles. Here we work with a subset of 384 genes that had non-negative normalized fluorescence readings across the time points and were assigned to one of the five phases of the cell cycle (namely early G1, late G1, S, G2, and M) given at the website `http://genome-www.stanford.edu/cellcycle/data/rawdata/` of the Cho data. The 17 time points were divided into two panels that correspond to two cell cycles and were normalized to have mean 0 and variance 1 within each panel; see Tamayo et al. (1999) and Yeung, Haynor, and Ruzzo (2001) for more details. This standardized dataset is available at `http://faculty.washington.edu/kayee/cluster/`.

With this time course data of $n = 384$ and $p = 17$, we take the design matrix W to be a 17×2 matrix where the lth row $(l = 1, \ldots, 17)$ is specified as in (8.37). In this study, we adopt the least squares estimate $\omega = 85$ obtained by Booth et al. (2004) in the analysis of the full data set of Cho data. The value of Φ is set to be 0, which leads to a better model based on BIC. For the design matrices of the random effects, we take $U = V = I_{17}$. That is, we assume an AR(1) autocorrelation variance structure for gene-specific random effects b_{hj} $(h = 1, \ldots, g;\ j = 1, \ldots, n)$. Model selection via BIC indicated that there are five clusters. The expression profiles for genes in each cluster, as well as the predicted average gene expression profile for each cluster, are presented in Figure 8.5. It can be seen that the five clusters are different in terms of either times to peak expression or the periods. Genes in the same cluster show certain similar functions. For example, genes that expressed at the M/G1 boundary such as *CDC*6, *CDC*46, and *MCM*3 involved in DNA replication (as shown in the CellCycle98 Excel worksheet given at the website of the Cho data) are clustered into the same group (Cluster 4). Also, genes that expressed in the G1 phase such as *CLB*5, *CLB*6, *CLN*1, *CLN*2, *PCL*1, *PCL*2, *HSL*1, *SWE*1, and *ZDS*2 involved in cell cycle regulation are clustered into the same group (Cluster 1). For the computation of the time-lagged correlation coefficients, we considered $\kappa = 1$ to 6. Significant correlation coefficients and the time lags for each pair of gene clusters are provided in Table 8.5, where the first column represents the leading gene clusters. It can be seen that genes in Cluster 2 (such as $HPR5$, $STU2$, $YKL052C$, and $YNR009W$) are potential activators, as discussed in Chen, Filkov, and Skiena (2001), for gene clusters 3 to 5 at different time lags. In addition, genes in Cluster 5 (such as the gene *CLB*2) are potential activators for gene cluster 4 (such as the *CDC*46 group) but are potential inhibitors for gene cluster 3 (such as *PCL*7 and $TOF2$).

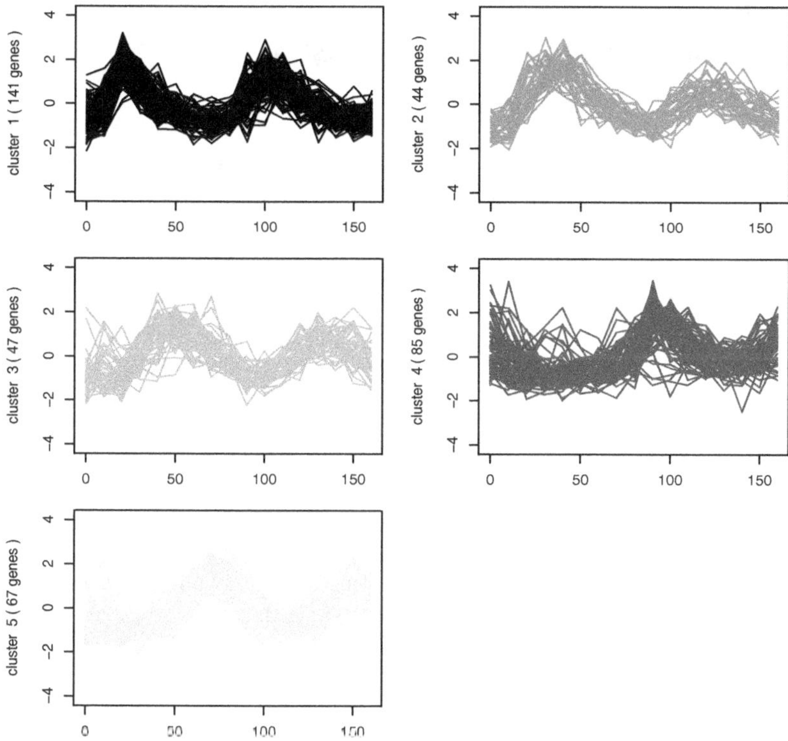

FIGURE 8.5
Clusters of co-expressed genes (Example 8.2) - the x-axis is the time point and the y-axis is the gene expression level. The gene expression profiles are presented in solid lines.

TABLE 8.5
Significant correlation coefficients and the time lags for each pair of co-expressed gene clusters (Example 8.2).

	Cluster 1	Cluster 2	Cluster 3	Cluster 4	Cluster 5
Cluster 1	—	—	—	—	—
Cluster 2	—	—	0.907 ($\kappa = 1$)	0.826 ($\kappa = 6$)	0.920 ($\kappa = 4$)
Cluster 3	0.826 ($\kappa = 6$)	—	—	0.835 ($\kappa = 5$)	0.906 ($\kappa = 2$)
Cluster 4	—	—	—	—	—
Cluster 5	—	-0.773 ($\kappa = 2$)	-0.767 ($\kappa = 3$)	0.917 ($\kappa = 2$)	—

8.5 Clustering of Multilevel Longitudinal Data

For clustering longitudinal data in the form of growth trajectories, growth mixture models (including those proposed by Muthén (2004), Vermunt (2007), and Muthén and Asparouhov (2009)) and mixture latent growth models developed by Vermunt (2003) and Vermunt (2007) have been adopted to identify different clusters of trajectory patterns and predictors of membership in these classes. Growth mixture models extend the conventional growth model (Raudenbush and Bryk, 2002) and the latent class growth analysis (LCGA) approach (Nagin and Land, 1993) to allow the presence of different clusters of trajectory patterns and heterogeneity in individual trajectories that vary around the mean trajectory within a cluster. The growth mixture models are thus flexible to model individual growth trajectories from unobserved sub-populations (latent trajectory clusters) with individual variation in growth parameters that are captured by random effects.

In contrast to the gene expression data presented in Sections 8.3 and 8.4, longitudinal data in the form of growth trajectories often involve a vector of covariates associated with the observed outcome variables over time for each individual and exhibit a data hierarchy structure as data are collected at multiple study units. For clustering this kind of multilevel longitudinal data, we thus extend the random-effects mixture modelling framework (8.2) by allowing the mixing proportions π_h to depend on covariate risk factors and incorporating correlation among growth trajectories within a hierarchical unit in the clustering process. The two sets of random effects (8.2) are now adopted to capture the individual level variation and the variation among higher-level study units simultaneously. Furthermore, the first extension facilitates the provision of a better clustering result where there exists additional information on an individual's risk factors that have an impact on membership of sub-populations. The extensions thus create a wider applicability of mixture model-based approaches for clustering hierarchically structured longitudinal trajectory data.

We assume that there are m higher level units, and within each unit there are n_i study individuals $(i = 1,\ldots,m)$. Thus, the total number of subjects is $n = \sum_{i=1}^{m} n_i$. The objectives are to identify the sub-population structure within the subjects and the risk factors that have impact on the trajectory patterns of an outcome measure; see, for example, Ng and McLachlan (2014a). We denote y_{ij} the observed p-dimensional trajectory for the jth individual in the ith unit and x_{ij} a vector of risk factors associated with Y_{ij} $(i = 1,\ldots,m;\ j = 1,\ldots,n_i)$. The risk factors x_{ij} are incorporated into the mixing proportions in (8.1) as

$$\pi_h(x_{ij};\alpha) = \frac{\exp(v_h^T x_{ij})}{1+\sum_{l=1}^{g-1}\exp(v_l^T x_{ij})} \qquad (h = 1,\ldots,g-1), \tag{8.41}$$

and $\pi_g(x_{ij}) = 1 - \sum_{l=1}^{g-1}\pi_l(x_{ij})$, where the first element of x_{ij} is assumed to be 1 to account for an intercept term and α contains the elements in v_h $(h = 1,\ldots,g-1)$, which are fixed effects (unknown constants) representing the log odds ratio of cluster membership for each risk factor in x_{ij}.

The linear mixed-effects model given in (8.2) is flexible to include different forms of W. For the clustering of multilevel longitudinal data, we adopt a polynomial regression form of $W\beta_h$ in (8.2) to fit a nonlinear relationship between the conditional mean element of y_{ij} at different time points. For example,

$$W = \begin{bmatrix} 1 & t_1 & t_1^2 & t_1^3 \\ 1 & t_2 & t_2^2 & t_2^3 \\ \vdots & \vdots & \vdots & \vdots \\ 1 & t_p & t_p^2 & t_p^3 \end{bmatrix} \tag{8.42}$$

represents a cubic polynomial regression form, which is commonly used in the analysis of growth trajectories and involves four parameters in β_h corresponding to the intercept, slope, quadratic, and cubic growth parameters; see also Muthén (2004). Another common choice for W in a clustering context is spline-based methods (James and Sugar, 2003), such as (cubic) B-splines, that have been used in the analysis of time-course gene expression data by Luan and Li (2003); see also Section 8.4. Because of their flexibility, both polynomial and spline basis functions are appropriate for clustering general growth trajectories that are not periodic. While splines are able to capture a lot of local information in the curvature of the trajectories via piecewise polynomials (Ramsay and Silverman, 2005), polynomial basis functions are capable of modelling global relationships between trajectory levels across time, without overfitting idiosyncratic noise from the observed data, especially when the number of measurements during the follow-up period, p, is small.

With (8.2), random effects are introduced to capture individual variation in trajectories that vary around the mean trajectory within a unit at s time periods, where $s \leq p$ and the total number of time points $p = \sum_{l=1}^{s} s_l$. In addition, it is assumed that the effects imposed by the higher-level units are random and shared among individuals in the same unit. The unobservable individual-specific and unit-specific random effects are denoted by $b_{hij} = (b_{1hij}, \ldots, b_{shij})^T$ and $c_{hi} = (c_{1hi}, \ldots, c_{phi})^T$, respectively, for $i = 1, \ldots, m$; $j = 1, \ldots, n_i$. Both the variance components B_h and C_h are assumed to be diagonal $(h = 1, \ldots, g)$.

8.5.1 EM-Based Estimation via Maximum Likelihood

We let $\Psi = (\alpha^T, \psi_1^T, \ldots, \psi_g^T)^T$ be the vector of all the unknown parameters, where ψ_h is the vector containing the unknown parameters in β_h, A_h, B_h, and C_h $(h = 1, \ldots, g)$. The ML estimation of Ψ can be undertaken by proceeding conditionally on the unit-specific random effects c_{hi}. In particular, the expected value of the complete-data log-likelihood can be decomposed into two terms with respect to the mixing proportions $\pi_h(x_{ij})$ and component densities. Thus separate maximizations can be performed independently for updating the estimates of α and ψ_h $(h = 1, \ldots, g)$. The E- and M-steps of the EM algorithm exist in closed form for ψ_h, as presented in Section 8.2 with the summation from $j = 1$ to n replaced by the summation from $i = 1$ to m for $j = 1$ to n_i, and π_h replaced by $\pi_h(x_{ij}; \alpha)$. As such, Equations (8.17) to

(8.20) in the M-step provided in Section 8.2 become:

$$\beta_h^{(k+1)} = \beta_h^{(k)} + \frac{[W^T W]^{(-1)} W M_h^{(k)} A_h^{(k)} \sum_{i=1}^m \sum_{j=1}^{n_i} \tau_{hij}^{(k)} (y_{ij} - W\beta_h^{(k)})}{\sum_{i=1}^m \sum_{j=1}^{n_i} \tau_{hij}^{(k)}}, \tag{8.43}$$

$$\theta_{bh}^{(k+1)} = \frac{E_{\Psi^{(k)}}(b_h^T b_h | y)}{(s \sum_{i=1}^m \sum_{j=1}^{n_i} \tau_{hij}^{(k)})}, \tag{8.44}$$

$$\theta_{ch}^{(k+1)} = \frac{E_{\Psi^{(k)}}(c_h^T c_h | y)}{pm}, \tag{8.45}$$

and

$$A_h^{(k+1)} = \frac{E_{\Psi^{(k)}}(\varepsilon_h^T \Omega_h \varepsilon_h | y)}{\sum_{i=1}^m \sum_{j=1}^{n_i} \tau_{hij}^{(k)}}, \tag{8.46}$$

where

$$E_{\Psi^{(k)}}(z_{hij}|y) = \tau_{hij}^{(k)} = \frac{\pi_h(x_{ij}|\alpha^{(k)}) f(y_{ij}|z_{hij} = 1; \psi_h^{(k)})}{\sum_{l=1}^g \pi_l(x_{ij}|\alpha^{(k)}) f(y_{ij}|z_{lij} = 1; \psi_l^{(k)})}, \tag{8.47}$$

with $f(y_{ij}|z_{hij} = 1; \psi_h^{(k)})$ given by (8.7),

$$E_{\Psi^{(k)}}(b_h^T b_h | y) = \sum_{i=1}^m \sum_{j=1}^{n_i} \tau_{hij}^{(k)} b_{hij}^{(k)^T} b_{hij}^{(k)} + F_h^{(k)}, \tag{8.48}$$

with

$$b_h^T = (b_{h11}^T, \ldots, b_{h1n_1}^T, \ldots, b_{hm1}^T, \ldots, b_{hmn_m}^T) \qquad (h = 1, \ldots, g), \tag{8.49}$$

$$b_{hij}^{(k)} = E_{\Psi^{(k)}}(b_{hij}|y) = \theta_{bh}^{(k)} U^T [A_h^{(k)} + \theta_{bh}^{(k)} UU^T]^{-1} \Bigg[(y_{ij} - W\beta_h^{(k)}) - \theta_{ch}^{(k)} VV^T M_h^{(k)} \sum_{l=1}^m \sum_{q=1}^{n_l} \tau_{hlq}^{(k)} (y_{lq} - W\beta_h^{(k)}) \Bigg], \tag{8.50}$$

and where

$$M_h^{(k)} = \Bigg[A_h^{(k)} + \theta_{bh}^{(k)} UU^T + \sum_{i=1}^m \sum_{j=1}^{n_i} \tau_{hij}^{(k)} \theta_{ch}^{(k)} VV^T \Bigg]^{-1}. \tag{8.51}$$

In (8.48),

$$F_h^{(k)} = \sum_{i=1}^m \sum_{j=1}^{n_i} \tau_{hij}^{(k)} \text{trace}(\theta_{bh}^{(k)} UU^T) - \text{trace}(\theta_{bh}^{(k)^2} U^T M_h^{(k)} U) - (\sum_{i=1}^m \sum_{j=1}^{n_i} \tau_{hij}^{(k)} - 1) \text{trace}(\theta_{bh}^{(k)^2} U^T [A_h^{(k)} + \theta_{bh}^{(k)} UU^T]^{-1} U). \tag{8.52}$$

$$E_{\Psi^{(k)}}(c_h^T c_h | y) = c_h^{(k)^T} c_h^{(k)} + G_h^{(k)}, \tag{8.53}$$

with

$$c_{hi}^{(k)} = E_{\Psi^{(k)}}(c_{hi}|y) = \theta_{ch}^{(k)} V^T M_h^{(k)} \sum_{l=1}^{m} \sum_{q=1}^{n_l} \tau_{hlq}^{(k)} (y_{lq} - W\beta_h^{(k)}) \tag{8.54}$$

$$G_h^{(k)} = \text{trace}(\theta_{ch}^{(k)2} VV^T) - \sum_{i=1}^{m} \sum_{j=1}^{n_i} \tau_{hij}^{(k)} \text{trace}(\theta_{ch}^{(k)2} V^T M_h^{(k)} V). \tag{8.55}$$

$$E_{\Psi^{(k)}}(\varepsilon_h^T \Omega_h \varepsilon_h|y) = \sum_{i=1}^{m} \sum_{j=1}^{n_i} \tau_{hij}^{(k)} \varepsilon_{hij}^{(k)T} A_h^{(k)-1} \varepsilon_{hij}^{(k)} + H_h^{(k)}, \tag{8.56}$$

with

$$\varepsilon_{hij}^{(k)} = y_{ij} - W\beta_h^{(k)} - Ub_{hij}^{(k)} - Vc_h^{(k)} \tag{8.57}$$

$$H_h^{(k)} = \sum_{i=1}^{m} \sum_{j=1}^{n_i} \tau_{hij}^{(k)} \text{trace}(A_h^{(k)}) - \text{trace}(A_h^{(k)T} M_h^{(k)} A_h^{(k)}) - \\ (\sum_{i=1}^{m} \sum_{j=1}^{n_i} \tau_{hij}^{(k)} - 1) \text{trace}(A_h^{(k)T} [A_h^{(k)} + \theta_{bh}^{(k)} UU^T]^{-1} A_h^{(k)}). \tag{8.58}$$

For α, the M-step involves solving the following system of nonlinear equations:

$$\sum_{i=1}^{m} \sum_{j=1}^{n_i} \left(\tau_{hij}^{(k)} - \frac{\exp(v_h^T x_{ij})}{1 + \sum_{l=1}^{g-1} \exp(v_l^T x_{ij})} \right) x_{ij} = 0 \qquad (h = 1, \ldots, g-1). \tag{8.59}$$

The MINPACK routine HYBRD1 is adopted to solve (8.59) for α.

8.5.2 Example 8.3: Cluster Analysis of a Multilevel Longitudinal Data Set

We consider the matched mother and child data files from the 1979 U.S. National Longitudinal Survey of Youth (NLSY), which are accessible at the website `https://www.nlsinfo.org/investigator/pages/search.jsp?s=NLSY79`. The original NLSY79 sample included 6,283 young women aged 14-22 years in 1979 at the first interview. Starting from 1986, a biennial supplement of interview questions was added to acquire information on the biological children of NLSY female respondents. These children are born to a sample of relatively young and disadvantaged mothers who are disproportionately Hispanic and African American (Carlson and Corcoran, 2001). In 1986, data on 5,255 children reported by 2,922 interviewed mothers were obtained. In this example, we focus on exploring the heterogeneity in trajectory of child behavioural problems from 4 to 14 years ($p = 6$ time points) and potential demographic, maternal, household characteristics that may explain the difference in behaviour-problem trajectory.

The behavioural problems of children were assessed using the Behaviour Problems Index (BPI), which is based on items from Zill and Peterson's adaption (Zill and Petersen, 1986) of the Child Behaviour Checklist developed by Achenbach and Edlebrock (1981) and is widely used in studies of behaviour problems for children aged

4 years or over. Higher BPI scores indicate a higher level of behaviour problems. The Center for Human Resource Research (2002) has standardized BPI scores by age and gender based on cross-sectional data of general population samples to quantify a child's tendency to internalize problems (such as anxiety and depression) or externalize behaviours (such as antisocial behaviour, peer problems, and hyperactivity); see Chang, Halpern, and Kaufman (2007) for more information. The NLSY79 cohort was selected in two phases to choose housing units for screening and then identify eligible respondents for interview (Moore et al., 2000). In this example, we consider the BPI data separately for internalized and externalized problems, from children who were born in 1988 and 1990 living with their biological mother as reported in 1992 and 1994 (i.e., when the children were 4 years old). The U.S. region of residence (Northeast, North Central, South, and West) for the mother reported at that year was adopted to form $m = 4$ higher-level units of collection districts. A key risk factor was maternal depressive symptoms measured using the seven-item Center for Epidemiologic Studies Depression Scale short form (CES-D-SF) when the children were 4 years old. The CES-D-SF assesses the frequency of depressive symptoms experienced in the previous week (Levine, 2013), ranging from 0 to 21. A higher score indicates more severe depressive symptoms. The CES-D-SF scale score was dichotomized using the previously reported cutoff score of 8 or higher to identify cases of maternal depressive symptoms, which showed good specificity of 0.97 and modest sensitivity of 0.69 in an evaluation study by Levine (2013).

Measurements of children's BPI were taken at $(p = 6)$ time points when they were aged 4 to 14 years, corresponding to the coding of $t = 0$ to $t = 5$ in (8.42). Children with less than 4 BPI measurements were excluded. Missing BPI values were then computed using the EM algorithm, assuming multivariate normal distribution. With this multilevel setting, we consider individual-specific random effects $b_{hij} = (b_{1hij}, b_{2hij})^T$ corresponding to individual variations in BPI standardized scores before and after 10 years of age $(s = 2)$. The second set of random effects is region-specific $c_{hi} = (c_{1hi}, \ldots, c_{6hi})^T$, capturing the correlation of individual trajectories of BPI standardized scores at the 6 time points from the same region of residence. The following covariates x_{ij}, which may be associated with children's behaviour problems, were included in the mixing proportions via a logistic function as in (8.41): Children's gender and race (3 categories: Hispanic, Black, Non-Hispanic/Non-Black), maternal depressive symptoms measured using CES-D-SF (2 categories: Case versus Normal), maternal age at child's birth, marital status (3 categories: Never Married; Married and Spouse Present; Other, including separated, divorced, widowed), maternal education level in terms of the highest grade completed, and the number of biological/step/adopted children in household. All maternal variables, except age at child's birth, were measured when the children were 4 years old.

In this example, we consider only 1 biological child (either the older or the one with less missing values in BPI measurements) from each mother. There are $n = 671$ children with completed data on all covariates. We first fit the mixture of LMMs without covariates x_{ij} to the data (separately for BPI internalized and externalized scores), where a quadratic regression form (as the cubic part is not significant) is adopted to capture the nonlinear relationship between the conditional mean at different time

TABLE 8.6
Model selection: mixture of LMMs without covariates (Example 8.3)

g	Mixing proportions	Log likelihood	AIC	BIC
(a) Internalize behaviour problems				
1	1.000	-16968.119	33958	34008
2*	0.675, 0.325	-15198.057	30442	30546
3	0.490, 0.398, 0.112	-15381.482	30833	30991
(b) Externalize behaviour problems				
1	1.000	-16816.082	33654	33704
2*	0.622, 0.378	-14856.670	29759	29863
3	0.469, 0.417, 0.113	-14897.307	29865	30022

* Number of components selected based on AIC and BIC.

points; see (8.42). Model selection was performed using the BIC and then covariates were entered into the mixture model to obtain the final clustering results.

The results for model selection presented in Table 8.6 indicated that there are two components of trajectories for both internalized and externalized behaviour problems. Covariates were then entered into the two-component mixture model. Non-significant factors were excluded from the final model except gender, which allows for exploring gender-specific behaviour problems. The trajectories patterns in the final models for internalized and externalized behaviour problems are displayed in Figure 8.6. For internalized behaviour problems, both clusters show a drop of BPI internalized scores after four years old ($t = 0$) and start to increase, but at different times, when the children reach twelve and ten years old (for the cluster with higher internalize problems), respectively. The clusters were named in terms of the overall level of behavioural problems based on the trajectories patterns. The "Better Behaviour" cluster indicates a group of children (67.5%) who have a generally lower BPI internalized score. The "Worse Behaviour" cluster represents a group of children (32.5%) whose BPI internalized scores are relatively larger. Variation in BPI internalized scores among the four U.S. regions of residence was larger for the worse behaviour cluster (estimated variance component $\hat{\theta}_{c2} = 38.1$; S.E. = 12.0) compared to the better behaviour cluster (estimated variance component $\hat{\theta}_{c1} = 6.6$; S.E. = 2.2).

The trajectory patterns for externalized behaviour problems are somewhat different, especially for children with greater behaviour problems. The "Better Behaviour" cluster represents a group of children (62.2%) who have a generally lower BPI externalized score, starting to increase at the age of eight years. The "Worse Behaviour" cluster indicates a group of children (37.8%) whose BPI externalized scores increase gradually after four years old and then start to decrease (improvement) slightly when they reach ten years old. Variation in BPI externalized scores among the four U.S. regions of residence was larger for the worse behaviour cluster (estimated variance component $\hat{\theta}_{c2} = 28.7$; S.E. = 12.1) compared to the better behaviour cluster (estimated variance component $\hat{\theta}_{c1} = 3.9$; S.E. = 1.6).

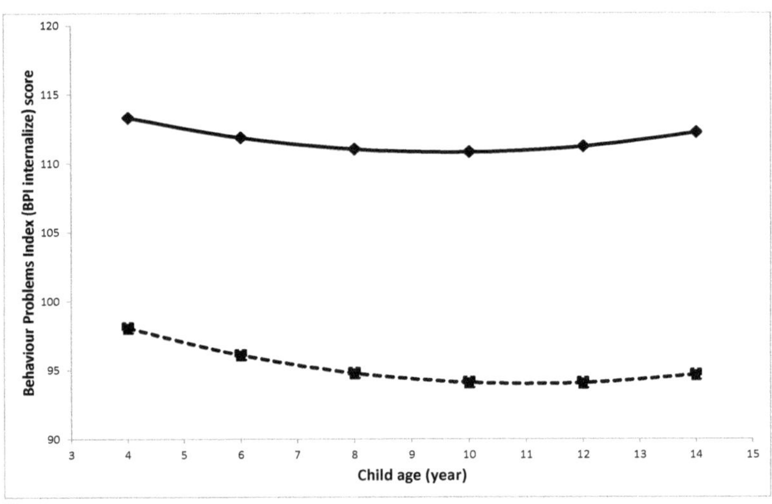

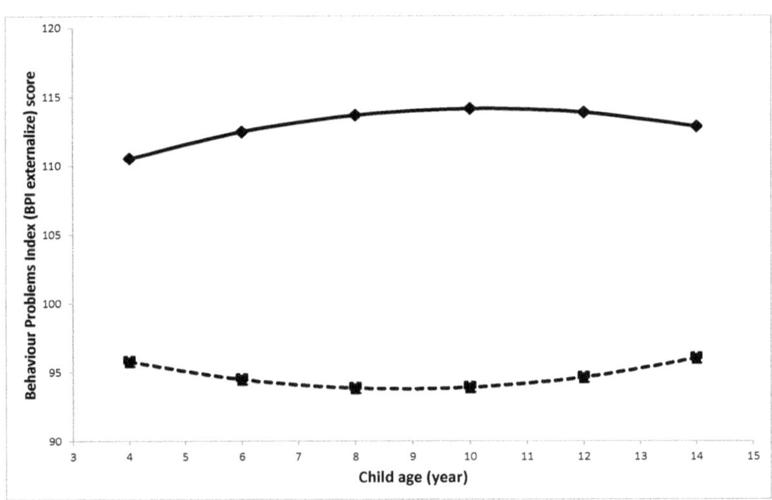

FIGURE 8.6
Estimated mean trajectories of BPI score in two clusters (Example 8.3): (Top) Internalized problems; (Bottom) Externalized problems ("Better Behaviour" - dotted line; "Worse Behaviour" - solid line; higher BPI scores indicate worse behaviour).

TABLE 8.7
Estimated adjusted odds ratios with a two-component mixture of linear mixed effects models (Example 8.3).

Risk factor	Adjusted OR (95% CI) relative to "Better Behaviour"	
	Internalized behaviour	Externalized behaviour
Gender (Female)	1.42 (0.99, 2.05)	0.72 (0.49, 1.04)
Maternal depressive symptoms		
Normal	Reference	Reference
Case	2.80* (1.76, 4.45)	3.16* (1.95, 5.12)
Maternal marital status		
Married and spouse present	Reference	Reference
Never married	1.54 (0.86, 2.76)	1.65 (0.86, 3.17)
Separated/divorced/widowed	1.95* (1.21, 3.15)	1.65* (1.04, 2.63)
Maternal educational level (higher)	0.89* (0.82, 0.97)	0.82* (0.74, 0.91)

* p-value < 0.05.

The estimated adjusted odds ratios (ORs) for the risk factors that differentiate "worse" versus "better" behaviour clusters were presented in Table 8.7 for both internalized and externalized behaviour problems. Although the gender effect was not statistically significant, the potential reverse of the gender effect on internalized and externalized behaviour problems was observed. Girls generally had a higher chance of having an internalized behaviour problem such as anxiety or depression, whereas boys had a higher chance of having an externalized behaviour problem such as antisocial behaviour, peer problems, or hyper-activity. Compared with the better behaviour cluster, children whose mothers had depressive symptoms were more likely to have internalized behaviour problems (adjusted OR = 2.80, 95% CI = 1.76-4.45) or externalized behaviour problems (adjusted OR = 3.16, 95% CI = 1.95-5.12). Also, children with mothers separated, divorced, or widowed, compared to those married and spouse present, had an elevated chance of having internalized behaviour problems (adjusted OR = 1.95, 95% CI = 1.21-3.15) or externalized behaviour problems (adjusted OR = 1.65, 95% CI = 1.04-2.63). On the other hand, children were less likely to have internalized behaviour problems (adjusted OR = 0.89, 95% CI = 0.82-0.97) or externalized behaviour problems (adjusted OR = 0.82, 95% CI = 0.74-0.91) when their mothers had a higher educational level. Children's race, maternal age at child's birth, and the number of children in the household were not significant risk factors. Whilst children's race was not a significant factor in the final models, there were discrepancies in children's race between those significant risk factors, as presented in Table 8.8. It shows that mothers with a Hispanic or Black race had a higher proportion of having depressive symptoms (p=0.062) or being separated, divorced, or widowed (p <0.001), but had a lower educational level (p <0.001).

TABLE 8.8
Discrepancy in children's race between the identified risk factors (Example 8.3).

Risk factor	Hispanic race	Black race	Others	p-value
Maternal depressive symptoms				0.062
Normal	103 (79.8%)	155 (78.3%)	295 (85.8%)	
Case	26 (20.2%)	43 (21.7%)	49 (14.2%)	
Maternal marital status				<0.001
Married and spouse present	90 (69.8%)	76 (38.4%)	295 (85.8%)	
Never married	14 (10.9%)	68 (34.3%)	10 (2.9%)	
Separated/divorced/widowed	25 (19.4%)	54 (27.3%)	39 (11.3%)	
Maternal educational level				<0.001
(highest grade completed)	12.1 (2.6)	12.6 (1.9)	13.6 (2.1)	

Note: Data are frequency (percentage) for categorical risk factors or mean (standard deviation) for the highest grade completed.

8.6 R and Fortran Programs for Fitting Mixtures of Linear Mixed Models

The R package "EMMIX-WIRE" (version 1.0.0) can be downloaded from the book's online webpage. It fits a mixture of linear mixed models with two sets of random effects terms via the EM algorithm. The R program codes for Examples 8.1 and 8.2 are provided in Figures 8.7 and 8.8, respectively. For fitting a mixture of linear mixed models where the mixing proportions π_h depend on covariate risk factors as given in (8.41), the Fortran program "emmix-wire_regr.f" can be downloaded from the book's online webpage, along with external Fortran programs "random.f" and "hybrid.f". The instructions to compile emmix-wire_regr.f with random.f and hybrid.f in Cygwin into an executable program EMMIXWIREregr and the program code for fitting the NLSY matched mother and child data (Example 8.3) using EMMIXWIREregr are provided in Figure 8.9.

Main Analysis for Example 8.1

```
setwd("c:/Example8.1/")                 # set the work directory
rm(list=ls())                           # remove list memory
data <- matrix(scan("gse36703.txt"),ncol=14,byrow=T)     # read the gene data
require(EMMIXcontrasts)                               # call the EMMIXcontrasts program
n1 <- 5; n2 <- 6; n3 <- 3                     # input the number of samples (3 classes)

# set up the design matrices
X <- U <- W <- cbind(rep(c(1,0,0),c(n1,n2,n3)),rep(c(0,1,0),c(n1,n2,n3)),
                                                    rep(c(0,0,1),c(n1,n2,n3)))
V <- diag(n1+n2+n3)

ng=9                                          # assume g=9 components
set.seed(123456)                              # set a seed for generating random numbers

# assume each component has their own covariance matrix of B_h (ncov=3)
# include the random effect term c_h in the model (nvcov=1)
ret <- emmixwire(data, g=ng, ncov=3, nvcov=1, X=X, U=U, W=W,
                                  debug=1, itmax=1000, epsilon=1e-4)
# print out the estimates
ret$lk; ret$pro; ret$beta; ret$sigma.e; ret$sigma.b; ret$sigma.c

# output the clustering result and the mixed contrast
wj <- scores.wire(ret, contrast=c(-0.5,-0.5,1))
mat <- cbind(ret$tau, ret$cluster, wj)
write.table(mat, file="cluster_wj", row.names=T, col.names=F)

# obtain replicated values of W_j (B=19) and fitting a 3-component mixture of t
wj0 <- wj2.permuted(data, ret, nB=19, contrast=c(-0.5,-0.5,1))
pv <- pvalue.wire(wj,wj0)
wjboot <- array(cbind(wj,wj0))
require(EMMIX)                                         # call the EMMIX program
set.seed(126)                                 # set a seed for generating random numbers

# initialization using 10 k-means and 10 random starts
# g=3 components of t-distribution with unequal standard deviation (ncov=3)
initobj <- init.mix(wjboot, g=3, distr="mvt", ncov=3, nkmeans=10, nrandom=10, nhclust=F)
obj <- EMMIX(wjboot, g=3, distr="mvt", ncov=3, clust=NULL, init=initobj, itmax=1000,
       epsilon=1e-6, debug=TRUE)

t_score <- (wjboot - 0.3719894)/sqrt(1.224055)        # reference to the null t
pv <- 2*(1-pt(abs(t_score),19.79059))                 # associated P-values
pv_wj <- pv[1:37628]
pv_wj[pv_wj==0] <- 1e-16
pv_wj[pv_wj==1] <- 1-(1e-16)
zz <- qnorm(1-pv_wj)                                  # convert to Z-scores
write.csv(cbind(wj,pv_wj,zz), "z_scores.csv")

# non-parametric clustering of highly ranked differentially-expressed genes
source("Nonparametric_clustering.R")                  # call the source program
data <- matrix(scan("gse36703_1_215.txt"),ncol=215,byrow=T)      # read Cluster 1 data

# with B=100 bootstrap replications and alpha=0.0002
n <- nrow(data)
nv <- ncol(data)
corr <- cal.measure(data, nv, measure="spearman")
pairnull <- permuteit(data, n, nv, B=100)
obj <- cutit(pairnull, corr, nv, alpha=0.0002)
Mmatrix <- cal.matrix(corr, nv, obj$cutcorr)
write.csv(Mmatrix, "matrix.csv", row.names=F)
ret <- clustit(corr, obj$pairrsort, Mmatrix, nv, obj$cutoff, obj$cutcorr)
obj[1:2]
write.csv(ret, "output.csv", row.names=F)             # use Excel to remove duplicates
```

FIGURE 8.7
The R program codes for Example 8.1.

Main Analysis for Example 8.2

```
setwd("c:/Example8.2/")             # set the work directory

require(emmixskew)                          # install EMMIXskew_1.0.1.zip and call it
source("Emmixwire_time_course.R")           # call the source program

# read the time-course data (1st column: gene name; 2nd column: true clustering)
datafile <- read.table("norm_cellcycle_384_17.txt", header=T)
orgCluster2 <- datafile[,2]                 # true clustering (optional)
data <- datafile[,-c(1,2)]                  # data of 17 time points

m <- 17
time <- seq(from=0, by=10, length=17)                    # 10 minutes intervals
period <- 85

# set up the design matrices
U <- W <- cbind(rep(1,m))
V <- diag(m)

ng=5                                        # assume g=5 components
set.seed(457)                               # set a seed for generating random numbers

# assume an AR(1) variance structure and use best seed for initialization
# include the random effect term c_h in the model (nvcov=1)
initclust <- sample(1:ng,384,replace=T)
obj5 <- emmixwire(data, ng, ncov="AR", nvcov=1, tmtype=T, one=0, common=0,
           clust=initclust, debug=0, itmax=1000, epsilon=1e-5,
           tmpoints=time, k=1, periods=85, initperiod=rep(period,ng))
initclust <- obj5$clust

# obtain the final model using clustering result from the best seed
obj5 <- emmixwire(data, ng, ncov="AR", nvcov=1, tmtype=T, one=0, common=0,
           clust=initclust, debug=0, itmax=1000, epsilon=1e-6,
           tmpoints=time, k=1, periods=85, initperiod=rep(period,ng))

# plot the clustering results and profiles
x11()
clust.plot.time(obj5$clust, data, base=4, nr=3)

# obtain the average gene profiles
profile <- array(0,c(384,17))
for (j in 1:384) {
  for (m in 1:17) {
    profile[j,m] <- obj5$mu[m,obj5$clust[j]] + obj5$eb1[j,m,obj5$clust[j]] +
                    obj5$ec1[m,obj5$clust[j]]   }}
mprof <- t(obj5$tau)%*%profile
for (i in 1:ng) {
  mprof[i,] <- mprof[i,]/sum(obj5$clust==i)   }

# obtain correlation coefficients with different time lags
# output format: (top) corr. coefficients at time lag 1 to 6
# output format: (down) adjusted p-values at time lag 1 to 6
datat <- mprof
source("corr_lag.R")                        # call the source program
mat1 <- cbind(corr1, corr2, corr3, corr4, corr5, corr6)
mat2 <- cbind(pv1, pv2, pv3, pv4, pv5, pv6)
mat <- rbind(mat1, mat2)
write.csv(mat,"corr_lag.csv")
```

FIGURE 8.8
The R program codes for Example 8.2.

Main Analysis for Example 8.3

```
$ cd c:/Chapter8                  # change to the work directory

$ gfortran -c random.f            # compile a Fortran file to obtain an .o file

$ gfortran -c hybrid.f

$ gfortran -c emmix-wire_regr.f   # compile the file to obtain emmix-wire_regr.o

# to obtain the executable program EMMIXWIREregr from the .o files
$ gfortran -o EMMIXWIREregr emmix-wire_regr.o random.o hybrid.o

$ ./EMMIXWIREregr                 # call the EMMIXWIREregr program
 Different residuals across groups? (1/0)
1                                           # assuming different residuals

 INPUT: no. of bootstrap (enter 1 for original data)
1                                           # for original data

# output file is "output"


# calculate standard errors using B=100 bootstrap replications

$ ./EMMIXWIREregr                 # call the EMMIXWIREregr program
 Different residuals across groups? (1/0)
1                                           # assuming different residuals

 INPUT: no. of bootstrap (enter 1 for original data)
100                                         # for 100 bootstrap replications

 Input the seeds for random number generation: input 3 integers between 1 and 30,000,
 e.g. 23 120 3411.
23 120 3411                                 # enter the seeds

 Do you want to print out the estimates
   for each bootstrap?     INPUT: 1(Yes); 0(No)
1                                           # to print out the estimates

# output file is "output"
```

FIGURE 8.9
The Fortran program codes for Example 8.3.

9

Miscellaneous: Handling of Missing Data

9.1 Introduction

In medical and health sciences research, we always inevitably face the problem of missing data. Due to the presence of missing data, studies are prone to bias and loss of statistical power, imposing an enormous challenge for interpretation and generalization of research findings. There are a number of approaches for handling missing data, and this chapter attempts to outline those approaches that are relevant in the applications of mixture models for cluster analysis of data with missing values. It is good practice to perform a sensitivity analysis with the use of different methods for handling missing data in order to assess the robustness of the findings obtained from each method.

Research studies in medical and health sciences, especially longitudinal studies, often encounter particular problems such as non-response, loss to follow-up, and withdrawal from the study; see, for example, Seaman and White (2013) and Ng, Scott, and Scuffham (2016), inevitably inducing the requirement of handling missing data in the data analysis (Sterne et al., 2009). In general, there are two key aspects regarding influences on findings due to missing data. The first concerns the accuracy and reliability of parameter estimates in the analytic results or whether the analytic methods are robust to handle the bias and the amount of variability in the data arisen from the existence of missing data. The second aspect relates to the representativeness of the study sample and hence the generalizability of findings when there are significant amounts of missing data (say, greater than 10%), as discussed in Atherton et al. (2008).

While the first aspect concerning methodological approaches to handle missing data will be the focus in this chapter, the second aspect about the interpretation and generalization of findings is also critical for assuring the quality of the research. In the latter, heavy selection bias could be induced due to non-response, where there are significant differences in characteristics and outcomes between the respondents and non-respondents as shown in Ng et al. (2010), Ng et al. (2014), Weuve et al. (2012), and Thompson and Arah (2014). As a result, an over-representation of relatively advantaged participants (such as individuals with higher education and socio-economic status) is often present in the sample. To this end, studies have been conducted to develop multivariate attrition models accounting for non-response of different types and adopt a (non-augmented) inverse probability weighting (IPW) method to adjust for non-response by re-balancing the set of complete cases; see Hofler et al. (2005), Ng, Scott, and Scuffham (2016), Sheikh (2007), and Thompson and Arah (2014). With the IPW adjustment, caution should be exercised in the estimation of standard errors of the weighted parameter estimates to account for additional variability resulting from estimation of weights (Rao et al., 2005). Alternatively, augmented IPW estimators in missing data models have been discussed in Scharfstein, Rotnitzky, and Robins (1999). In a related context, use of record-linkage has been proposed to adjust for survey non-representativeness and handle non-response (Gorman et al., 2017; Gray et al., 2013).

Concerning methodological approaches to handle missing data, the difficulty in obtaining valid and reliable estimates of model parameters depends on the mechanism that induces missing data. A useful account of missing-value mechanisms was provided by Rubin (1976) and in the monograph by Little and Rubin (2002). Let the complete data be $Y = (Y_{obs}, Y_{mis})$, where the variables corresponding to observed and missing values are denoted by Y_{obs} and Y_{mis}, respectively. Denote "missingness" as the probability that a data value is missing rather than observed and we summarize missingness of the data by a variable M, indicating the data values are missing (Y_{mis}) or not. In general, missingness can depend on both observed and missing values such that the distribution of M is expressed as

$$\text{pr}(M|Y_{obs}, Y_{mis}; \phi), \tag{9.1}$$

specified up to a vector ϕ of unknown parameters, which is assumed to be distinct from the model parameters denoted by Ψ.

If missingness is independent of both observed and missing values, then the mechanism is considered as "missing completely at random" (MCAR). That is, the distribution of M is given by

$$\text{pr}(M|Y_{obs}, Y_{mis}; \phi) = \text{pr}(M|\phi), \tag{9.2}$$

independent of both Y_{obs} and Y_{mis}. Practically, the MCAR mechanism is equivalent to random discard of some values in Y independently and that the missing values Y_{mis} can be considered as a random sample from the population. That is, Y_{obs} can be used unbiasedly to estimate Ψ, only that the sample size is smaller due to the missing values.

The second type of missing-value mechanism corresponds to the case where missingness depends only on observed values but not on missing values. This mechanism is referred to as "missing at random" (MAR) and the distribution of M is given by

$$\text{pr}(M|Y_{obs}, Y_{mis}; \phi) = \text{pr}(M|Y_{obs}; \phi). \tag{9.3}$$

In practice, it is anticipated that individuals with complete data differ from individuals with missing data, but the pattern of missingness is "traceable" from observed values Y_{obs} in the data set, rather than being due to the specific variables with missing values Y_{mis}.

Another type of missing-value mechanism is known as "not missing at random" (NMAR) when missingness depends on both observed and missing values. With the distribution of M given by (9.1), missingness is non-random and "non-ignorable" as the inference about Ψ will depend on the mechanism that causes data values to be missing.

It is possible to test the MCAR assumption against the alternative hypothesis that missingness is MAR. However, testing the MAR assumption against a NMAR alternative is impossible without additional information, as discussed in Horton and Lipsitz (2001). Certain statistical modelling approaches work well with missing values under the MCAR mechanism, such as the generalized estimating equations (GEE) in the analysis of longitudinal data; see, for example, Hogan, Roy, and Korkontzelou (2004) and Yelland et al. (2011).

9.2 Mixture Model-Based Clustering of Data with Missing Values

There are a number of approaches to handling missing values in statistical analyses of data acquired from medical and health sciences research, as summarized in Bennett (2001). Most of these existing methods for handling missing data focus on the case of ignorable MAR mechanism and continuous variables under the normality assumption. Whereas methods that ignore missing values (such as complete-case analysis) or that fill in the missing values using single imputation methods (such as mean substitution) have been proved to be biased and underestimate the variability under the MAR mechanism, the multiple imputation method and likelihood-based methods such as the EM method have been widely used to provide robust estimation of parameters and their standard errors. In particular, robust estimation of the mean and covariance matrix from data with potential outliers and ignorable missing values has been developed using a general mixture model that includes multivariate t and contaminated multivariate normal distributions as special cases; see Little and Rubin (2002, Chapter 12). The mixture model is expressed as

$$\int \phi(y_j;\mu,\Sigma/u_j)dH(u_j), \tag{9.4}$$

where $H(u_j)$ is a known probability distribution; see also McLachlan and Peel (2000, Sections 7.4-7.5). It can be seen that (9.4) is a multivariate t distribution if $H(u)$ is a chi-squared distribution with its degrees of freedom ν. It is a contaminated multivariate normal distribution when $H(u)$ places mass $(1-\pi)$ at the point $u=1$ and mass π at the point $u=\lambda$. That is, (9.4) becomes

$$(1-\pi)\phi(y_j;\mu,\Sigma)+\pi\phi(y_j;\mu,\Sigma/\lambda), \tag{9.5}$$

where both π and λ are small, representing the small proportion of observations that have a relatively large variance. ML estimates of μ and Σ are obtained by applying the EM algorithm, treating the values of Y_{mis} and U as missing data, as described in Little and Rubin (2002, Chapter 12).

Attention has also been given to doubly robust (DR) estimators, which offer a desirable property in handling missing data to account for model misspecification; see, for example, Bang and Robins (2005) and Rotnitzky et al. (2012). Under a MAR mechanism, a DR estimator remains "consistent" when either a model for the missingness mechanism or a model for the distribution of the complete data is correctly specified. Further development to allow for consistency against model misspecifications in any of the multiple models for both the missingness mechanism and complete-data distribution is available in Han (2014). A comparison of multiple imputation and doubly robust estimation for analyses of missing data is provided by Carpenter, Kenward, and Vansteelandt (2006). Harkanen et al. (2014) compared IPW with DR methods in correction of non-response effects. The influences of imputation methods on the accuracy of classifiers in the context of supervised learning have been studied recently by Hunt (2017).

In this chapter, we focus on the multiple imputation methods and the EM algorithm (as described in the following sections) for handling missing data when mixture models are applied for cluster analysis. Methods that use data augmentation/imputation via multiple imputation and parameter estimation using the EM algorithm have been discussed in Di Zio, Guarnera, and Luzi (2007) and Lehmann and Schlattmann (2017). The capacity of clustering methods in handling missing data is fundamentally important as inappropriate treatment of missing data may induce biases in parameter estimates and/or errors in classification (García-Laencina, Sancho-Gómes, and Figueiras-Vidal, 2010). We will consider two types of problems in the framework of mixture modelling, concerning the modelling of a mixture of multivariate normal distributions with missing data (Example 9.1) and missing data in a longitudinal study (Example 9.2).

9.3 Multiple Imputation Approach

Multiple imputation involves a three-step process (Rubin, 1996), in which the first step replaces each missing value by M sets of plausible values to form M complete data sets. Each of these complete data sets is then analyzed in the second step, and the results are finally combined during the third step into a single summary finding that takes into account the extra variation due to the missing values and the uncertainties of imputation.

The imputation step is the most critical, where it relies upon the missingness mechanism assumed and the relationship between the missing and observed data; see Horton and Lipsitz (2001). For example, under the MAR assumption (9.3), the imputation step can be undertaken by generating sets of imputations from the distribution $\mathrm{pr}(M|Y_{obs};\phi)$. A variety of imputation models can be adopted, such as regression and propensity methods, taking into account the uncertainty of the imputation. Some models, such as the predictive mean matching method discussed in Little and Rubin (2002, Section 4.3), ensures that imputed values are plausible (or are not out of range) by replacing the imputation model estimates with the "closest" values in Y_{obs}. With the assumption of multivariate normality, a Markov Chain Monte Carlo (MCMC) method can be used to impute the missing values, where a Markov chain is constructed to simulate random numbers drawn from the posterior distribution of $\text{pr}(M|Y_{obs};\phi^{(k)})$ in the $(k+1)$th iteration. In the first iteration, $\phi^{(0)}$ are obtained using Y_{obs} only and in subsequent iterations, $\phi^{(k)}$ $(k \geq 1)$ are obtained using Y_{obs} and the current imputed data. The iterative process is terminated when the model estimates in $\phi^{(k+1)}$ relative to $\phi^{(k)}$ change by an amount smaller than a certain threshold. An Imputation-Posterior (IP) algorithm has also been developed by Schafer (1997), where in each iteration there is an additional step to draw random parameter values $\phi^{(k)}$ from the posterior distribution of $\text{pr}(\phi|Y_{obs})$ using the current imputed data. It may be preferable to draw $\phi^{(k)}$ from the (current) conditional distribution, as it better preserves data covariance.

In the second step, the analysis of interest is then performed for each of the M imputed (complete) data sets to obtain the parameter estimates and their standard errors.

Given the M sets of estimates and their standard errors, the final combination step calculates estimates of the within- and between-imputation variability to provide consistent estimates of the parameters and their standard errors. Let $\hat{\Psi}^*_m$ and $\text{Cov}(\hat{\Psi}^*_m)$ be the vector of parameter estimates and the covariance matrix for the mth imputed complete data $(m = 1,\ldots,M)$. With the combination step, the combined estimates of the parameters are the average parameter values across all M sets of imputed data:

$$\overline{\hat{\Psi}} = \sum_{m=1}^{M} \hat{\Psi}^*_m/M. \tag{9.6}$$

The variability associated with $\overline{\hat{\Psi}}$ has two components, where the within-imputation variation is given by the averaged $\text{Cov}(\hat{\Psi}^*_m)$ for all M imputed data sets as

$$\overline{\text{Cov}(\hat{\Psi})} = \sum_{m=1}^{M} \text{Cov}(\hat{\Psi}^*_m)/M \tag{9.7}$$

and the between-imputation variation is calculated as

$$\text{Cov}_{\text{between}} = \sum_{m=1}^{M} (\hat{\Psi}^*_m - \overline{\hat{\Psi}})(\hat{\Psi}^*_m - \overline{\hat{\Psi}})^T/(M-1). \tag{9.8}$$

The covariance matrix of $\hat{\Psi}$ is then approximated by the total variability associated with $\overline{\hat{\Psi}}$. That is,

$$\text{Cov}(\hat{\Psi}) \approx \overline{\text{Cov}(\hat{\Psi})} + \frac{M+1}{M}\text{Cov}_{\text{between}} \tag{9.9}$$

and $(1+1/M)$ is an adjustment for finite M; see Little and Rubin (2002, Section 5.4). The standard error of the ith element of $\hat{\Psi}$ can be estimated by the positive square root of the ith diagonal element of $\text{Cov}(\hat{\Psi})$ in (9.9).

More detailed descriptions of the multiple imputation techniques, including the asymptotic behaviour of multiple imputation methods, can be found in Rubin (1996) and the monographs by Little and Rubin (2002) and Schafer (1997).

The three-step multiple imputation method can be readily applied in a mixture model-based approach for cluster analysis of data with missing values, such as using the Multiple Imputation by Chained Equations (MICE) in R developed by van Buuren and Groothuis-Oudshoorn (2011) with a variety of imputation approaches including regression and predictive mean matching. However, the imputation step does not account for the group structure of the data. In the second step, the mixture model of interest is performed for each of the M imputed data sets. The combined parameter estimates and their standard errors are obtained in the final step.

9.4 EM Algorithm

With the incomplete-data framework, the EM algorithm described in Section 1.3 is directly applicable for handling missing data. The E-step computes the Q-function, which is equivalent to obtaining the distribution of the missing values Y_{mis} on the basis of the observed data Y_{obs} and the current estimate of the model parameters Ψ, in order to provide the "best" guess on Y_{mis} via expectation of the conditional distribution. The M-step updates parameter estimates by maximizing the conditional expectation of the (complete-data) likelihood function as if no data were missing. The E- and M-steps are alternated repeatedly until convergence is obtained; see Section 1.3 for more details.

With the EM approach, a simple way to handle missing values in a mixture model-based cluster analysis application is to obtain the guess on Y_{mis} by ignoring the group structure of the data in forming the Q-function. With the MAR assumption, the R package AMELIA II developed by Honaker, King, and Blackwell (2015) combines the EM algorithm with a bootstrap approach to take draws from the posterior distribution of $\text{pr}(\phi|Y_{obs})$. With this EM Bootstrap (EMB) algorithm, uncertainty in parameter estimation in each draw is simulated by bootstrapping the data and the EM algorithm is then performed to obtain the mode of the posterior for the bootstrap sample; see Honaker and King (2010). Imputation of the missing values can be obtained by drawing values of Y_{mis} from its distribution conditional on Y_{obs} and the draws of the posterior of the complete-data parameters, as described in Honaker, King, and Blackwell (2015). Similar to the multiple imputation method, this procedure can be repeated M times and the estimates of parameters and their standard errors obtained from mixture modelling of each imputed data set can be combined in the final step, as given in (9.6) to (9.9) in Section 9.3.

An alternative approach is to establish the Q-function within the mixture model framework. Here, a mixture of multivariate normal distributions is considered, as shown in Ghahramani and Jordan (1994); see also Sovilj et al. (2016) who consider implementing the EM algorithm with a multiple imputation strategy. Let

$y_j^c = (y_j^{oT}, y_j^{mT})^T$ denote the complete data for the jth entity ($j = 1, \ldots, n$) containing, respectively, the sets of observed variables and missing variables such that Y_{obs} and Y_{mis} are given by

$$\begin{aligned} Y_{obs} &= (y_1^{oT}, \ldots, y_n^{oT})^T \\ Y_{mis} &= (y_1^{mT}, \ldots, y_n^{mT})^T. \end{aligned} \tag{9.10}$$

In the E-step, the following complete-data sufficient statistics in an analogue form presented in Section 2.2 are required:

$$\begin{aligned} T_{h1}^{(k)} &= \sum_{j=1}^{n} \tau_h(y_j^o; \Psi^{(k)}) \\ T_{h2}^{(k)} &= \sum_{j=1}^{n} \tau_h(y_j^o; \Psi^{(k)}) E_{\Psi^{(k)}}(y_j^c | Z_{hj} = 1, y_{obs}) \\ T_{h3}^{(k)} &= \sum_{j=1}^{n} \tau_h(y_j^o; \Psi^{(k)}) E_{\Psi^{(k)}}(y_j^c y_j^{cT} | Z_{hj} = 1, y_{obs}), \end{aligned} \tag{9.11}$$

where

$$\tau_h(y_j^o; \Psi^{(k)}) = E_{\Psi^{(k)}}(Z_{hj} | y_{obs}) = \frac{\pi_h^{(k)} \phi(y_j^o; \mu_h^{o(k)}, \Sigma_h^{oo(k)})}{\sum_{l=1}^{g} \pi_l^{(k)} \phi(y_j^o; \mu_l^{o(k)}, \Sigma_l^{oo(k)})} \tag{9.12}$$

is the estimated posterior probability of component membership based only on the observed variables y_j^o for each jth entity using the current parameter estimates in $\Psi^{(k)}$ ($h = 1, \ldots, g$; $j = 1, \ldots, n$), which are split into the observed and missing parts according to the missingness of the jth entity. That is, the current estimates of the mean vector $\mu_h^{(k)}$ and the covariance matrix $\Sigma_h^{(k)}$ are partitioned as

$$\mu_h^{(k)} = \begin{pmatrix} \mu_h^{o(k)} \\ \mu_h^{m(k)} \end{pmatrix} \tag{9.13}$$

and

$$\Sigma_h^{(k)} = \begin{bmatrix} \Sigma_h^{oo(k)} & \Sigma_h^{om(k)} \\ \Sigma_h^{mo(k)} & \Sigma_h^{mm(k)} \end{bmatrix}, \tag{9.14}$$

where, to keep notations simple, we drop the subscript j that indicates the matching of missingness structure of the jth entity. In (9.11), $E_{\Psi^{(k)}}(y_j^c | Z_{hj} = 1, y_{obs})$ is equal to y_j^o for those observed values. The missing part corresponding to y_j^m in y_j^c is replaced by the expectation conditioned on y_j^o, $z_{hj} = 1$, and $\Psi^{(k)}$ as

$$y_j^m = \mu_h^{m(k)} + \Sigma_h^{mo(k)} {\Sigma_h^{oo(k)}}^{-1} (y_j^o - \mu_h^{o(k)}). \tag{9.15}$$

Using the matrix manipulations given in Lin, Lee, and Ho (2006) and denoting the dimension of y_j^c as p, it follows that

$$E_{\Psi^{(k)}}(y_j^c | Z_{hj} = 1, y_{obs}) = \mu_h^{(k)} + \Sigma_h^{(k)} S_{hj}^{(k)} (y_j^c - \mu_h^{(k)}), \tag{9.16}$$

where $S_{hj}^{(k)}$ is a $p \times p$ matrix obtained by

$$S_{hj}^{(k)} = O_j^T \Sigma_h^{oo(k)^{-1}} O_j = O_j^T (O_j \Sigma_h^{(k)} O_j^T)^T O_j, \tag{9.17}$$

with O_j a binary indicator matrix extracted from a p-dimensional identity matrix I_p corresponding to row-positions of y_j^o in y_j^c; see Lin, Lee, and Ho (2006).

For the conditional expectation matrix $E_{\Psi^{(k)}}(y_j^c y_j^{cT} | Z_{hj} = 1, y_{obs})$, the sub-matrix corresponding to the missing part is replaced by the conditional expectation as

$$y_j^m y_j^{mT} + \Sigma_h^{mm(k)} - \Sigma_h^{mo(k)} \Sigma_h^{oo(k)^{-1}} \Sigma_h^{om(k)}; \tag{9.18}$$

see Ghahramani and Jordan (1994). With (9.17), it follows that

$$E_{\Psi^{(k)}}(y_j^c y_j^{cT} | Z_{hj} = 1, y_{obs}) = y_j^c y_j^{cT} + \Sigma_h^{(k)} - \Sigma_h^{(k)} S_{hj}^{(k)} \Sigma_h^{(k)}; \tag{9.19}$$

see Lin, Lee, and Ho (2006). On the basis of the sufficient statistics in (9.11), the M-step is given by

$$\begin{aligned}
\pi_h^{(k+1)} &= T_{h1}^{(k)} / n \\
\mu_h^{(k+1)} &= T_{h2}^{(k)} / T_{h1}^{(k)} \\
\Sigma_h^{(k+1)} &= \{T_{h3}^{(k)} - T_{h1}^{(k)^{-1}} T_{h2}^{(k)} T_{h2}^{(k)^T}\} / T_{h1}^{(k)}.
\end{aligned} \tag{9.20}$$

9.5 Example 9.1: Multivariate Normal Mixture Model

We consider the well-known set of *Iris* data first analyzed by Fisher (1936). The data set is available from the UCI Repository of machine learning databases (Bache and Lichman, 2013), containing the measurements of the length and width of both sepals and petals of 50 plants for each of three types of *Iris* species: *setosa*, *versicolor*, and *virginica*. The data set has been analyzed using a wide variety of techniques, such as the use of mixture model-based approaches by McLachlan and Peel (2000, Section 3.11), Ng and McLachlan (2004b), and McLachlan and Ng (2009). Here we adopt a 3-component mixture of multivariate normal distributions with unrestricted component-covariance matrices to study the relative performance of the aforementioned methods for handling missing data imposed on the *Iris* data under various missing-value mechanisms. The comparison is based on the bias of the estimates of model parameters and their standard errors, and the misclassification rate.

In this example, we consider four different missing-value mechanisms. The first corresponds to MCAR, where missingness is completely random at a probability of 10% for all types of *Iris* with their four measurements. The second mechanism is similar to the first one, except that missing values only happen for *Iris versicolor*. The other two types of species do not have missing values. The third mechanism corresponds to MAR, where missingness depends only on observed values but not on missing values. Specifically, measurements of the length and the width of petals for all types of *Iris* are missing with a probability of 30% and 20%, respectively, if the length of sepals is greater than 5 or the width of sepals is greater than 3. Finally, the fourth mechanism is similar to the third one, except that missing values only happen for *Iris versicolor* and the probabilities of missingness are replaced by 40% and 30% for the length and the width of petals, respectively. Ten sets of data with missing values were generated for each of the four missing-value mechanisms. For Cases 1 to 4 of missing-value mechanisms, the averaged proportions of missingness are 9.6%, 3.3%, 11.1%, and 5.4%, respectively.

We consider three different approaches in handling missing data within the application of multivariate normal mixture models with unrestricted component covariance matrices to each missing-value mechanism of the *Iris* data. The first approach makes use of multiple imputation via the MICE R-package version 2.30 (van Buuren and Groothuis-Oudshoorn, 2011). With this approach, the imputation step is performed according to the predictive mean matching (pmm) method with $M = 10$ multiple imputations, followed by the fitting of multivariate normal mixture models and, in the final step, the pooling of M sets of findings to obtain the parameter estimates. The second approach adopts Amelia II program version 1.7.4 (Honaker, King, and Blackwell, 2015), which assumes multivariate normally distributed data and uses the EMB algorithm (a bootstrap-based EM approach) to impute missing values to create imputed datasets for the fitting of multivariate normal mixture models. Similar to the multiple imputation method above, $M = 10$ complete datasets are produced by combining the EM algorithm with a bootstrap approach to draw missing values from their posterior distributions, without any priors specified; see Honaker, King, and Blackwell (2015) for details. With the MICE and Amelia II programs, all four random variates are entered into the imputation step. Including as many predictors as possible in imputation tends to make the MAR assumption more plausible in practice; see the discussion in van Buuren and Groothuis-Oudshoorn (2011). In the third approach, the EM procedure for a mixture of multivariate normal distributions with missing data described in Section 9.4 is used. This "EM-mix" method accounts for the group structure of the data in forming the Q-function for computing Y_{mis}; see (9.11) to (9.20).

The results for Case 1 (MCAR mechanism) are given in Table 9.1. By fitting a three-component multivariate normal mixture model with unrestricted component-covariance matrices (non-diagonal estimates not reported) to the complete *Iris* data, the subjects were well classified, with only 5 (3% of 150) misclassification errors (mainly between *versicolor* and *virginica*). No appreciable bias is observed, but the ratios of estimated standard error over true value are relatively large for parameters σ^2_{233}, σ^2_{333}, and σ^2_{344}, indicating the uncertainty involved in estimation of these parameters in the component covariance matrices. With the multiple imputation (MI) method, the averaged misclassification rate increases to 25.3 (17% of 150). Appreciable bias is observed for the estimated mean vector of *virginica* ($\mu_{3l}, l = 1, \ldots, 4$). The estimates of σ^2_{333} and σ^2_{344} are also biased. With the bootstrap-based EM (EMB) algorithm, the averaged misclassification rate further increases to

TABLE 9.1
Estimated biases and standard errors of the ML estimators for multivariate normal mixture models with missing data (Example 9.1: Case 1).

Parameter	Completed data[a]		MI[b]	EMB[c]	EM-mix[d]
True value	Est (SE)	Bias	Bias (SE)	Bias (SE)	Bias (SE)
$\pi_1 = 0.33$	0.33 (0.04)	0.000	-0.003 (0.01)	-0.008 (0.01)	-0.000 (0.00)
$\pi_2 = 0.33$	0.30 (0.05)	-0.034	-0.009 (0.02)	0.036 (0.04)	-0.037 (0.01)
$\pi_3 = 0.33$	0.37 (0.05)	0.034	0.011 (0.02)	-0.027 (0.03)	0.037 (0.01)
$\mu_{11} = 5.01$	5.01 (0.05)	0.000	0.006 (0.01)	-0.003 (0.02)	0.006 (0.01)
$\mu_{12} = 3.42$	3.42 (0.05)	0.000	-0.005 (0.02)	-0.004 (0.01)	0.004 (0.02)
$\mu_{13} = 1.46$	1.46 (0.03)	0.000	0.006 (0.01)	0.012 (0.01)	0.005 (0.01)
$\mu_{14} = 0.24$	0.24 (0.02)	0.000	0.001 (0.00)	-0.001 (0.00)	0.001 (0.00)
$\mu_{21} = 5.94$	5.92 (0.09)	-0.021	0.081 (0.08)	0.114 (0.08)	-0.017 (0.02)
$\mu_{22} = 2.77$	2.78 (0.06)	0.008	0.082 (0.06)	0.100 (0.07)	0.015 (0.02)
$\mu_{23} = 4.26$	4.20 (0.08)	-0.058	0.161 (0.23)	0.227 (0.24)	-0.043 (0.02)
$\mu_{24} = 1.33$	1.30 (0.04)	-0.029	0.112 (0.15)	0.166 (0.14)	-0.032 (0.01)
$\mu_{31} = 6.59$	6.54 (0.10)	-0.043	-0.131 (0.08)	-0.102 (0.12)	-0.052 (0.02)
$\mu_{32} = 2.97$	2.95 (0.05)	-0.025	-0.082 (0.05)	-0.066 (0.05)	-0.029 (0.01)
$\mu_{33} = 5.55$	5.48 (0.10)	-0.072	-0.312 (0.21)	-0.365 (0.26)	-0.094 (0.03)
$\mu_{34} = 2.03$	1.98 (0.06)	-0.041	-0.199 (0.14)	-0.254 (0.16)	-0.049 (0.02)
$\sigma^2_{111} = 0.12$	0.12 (0.02)	-0.002	-0.002 (0.01)	0.004 (0.01)	0.000 (0.00)
$\sigma^2_{122} = 0.15$	0.14 (0.03)	-0.003	-0.001 (0.01)	-0.009 (0.01)	-0.004 (0.01)
$\sigma^2_{133} = 0.03$	0.03 (0.01)	-0.001	0.000 (0.00)	0.013 (0.01)	0.000 (0.00)
$\sigma^2_{144} = 0.01$	0.01 (0.00)	-0.000	0.000 (0.00)	0.003 (0.00)	0.000 (0.00)
$\sigma^2_{211} = 0.27$	0.28 (0.06)	0.009	0.038 (0.03)	0.058 (0.04)	0.016 (0.01)
$\sigma^2_{222} = 0.10$	0.09 (0.02)	-0.006	-0.003 (0.01)	0.004 (0.01)	-0.005 (0.01)
$\sigma^2_{233} = 0.22$	0.20 (0.06)	-0.020	0.081 (0.08)	0.191 (0.12)	-0.014 (0.02)
$\sigma^2_{244} = 0.04$	0.03 (0.01)	-0.007	0.042 (0.04)	0.073 (0.04)	-0.006 (0.00)
$\sigma^2_{311} = 0.40$	0.39 (0.07)	-0.017	0.058 (0.08)	0.079 (0.11)	-0.006 (0.02)
$\sigma^2_{322} = 0.10$	0.11 (0.02)	0.006	0.019 (0.01)	0.018 (0.02)	0.010 (0.01)
$\sigma^2_{333} = 0.30$	0.33 (0.08)	0.023	0.268 (0.23)	0.503 (0.35)	0.043 (0.03)
$\sigma^2_{344} = 0.08$	0.09 (0.02)	0.010	0.032 (0.02)	0.056 (0.04)	0.015 (0.01)
Misclassified error (average)	5		25.3	33.8	5.3

[a] Fitting a 3-component mixture model of multivariate normal distributions to the complete data without any missing values.
[b] Multiple imputation via the MICE R-package.
[c] Amelia II R-package using a bootstrap-based EM algorithm to impute missing values.
[d] EM algorithm for a multivariate normal mixture model with missing values.

33.8 (23% of 150). Appreciable bias is observed for the estimated mean vector of *versicolor* ($\mu_{2l}, l = 1,\ldots,4$) as well as the estimated mean vector of *virginica* (noticeably μ_{33} and μ_{34}). The estimates of σ^2_{233}, σ^2_{244}, σ^2_{333}, and σ^2_{344} are also biased. It appears that the predictive mean matching (pmm) technique adopted in the MI method improves the clustering results by ensuring imputations are restricted to the observed values and that the method can preserve non-linear relations even if the structural part of the imputation model is incorrect; as discussed in van Buuren and Groothuis-Oudshoorn (2011). For the Amelia II program with the EMB algorithm, logical bounds on the imputations can be set up using the "bounds" option, where draws from truncated distributions are taken to obtain imputed values that satisfy the bounds (Honaker, King, and Blackwell, 2015). Using the EM procedure for a mixture of multivariate normal distributions with missing data, the EM-mix algorithm produces results that are comparable to those obtained by the normal mixture model fitted to the complete data. The average misclassification is 5.3 (4% of 150) and no appreciable bias is observed.

The results for Case 2 (MCAR mechanism, missingness only for *Iris versicolor*) are given in Table 9.2. It is observed that there are fewer misclassification errors when the missingness occurs only in one of the three clusters, compared to Case 1 where missingness occurs in all clusters. Not only is the estimation of parameters for the two components with complete data generally well taken, the bias in estimates for the component with missing values is not appreciable under the MCAR mechanism, as shown in the results using the MI method (the averaged misclassification rate is 6.8 (5% of 150)). With the EMB algorithm, the average misclassification rate is 10.4 (7% of 150). No appreciable bias is observed except for the estimates of mean μ_{33} and its variance σ^2_{333}. The average misclassification rate for the EM-mix algorithm is 6.5 (4% of 150).

The results for Case 3 (MAR mechanism, missingness occurs in the last two dimensions) are given in Table 9.3. With the MI method, the averaged misclassification rate is 17.6 (12% of 150). No appreciable bias is observed except for the estimates of mean μ_{33} and its variance σ^2_{333}. With the EMB algorithm, the average misclassification rate is 24.8 (17% of 150). Appreciable bias is observed for the estimated mean vector of *virginica* (noticeably μ_{33} and μ_{34}). The estimates of σ^2_{233}, σ^2_{244}, σ^2_{333}, and σ^2_{344} are also heavily biased. The average misclassification rate for the EM-mix algorithm is 6.5 (4% of 150) and no appreciable bias is observed.

The results for Case 4 (MAR mechanism, missingness only for *Iris versicolor* and occurs in the last two dimensions) are given in Table 9.4. Using the MI method, the average misclassification rate is 15.2 (10% of 150). Appreciable bias is observed for the estimates of mean μ_{33} and μ_{34} and their variances σ^2_{333} and σ^2_{344}. With the EMB algorithm, the average misclassification rate is 17.6 (12% of 150). Appreciable bias is observed for the estimates of mean μ_{33} and μ_{34}. The estimates of σ^2_{233}, σ^2_{244}, σ^2_{333}, and σ^2_{344} are also biased. The average misclassification rate for the EM-mix algorithm is 6.4 (4% of 150).

To summarize, with multivariate normal mixture models involving missing data, the EM procedure described in Section 9.4 performs better than the MI method and the EMB algorithm. This is because the EM-mix procedure accounts for the group

TABLE 9.2
Estimated biases and standard errors of the ML estimators for multivariate normal mixture models with missing data (Example 9.1: Case 2).

Parameter	Completed data[a]		MI[b]	EMB[c]	EM-mix[d]
True value	Est (SE)	Bias	Bias (SE)	Bias (SE)	Bias (SE)
π_1= 0.33	0.33 (0.04)	0.000	0.001 (0.00)	0.000 (0.00)	0.000 (0.00)
π_2= 0.33	0.30 (0.05)	-0.034	-0.031 (0.01)	-0.028 (0.03)	-0.049 (0.03)
π_3= 0.33	0.37 (0.05)	0.034	0.030 (0.01)	0.028 (0.03)	0.049 (0.03)
μ_{11}= 5.01	5.01 (0.05)	0.000	0.000 (0.00)	0.000 (0.00)	-0.000 (0.00)
μ_{12}= 3.42	3.42 (0.05)	0.000	-0.001 (0.00)	0.000 (0.00)	-0.000 (0.00)
μ_{13}= 1.46	1.46 (0.03)	0.000	0.000 (0.00)	0.000 (0.00)	0.000 (0.00)
μ_{14}= 0.24	0.24 (0.02)	0.000	0.000 (0.00)	0.000 (0.00)	0.000 (0.00)
μ_{21}= 5.94	5.92 (0.09)	-0.021	0.000 (0.03)	0.017 (0.05)	-0.020 (0.02)
μ_{22}= 2.77	2.78 (0.06)	0.008	0.030 (0.01)	0.043 (0.02)	0.012 (0.01)
μ_{23}= 4.26	4.20 (0.08)	-0.058	-0.019 (0.07)	-0.005 (0.11)	-0.072 (0.04)
μ_{24}= 1.33	1.30 (0.04)	-0.029	-0.008 (0.04)	0.015 (0.07)	-0.032 (0.01)
μ_{31}= 6.59	6.54 (0.10)	-0.043	-0.056 (0.01)	-0.077 (0.03)	-0.065 (0.04)
μ_{32}= 2.97	2.95 (0.05)	-0.025	-0.027 (0.01)	-0.038 (0.02)	-0.034 (0.02)
μ_{33}= 5.55	5.48 (0.10)	-0.072	-0.092 (0.05)	-0.138 (0.07)	-0.102 (0.08)
μ_{34}= 2.03	1.98 (0.06)	-0.041	-0.056 (0.03)	-0.086 (0.06)	-0.059 (0.04)
σ^2_{111}= 0.12	0.12 (0.02)	-0.002	-0.003 (0.00)	-0.002 (0.00)	-0.003 (0.00)
σ^2_{122}= 0.15	0.14 (0.03)	-0.003	-0.002 (0.00)	-0.003 (0.00)	-0.003 (0.00)
σ^2_{133}= 0.03	0.03 (0.01)	-0.001	-0.001 (0.00)	0.000 (0.00)	-0.001 (0.00)
σ^2_{144}= 0.01	0.01 (0.00)	-0.000	0.000 (0.00)	0.000 (0.00)	-0.000 (0.00)
σ^2_{211}= 0.27	0.28 (0.06)	0.009	0.008 (0.01)	0.014 (0.02)	0.013 (0.02)
σ^2_{222}= 0.10	0.09 (0.02)	-0.006	-0.005 (0.01)	-0.005 (0.01)	-0.009 (0.01)
σ^2_{233}= 0.22	0.20 (0.06)	-0.020	-0.010 (0.02)	0.068 (0.15)	-0.018 (0.01)
σ^2_{244}= 0.04	0.03 (0.01)	-0.007	-0.001 (0.01)	0.021 (0.03)	-0.007 (0.00)
σ^2_{311}= 0.40	0.39 (0.07)	-0.017	0.006 (0.02)	0.028 (0.05)	-0.008 (0.01)
σ^2_{322}= 0.10	0.11 (0.02)	0.006	0.007 (0.00)	0.009 (0.01)	0.007 (0.00)
σ^2_{333}= 0.30	0.33 (0.08)	0.023	0.054 (0.05)	0.116 (0.11)	0.043 (0.05)
σ^2_{344}= 0.08	0.09 (0.02)	0.010	0.014 (0.01)	0.018 (0.01)	0.018 (0.02)
Misclassified error (average)	5		6.8	10.4	6.5

[a] Fitting a 3-component mixture model of multivariate normal distributions to the complete data without any missing values.
[b] Multiple imputation via the MICE R-package.
[c] Amelia II R-package using a bootstrap-based EM algorithm to impute missing values.
[d] EM algorithm for a multivariate normal mixture model with missing values.

TABLE 9.3

Estimated biases and standard errors of the ML estimators for multivariate normal mixture models with missing data (Example 9.1: Case 3).

Parameter	Completed data[a]		MI[b]	EMB[c]	EM-mix[d]
True value	Est (SE)	Bias	Bias (SE)	Bias (SE)	Bias (SE)
$\pi_1 = 0.33$	0.33 (0.04)	0.000	-0.006 (0.01)	-0.011 (0.01)	0.000 (0.00)
$\pi_2 = 0.33$	0.30 (0.05)	-0.034	-0.016 (0.04)	-0.007 (0.05)	-0.034 (0.02)
$\pi_3 = 0.33$	0.37 (0.05)	0.034	0.022 (0.04)	0.018 (0.05)	0.034 (0.02)
$\mu_{11} = 5.01$	5.01 (0.05)	0.000	-0.051 (0.16)	-0.099 (0.21)	0.000 (0.00)
$\mu_{12} = 3.42$	3.42 (0.05)	0.000	-0.036 (0.11)	-0.069 (0.14)	0.000 (0.00)
$\mu_{13} = 1.46$	1.46 (0.03)	0.000	-0.018 (0.05)	-0.025 (0.05)	-0.003 (0.01)
$\mu_{14} = 0.24$	0.24 (0.02)	0.000	-0.002 (0.01)	-0.001 (0.01)	0.003 (0.00)
$\mu_{21} = 5.94$	5.92 (0.09)	-0.021	0.019 (0.07)	0.027 (0.08)	0.006 (0.02)
$\mu_{22} = 2.77$	2.78 (0.06)	0.008	0.048 (0.05)	0.088 (0.07)	0.008 (0.02)
$\mu_{23} = 4.26$	4.20 (0.08)	-0.058	0.014 (0.16)	-0.040 (0.19)	-0.040 (0.04)
$\mu_{24} = 1.33$	1.30 (0.04)	-0.029	0.024 (0.10)	0.039 (0.11)	-0.023 (0.01)
$\mu_{31} = 6.59$	6.54 (0.10)	-0.043	-0.076 (0.05)	-0.076 (0.08)	-0.065 (0.03)
$\mu_{32} = 2.97$	2.95 (0.05)	-0.025	-0.037 (0.02)	-0.053 (0.02)	-0.024 (0.02)
$\mu_{33} = 5.55$	5.48 (0.10)	-0.072	-0.143 (0.14)	-0.223 (0.14)	-0.067 (0.04)
$\mu_{34} = 2.03$	1.98 (0.06)	-0.041	-0.109 (0.08)	-0.172 (0.06)	-0.048 (0.03)
$\sigma^2_{111} = 0.12$	0.12 (0.02)	-0.002	-0.003 (0.00)	-0.005 (0.00)	-0.002 (0.00)
$\sigma^2_{122} = 0.15$	0.14 (0.03)	-0.003	-0.006 (0.01)	-0.006 (0.01)	-0.003 (0.00)
$\sigma^2_{133} = 0.03$	0.03 (0.01)	-0.001	0.002 (0.00)	0.023 (0.01)	-0.001 (0.00)
$\sigma^2_{144} = 0.01$	0.01 (0.00)	-0.000	0.000 (0.00)	0.004 (0.00)	-0.000 (0.00)
$\sigma^2_{211} = 0.27$	0.28 (0.06)	0.009	0.026 (0.02)	0.066 (0.10)	0.024 (0.03)
$\sigma^2_{222} = 0.10$	0.09 (0.02)	-0.006	-0.001 (0.01)	0.003 (0.01)	-0.003 (0.01)
$\sigma^2_{233} = 0.22$	0.20 (0.06)	-0.020	0.067 (0.07)	0.258 (0.34)	0.008 (0.04)
$\sigma^2_{244} = 0.04$	0.03 (0.01)	-0.007	0.018 (0.03)	0.052 (0.05)	-0.005 (0.01)
$\sigma^2_{311} = 0.40$	0.39 (0.07)	-0.017	0.014 (0.04)	0.050 (0.04)	-0.002 (0.02)
$\sigma^2_{322} = 0.10$	0.11 (0.02)	0.006	0.006 (0.01)	0.004 (0.01)	0.003 (0.01)
$\sigma^2_{333} = 0.30$	0.33 (0.08)	0.023	0.141 (0.12)	0.378 (0.18)	0.024 (0.04)
$\sigma^2_{344} = 0.08$	0.09 (0.02)	0.010	0.026 (0.01)	0.051 (0.02)	0.016 (0.02)
Misclassified error (average)	5		17.6	24.8	6.5

[a] Fitting a 3-component mixture model of multivariate normal distributions to the complete data without any missing values.

[b] Multiple imputation via the MICE R-package.

[c] Amelia II R-package using a bootstrap-based EM algorithm to impute missing values.

[d] EM algorithm for a multivariate normal mixture model with missing values.

TABLE 9.4
Estimated biases and standard errors of the ML estimators for multivariate normal mixture models with missing data (Example 9.1: Case 4).

Parameter	**Completed data**[a]		**MI**[b]	**EMB**[c]	**EM-mix**[d]
True value	**Est (SE)**	**Bias**	**Bias (SE)**	**Bias (SE)**	**Bias (SE)**
$\pi_1 = 0.33$	0.33 (0.04)	0.000	0.000 (0.00)	-0.001 (0.00)	0.000 (0.00)
$\pi_2 = 0.33$	0.30 (0.05)	-0.034	-0.020 (0.03)	-0.009 (0.02)	-0.051 (0.01)
$\pi_3 = 0.33$	0.37 (0.05)	0.034	0.020 (0.03)	0.011 (0.03)	0.051 (0.01)
$\mu_{11} = 5.01$	5.01 (0.05)	0.000	0.000 (0.00)	-0.001 (0.00)	0.000 (0.00)
$\mu_{12} = 3.42$	3.42 (0.05)	0.000	0.000 (0.00)	-0.001 (0.00)	0.000 (0.00)
$\mu_{13} = 1.46$	1.46 (0.03)	0.000	0.000 (0.00)	0.000 (0.00)	0.000 (0.00)
$\mu_{14} = 0.24$	0.24 (0.02)	0.000	0.000 (0.00)	-0.001 (0.00)	0.000 (0.00)
$\mu_{21} = 5.94$	5.92 (0.09)	-0.021	0.022 (0.04)	0.044 (0.05)	-0.023 (0.01)
$\mu_{22} = 2.77$	2.78 (0.06)	0.008	0.026 (0.02)	0.043 (0.03)	0.006 (0.01)
$\mu_{23} = 4.26$	4.20 (0.08)	-0.058	0.062 (0.12)	0.031 (0.13)	-0.078 (0.05)
$\mu_{24} = 1.33$	1.30 (0.04)	-0.029	0.034 (0.07)	0.062 (0.07)	-0.040 (0.01)
$\mu_{31} = 6.59$	6.54 (0.10)	-0.043	-0.069 (0.04)	-0.081 (0.06)	-0.069 (0.02)
$\mu_{32} = 2.97$	2.95 (0.05)	-0.025	-0.055 (0.03)	-0.057 (0.03)	-0.031 (0.01)
$\mu_{33} = 5.55$	5.48 (0.10)	-0.072	-0.149 (0.07)	-0.193 (0.09)	-0.091 (0.03)
$\mu_{34} = 2.03$	1.98 (0.06)	-0.041	-0.124 (0.07)	-0.133 (0.07)	-0.052 (0.02)
$\sigma^2_{111} = 0.12$	0.12 (0.02)	-0.002	-0.002 (0.00)	-0.002 (0.00)	-0.002 (0.00)
$\sigma^2_{122} = 0.15$	0.14 (0.03)	-0.003	-0.003 (0.00)	-0.004 (0.00)	-0.003 (0.00)
$\sigma^2_{133} = 0.03$	0.03 (0.01)	-0.001	-0.001 (0.00)	-0.001 (0.00)	-0.001 (0.00)
$\sigma^2_{144} = 0.01$	0.01 (0.00)	-0.000	0.000 (0.00)	0.000 (0.00)	-0.000 (0.00)
$\sigma^2_{211} = 0.27$	0.28 (0.06)	0.009	0.019 (0.01)	0.028 (0.03)	0.014 (0.01)
$\sigma^2_{222} = 0.10$	0.09 (0.02)	-0.006	-0.009 (0.00)	-0.005 (0.00)	-0.005 (0.00)
$\sigma^2_{233} = 0.22$	0.20 (0.06)	-0.020	0.034 (0.04)	0.099 (0.06)	-0.019 (0.02)
$\sigma^2_{244} = 0.04$	0.03 (0.01)	-0.007	0.020 (0.02)	0.031 (0.02)	-0.008 (0.00)
$\sigma^2_{311} = 0.40$	0.39 (0.07)	-0.017	0.032 (0.07)	0.032 (0.05)	-0.011 (0.01)
$\sigma^2_{322} = 0.10$	0.11 (0.02)	0.006	0.015 (0.01)	0.013 (0.01)	0.006 (0.00)
$\sigma^2_{333} = 0.30$	0.33 (0.08)	0.023	0.150 (0.13)	0.191 (0.13)	0.028 (0.01)
$\sigma^2_{344} = 0.08$	0.09 (0.02)	0.010	0.025 (0.01)	0.024 (0.01)	0.014 (0.01)
Misclassified error (average)	5		15.2	17.6	6.4

[a] Fitting a 3-component mixture model of multivariate normal distributions to the complete data without any missing values.
[b] Multiple imputation via the MICE R-package.
[c] Amelia II R-package using a bootstrap-based EM algorithm to impute missing values.
[d] EM algorithm for a multivariate normal mixture model with missing values.

structure of the data in forming the Q-function for computing Y_{mis}. In general, the results in estimation of parameters and classification are better for all three missing-data handling methods considered in Example 9.1 when missingness occurs only in one cluster. In particular, the MI method and the EMB algorithm provide comparable results with the EM-mix procedure in Case 2 under the MCAR mechanism and that missingness occurs in one cluster only; see Table 9.2. Comparison of methods for handling missing data in fitting normal mixture models using the *Iris* data and other multivariate data sets under various missingness mechanisms can also be found in Lin, Lee, and Ho (2006) and Eirola et al. (2014).

9.6 Missing Data in Longitudinal Studies

Longitudinal studies often encounter particular problems such as non-response, loss to follow-up, and withdrawal from the study. Findings based on data provided by respondents may be biased due to potential over-representation of relatively advantaged participants who respond, such as people with higher education or socio-economic status; see, for example, Ng et al. (2014) and Cameron et al. (2014). As described in Section 9.1, attrition studies attempt to investigate specific predictors of attrition from a wide variety of exposures, under different types of attrition including sub-types such as "contactable" non-respondents versus lost to follow-up participants (Ng, Scott, and Scuffham, 2016). Under the assumption of MAR mechanism, the attrition model derived may be adopted to calculate IPW adjustment for re-balancing the set of complete cases in order to improve generalizability of research findings from longitudinal studies due to non-response bias.

Longitudinal studies are prone to selection bias and loss of statistical power due to attrition, as discussed in Atherton et al. (2008), Bambs et al. (2013), and Weuve et al. (2012). For longitudinal data, an important concept about missingness concerns the pattern of missing data. The missingness is defined as monotone (or dropout) when the data matrix can be rearranged in a way that an outcome measure Y_{t_i} for an individual missing at certain time t_i implies that outcome measures at time $t \geq t_i$ are all missing; see, for example, Horton and Lipsitz (2001) and Ibrahim and Molenberghs (2009). If the pattern of missing data is monotone, simpler imputation methods can be adopted to handle missing data in the analysis. In practice, a common method in longitudinal data analysis for handling monotone missing values is Last Observation Carried Forward (LOCF), assuming that the outcome measure of an individual is unchanged after dropout. However, as a consequence, imputation uncertainty has not been accounted for in the analysis of LOCF imputed data.

Estimation of parameters with non-ignorable missing data mechanism is complex. Likelihood-based methods require specification of the joint distribution of the data and the missingness $f(Y,M)$, which can be classified into three types of models, namely, selection, pattern-mixture, and shared-parameter models; see Little and Rubin (2002, Chapter 15) and Ibrahim and Molenberghs (2009). Selection models write the joint distribution in the form of $f(M|Y)f(Y)$, where the first factor attempts to model the missingness conditional on hypothetical complete data; see also Diggle and Kenward (1994), Kenward (1998), and Little (2009). The pattern-mixture models partition the joint distribution as $f(Y|M)f(M)$ and that $f(Y|M)$ characterizes the distribution of Y in the strata defined by different patterns of missing data, as described in Little and Rubin (2002, Sections 15.1 and 15.5). In particular, the marginal distribution of Y will have a mixture model form, where the mixing proportion π_h represents the probability of the hth pattern of missing data. With shared-parameter models, both $f(Y)$ and $f(M)$ account for the dependence on latent variables U such as random effects. By assuming conditionally independence given U, the joint distribution is then given by

$$f(Y,M) = \int f(Y|U)f(M|U)f(U)dU;$$

see, for example, Kenward and Rosenkranz (2011).

Pattern mixture models attempt to stratify individuals by their missing data patterns. Strategies to fit pattern mixture models via multiple imputation have been considered in Thijs et al. (2002) and Tang (2017). However, pattern mixture models are often under-identified as the observed data do not provide direct information for each pattern of dropout to identify the distributions for the incomplete patterns. It is thus necessary to identify restrictions in order to achieve mixture model identification, as described in Little and Rubin (2002, Section 15.5) and Wang and Daniels (2011). In practice, the missing data patterns can be represented by (between-subjects) indicator variables in longitudinal data analysis. These pattern indicator variables can then be incorporated into the statistical model of interest in order to determine if the missing data pattern has any power in predicting outcome measures and any interactions with other model covariates such as intervention or time. More importantly, the conditional distribution $f(Y|M)$ is flexible to take various forms, such as mixed-effects models as shown in Hedeker and Gibbons (2006, Chapter 14). However, the effective use of non-ignorable models described above requires knowledge of the missing data mechanism that is often lacking in general clustering applications; see the discussion in Hunt and Jorgensen (2003).

9.7 Example 9.2: Clustering Longitudinal Data with Missing Values

We consider a synthetic growth trajectory data set, which is based on the longitudinal cortisol data described in Ram and Grimm (2009) as part of the MacArthur Successful Aging Studies, which examined the time course of cortisol production and dispersion in response to a "naturalistic" driving simulation challenge (Seeman, Singer, and Charpentier, 1995). Nine longitudinal measurements of cortisol level were obtained over the course of a few hours from 34 participants. The first two measurements were taken as baseline measurements, after which individuals began a driving simulation challenge task. Three measurements were taken during and directly after the challenge as indicators of cortisol production or response. The last four measurements were taken as indicators of cortisol dissipation during a post-challenge rest period; see Ram and Grimm (2009) and Ng and McLachlan (2014a) for more information. A growth mixture model (GMM) was taken by Ram and Grimm (2009) to fit the longitudinal cortisol data and identify two clusters ($g = 2$) of participants, corresponding to a "typical" response pattern and a "chronic-stress" pattern where cortisol levels did not dissipate following the stressor; see, for example, Ram and Grimm (2009) and Miller, Chen, and Zhou (2007). Figure 9.1 displays simulated trajectories for these two clusters of participants.

In this example, we consider a GMM where a random effect b_{hj} is introduced to capture individual variation in cortisol levels at various time points. With reference to the framework of mixtures of linear mixed models described in Section 8.1, the measurement vector of cortisol levels y_j for the jth individual follows the model

$$y_j = W\beta_h + 1b_{hj} + \varepsilon_{hj}, \tag{9.21}$$

given that y_j belongs to the hth component of the mixture model and where elements of β_h are fixed effects (unknown constants) modelling the conditional mean of y_j in the hth component ($h = 1,\ldots,g;\ j = 1,\ldots,n$). In (9.21), 1 is a vector of ones representing the known design matrix for the corresponding random effect b_{hj} and ε_{hj} is the measurement error vector. With the LMM framework, ε_{hj} and b_{hj} are taken to be a normal distribution, $N(0, \sigma_{ah}^2 I)$ and $N(0, \sigma_{bh}^2)$, respectively. We also consider the situation where risk factors x_j are incorporated into the mixing proportions $\pi_h(x_j)$ to model individual probabilities of membership of the g components, where the first element of x_j is assumed to be 1 to account for an intercept term. That is,

$$\pi_h(x_j;\alpha) = \frac{\exp(v_h^T x_j)}{1 + \sum_{l=1}^{g-1} \exp(v_l^T x_j)} \qquad (h = 1,\ldots,g-1), \tag{9.22}$$

and $\pi_g(x_j) = 1 - \sum_{l=1}^{g-1} \pi_l(x_j)$, where α contains the elements in v_h ($h = 1,\ldots,g-1$), which are fixed effects representing the log odds ratio of component membership for each risk factor in x_j; see, for example Ng and McLachlan (2014a). In addition, a cubic polynomial regression form of $W\beta_h$ in (9.21) is adopted to fit a nonlinear relationship between the conditional mean element of y_j at different time points, where

$$W = \begin{bmatrix} 1 & t_0 & t_0^2 & t_0^3 \\ 1 & t_1 & t_1^2 & t_1^3 \\ \vdots & \vdots & \vdots & \vdots \\ 1 & t_8 & t_8^2 & t_8^3 \end{bmatrix}, \tag{9.23}$$

and thus β_h involves four coefficients corresponding to the intercept, slope, quadratic, and cubic growth parameters; see Muthén (2004) for details.

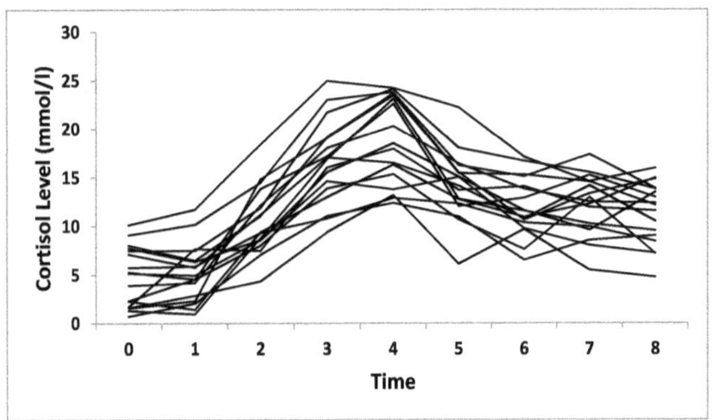

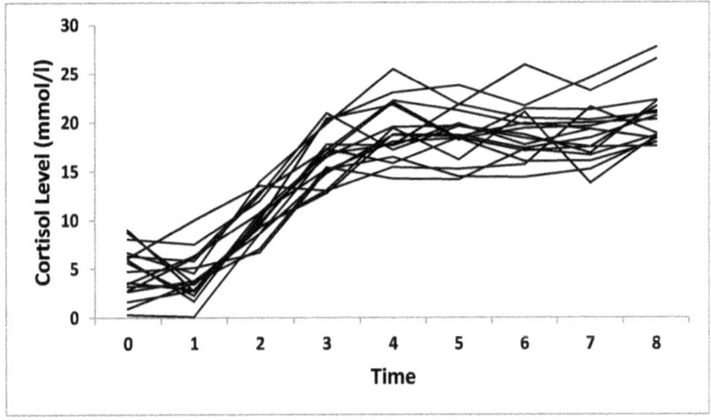

FIGURE 9.1
Simulated longitudinal cortisol data in two groups: (Top) - "typical" individuals; (Bottom) - "chronic-stress" individuals. Times 0 and 1: baseline phase; Times 2 to 4: response phase; Times 5 to 8: dissipation phase. Adapted from Ng and McLachlan (2014a).

We evaluate for the situation where there are missing outcomes as well as missing predictors. We generated longitudinal data of cortisol levels on the basis of the estimated parameters for the 2-component GMM given in Table 2 of Ram and Grimm (2009). Assuming x_j $(j = 1,\ldots,n)$ are three-dimensional vectors with the first element equal to one, we generate continuous two-dimensional vectors independently from $N(0, I_2)$ to form the last two elements of x_j. Realizations of the component indicator variable Z were simulated in which an individual has a probability of $\pi_1(x_j;\alpha)$ to be from the first component and has a probability of $(1-\pi_1(x_j;\alpha))$ to be from the second component, with $v_1 = (-0.5, 0.2, -0.2)^T$; see Equation (9.22). Suppose the jth individual belongs to the first component, then a synthetic growth trajectory is generated from the conditional probability density function corresponding to (9.21), where $\beta_1 = (4.0, 2.0, 0.25, -0.05)^T$, with b_{1j} generated from $N(0, \sigma^2_{b1})$. A measurement error vector generated from $N(0, \sigma^2_{a1} I)$ was then added to the synthetic growth trajectory. Similarly, for an individual belonging to the second component, a synthetic growth trajectory was generated using $\beta_2 = (4.5, 2.5, 0.4, -0.05)^T$, with b_{2j} generated from $N(0, \sigma^2_{b2})$.

A nonlinear GMM (with intercept, slope, quadratic and cubic growth parameters) was fitted to the synthetic growth trajectories with $g = 2$ "latent classes" using Mplus Version 7.4 (Muthén and Muthén, 1998-2015), where all within-class variances except for intercept (corresponding to σ^2_{bh}) were fixed to zero. The categorical latent class variable was set to relate the covariates in x_j via a logistic model. The GMM analyses were implemented with 200 random sets of starting values and 20 final optimizations; see Muthén and Muthén (1998-2015, Chapter 8) for methodological details and Dunn, Ng, Holland, et al. (2013) and Dunn, Ng, Breithart, et al. (2013) for applications.

In this example, we consider four different missing-value mechanisms. The first corresponds to MCAR with a monotone pattern of missing data, where missingness occurs only in outcome cortisol measures. Measurements of cortisol at Times 2 to 4 (response phase) are missing with a probability of 10%, 15%, 20%, respectively, whereas measurements at Times 5 to 8 (dissipation phase) are missing with a probability of 25%, 30%, 35%, and 40%, respectively. To ensure a monotone pattern, when a cortisol measure is missing at certain time t_i, measures at $t \geq t_i$ will all be missing. The second mechanism is similar to the first one, except that the last step for ensuring a monotone pattern is omitted and the probabilities of missingness are 10%, 20%, 35%, 50%, 65%, 80%, and 90%, respectively. The second mechanism is therefore non-monotone MCAR. In the third mechanism, we consider a MAR with a monotone pattern of missing data for outcome cortisol measures. If both x_{2j} and x_{3j} are greater than 0.5, measurements of cortisol at Times 5 to 8 (dissipation phase) are missing with a probability of 50%, 60%, 70%, and 80%, respectively. Also, we randomly add missing values for covariates b_{2j} and b_{3j}, with a probability of 10%. Finally, the fourth mechanism is similar to the third one, except that missing values of b_{2j} and b_{3j} only happen for the second latent class corresponding to the chronic-stress pattern, with a probability of 15%. Ten sets of data with missing values were generated for each of the four missing-value mechanisms. For Cases 1 to

4 of missing-value mechanisms, the average proportions of missingness are 33.2%, 31.8%, 9.3%, and 9.0%, respectively.

Three different approaches in handling longitudinal data with missing values in the application of GMMs were considered. The first approach is the MI method with pmm and $M = 5$, via the MICE R-package version 2.30 (van Buuren and Groothuis-Oudshoorn, 2011). The second approach is the bootstrap-based EM approach (EMB) method with $M = 5$, using Amelia II version 1.7.4 (Honaker, King, and Blackwell, 2015). With the MICE and Amelia II programs, all simulated cortisol measurements at nine different time points and two simulated covariates are entered into the imputation step. While including as many predictors as possible makes the MAR assumption more plausible (van Buuren and Groothuis-Oudshoorn, 2011), no consideration of specific data structure has been taken in the imputation step to account for the longitudinal nature of repeated cortisol measurements. In the third approach, missing values in the outcome cortisol measures were handled in Mplus using a robust full information ML (FIML) estimation procedure, which makes use of all available data to generate ML-based sufficient statistics with the assumption of MAR; see Muthén and Muthén (1998-2015) and Graham (2009). In the cases of missing covariate values (the third and fourth mechanisms), the IMPUTE option in Mplus was used to impute the missing covariate values with $M = 5$ imputed data sets as described in Muthén and Muthén (1998-2015, Chapter 11); see also Chambers et al. (2017). The comparison of these three approaches is based on the bias of the estimates of model parameters and their standard errors.

The results for Case 1 are given in Table 9.5. By fitting a GMM with two latent classes to the complete synthetic growth trajectory data, the subjects were perfectly classified (zero misclassified errors), with no appreciable bias observed. Using the multiple imputation (MI) method, bias is observed for the estimation of β_{11} and β_{12}, as well as variation in measurement errors (σ_a^2) and the variance components (σ_{b1}^2 and σ_{b2}^2). The averaged standard errors of estimates are, in general, larger compared to those based on complete data. With the bootstrap-based EM (EMB) algorithm, appreciable bias is observed for the estimation of σ_a^2, σ_{b1}^2, σ_{b2}^2, and the estimate of ν_{11}. Using the FIML estimation procedure in Mplus, estimates and their standard errors obtained are comparable to those obtained with complete data.

The results for Case 2 are given in Table 9.6. Under a non-monotone missing value mechanism, the MI method produces biased estimates for β_{11}, σ_a^2, σ_{b1}^2, and σ_{b2}^2. The average standard errors of estimates are generally larger compared to those based on complete data. The EMB algorithm again obtains heavily biased estimates for σ_a^2, σ_{b1}^2, and σ_{b2}^2. With the FIML procedure in Mplus, no appreciable bias is observed.

For Case 3 (Table 9.7), missingness in covariates has no appreciable effects on the estimates of ν_{12} and ν_{13} in the mixing proportions for all three methods. The estimations for σ_a^2, σ_{b1}^2, and σ_{b2}^2 are better because the proportion of missingness is smaller compared to Cases 1 and 2. However, bias is still observed in estimation of σ_a^2 using the MI method and all σ_a^2, σ_{b1}^2, and σ_{b2}^2 with the EMB method. The FIML procedure in Mplus, with the IMPUTE option to impute missing values in covariates, produces comparable estimates and standard errors against those obtained with

TABLE 9.5
Estimated biases and standard errors of the ML estimators for growth mixture models with missing data (Example 9.2: Case 1).

Parameter	Completed data[a]		MI[b]	EMB[c]	Mplus[d]
True value	Est (SE)	Bias	Bias (SE)	Bias (SE)	Bias (SE)
ν_{11}= -0.5	-0.50 (0.09)	0.003	0.006 (0.16)	0.113 (0.13)	0.007 (0.11)
ν_{12}= 0.2	0.21 (0.09)	0.011	0.017 (0.14)	0.024 (0.11)	0.047 (0.10)
ν_{13}= -0.2	-0.20 (0.09)	-0.001	0.066 (0.14)	0.016 (0.11)	-0.005 (0.10)
β_{11}= 4.0	3.95 (0.10)	-0.051	-0.163 (0.15)	-0.008 (0.12)	-0.059 (0.11)
β_{12}= 2.0	2.05 (0.08)	0.048	0.115 (0.14)	0.070 (0.12)	0.046 (0.10)
β_{13}= 0.25	0.23 (0.02)	-0.017	-0.027 (0.05)	0.005 (0.04)	-0.010 (0.03)
β_{14}= -0.05	-0.05 (0.00)	0.002	0.002 (0.00)	0.001 (0.00)	0.001 (0.00)
β_{21}= 4.5	4.55 (0.08)	0.047	0.073 (0.12)	0.041 (0.09)	0.056 (0.08)
β_{22}= 2.5	2.48 (0.06)	-0.016	-0.083 (0.11)	-0.004 (0.10)	-0.024 (0.08)
β_{23}= 0.4	0.40 (0.02)	0.000	0.012 (0.03)	-0.012 (0.03)	-0.001 (0.03)
β_{24}= -0.05	-0.05 (0.00)	0.000	0.000 (0.00)	0.001 (0.00)	0.001 (0.00)
σ_a^2= 1.0	1.00 (0.02)	0.003	0.201 (0.06)	1.158 (0.16)	0.009 (0.03)
σ_{b1}^2= 1.0	1.02 (0.11)	0.023	-0.202 (0.16)	1.730 (0.42)	0.035 (0.13)
σ_{b2}^2= 1.0	1.00 (0.09)	0.000	0.215 (0.12)	1.409 (0.33)	-0.001 (0.10)

[a] Fitting a GMM with 2 latent classes using Mplus to the complete data without any missing values.
[b] Multiple imputation via the MICE R-package.
[c] Amelia II R-package using a bootstrap-based EM algorithm to impute missing values.
[d] Robust full information ML (FIML) estimation via Mplus.

TABLE 9.6
Estimated biases and standard errors of the ML estimators for growth mixture models with missing data (Example 9.2: Case 2).

Parameter	Completed data[a]		MI[b]	EMB[c]	Mplus[d]
True value	Est (SE)	Bias	Bias (SE)	Bias (SE)	Bias (SE)
ν_{11}= -0.5	-0.50 (0.09)	0.003	0.051 (0.14)	0.027 (0.10)	0.002 (0.10)
ν_{12}= 0.2	0.21 (0.09)	0.011	-0.007 (0.13)	0.009 (0.09)	0.014 (0.09)
ν_{13}= -0.2	-0.20 (0.09)	-0.001	0.033 (0.12)	0.005 (0.09)	-0.002 (0.09)
β_{11}= 4.0	3.95 (0.10)	-0.051	-0.127 (0.15)	-0.011 (0.12)	-0.049 (0.10)
β_{12}= 2.0	2.05 (0.08)	0.048	0.075 (0.15)	0.001 (0.21)	0.047 (0.09)
β_{13}= 0.25	0.23 (0.02)	-0.017	-0.022 (0.05)	0.007 (0.09)	-0.016 (0.03)
β_{14}= -0.05	-0.05 (0.00)	0.002	0.002 (0.00)	0.000 (0.01)	0.001 (0.00)
β_{21}= 4.5	4.55 (0.08)	0.047	0.065 (0.12)	0.057 (0.09)	0.053 (0.08)
β_{22}= 2.5	2.48 (0.06)	-0.016	-0.062 (0.11)	-0.087 (0.16)	-0.044 (0.08)
β_{23}= 0.4	0.40 (0.02)	0.000	0.017 (0.04)	0.034 (0.07)	0.014 (0.03)
β_{24}= -0.05	-0.05 (0.00)	0.000	-0.002 (0.00)	-0.004 (0.01)	-0.001 (0.00)
σ_a^2= 1.0	1.00 (0.02)	0.003	0.147 (0.07)	0.900 (0.21)	0.002 (0.03)
σ_{b1}^2= 1.0	1.02 (0.11)	0.023	-0.148 (0.15)	0.913 (0.29)	0.042 (0.13)
σ_{b2}^2= 1.0	1.00 (0.09)	0.000	-0.169 (0.12)	0.768 (0.22)	0.004 (0.09)

[a] Fitting a GMM with 2 latent classes using Mplus to the complete data without any missing values.
[b] Multiple imputation via the MICE R-package.
[c] Amelia II R-package using a bootstrap-based EM algorithm to impute missing values.
[d] Robust full information ML (FIML) estimation via Mplus.

TABLE 9.7

Estimated biases and standard errors of the ML estimators for growth mixture models with missing data (Example 9.2: Case 3).

Parameter	Completed data[a]		MI[b]	EMB[c]	Mplus[d]
True value	Est (SE)	Bias	Bias (SE)	Bias (SE)	Bias (SE)
ν_{11}= -0.5	-0.50 (0.09)	0.003	0.029 (0.14)	0.012 (0.10)	0.005 (0.09)
ν_{12}= 0.2	0.21 (0.09)	0.011	-0.032 (0.13)	-0.004 (0.10)	-0.010 (0.09)
ν_{13}= -0.2	-0.20 (0.09)	-0.001	0.024 (0.13)	0.016 (0.09)	0.019 (0.09)
β_{11}= 4.0	3.95 (0.10)	-0.051	-0.071 (0.14)	-0.036 (0.10)	-0.049 (0.10)
β_{12}= 2.0	2.05 (0.08)	0.048	0.079 (0.12)	0.038 (0.08)	0.050 (0.08)
β_{13}= 0.25	0.23 (0.02)	-0.017	-0.024 (0.04)	-0.011 (0.03)	-0.017 (0.02)
β_{14}= -0.05	-0.05 (0.00)	0.002	0.002 (0.00)	0.001 (0.00)	0.002 (0.00)
β_{21}= 4.5	4.55 (0.08)	0.047	0.027 (0.11)	0.041 (0.08)	0.047 (0.08)
β_{22}= 2.5	2.48 (0.06)	-0.016	-0.067 (0.10)	-0.015 (0.07)	-0.016 (0.06)
β_{23}= 0.4	0.40 (0.02)	0.000	0.011 (0.03)	-0.002 (0.02)	0.000 (0.02)
β_{24}= -0.05	-0.05 (0.00)	0.000	0.000 (0.00)	0.000 (0.00)	0.000 (0.00)
σ_a^2= 1.0	1.00 (0.02)	0.003	0.104 (0.05)	0.281 (0.06)	0.006 (0.02)
σ_{b1}^2= 1.0	1.02 (0.11)	0.023	0.022 (0.17)	0.366 (0.21)	0.036 (0.11)
σ_{b2}^2= 1.0	1.00 (0.09)	0.000	-0.004 (0.15)	0.240 (0.15)	-0.006 (0.09)

[a] Fitting a GMM with 2 latent classes using Mplus to the complete data without any missing values.

[b] Multiple imputation via the MICE R-package.

[c] Amelia II R-package using a bootstrap-based EM algorithm to impute missing values.

[d] Robust full information ML (FIML) estimation via Mplus.

complete data. The same observation about the performance of the three methods is found for Case 4 (Table 9.8), where missingness in covariates occurs only for a certain class.

Overall, the FIML procedure with the IMPUTE option in Mplus handles missing values in outcome longitudinal measures and covariates very well in all four different missingness mechanisms, in terms of comparability in both parameter estimates and their standard errors compared to those obtained with complete data. The MI method and the EMB algorithm do not account for the group structure of the data nor the longitudinal nature of repeated cortisol measurements in the imputation step. These two methods lead to biased estimates for variation in measurement errors and the variance components, especially when the proportion of missing values is large (such as in Cases 1 and 2). Accounting for the longitudinal nature of missing data in multiple imputation is available in Zhang (2016).

TABLE 9.8
Estimated biases and standard errors of the ML estimators for growth mixture models with missing data (Example 9.2: Case 4).

Parameter	Completed data[a]		MI[b]	EMB[c]	Mplus[d]
True value	Est (SE)	Bias	Bias (SE)	Bias (SE)	Bias (SE)
v_{11}= -0.5	-0.50 (0.09)	0.003	0.027 (0.14)	0.009 (0.10)	0.006 (0.09)
v_{12}= 0.2	0.21 (0.09)	0.011	-0.026 (0.13)	0.001 (0.09)	-0.003 (0.09)
v_{13}= -0.2	-0.20 (0.09)	-0.001	-0.012 (0.13)	-0.003 (0.09)	0.003 (0.09)
β_{11}= 4.0	3.95 (0.10)	-0.051	-0.078 (0.14)	-0.039 (0.10)	-0.048 (0.10)
β_{12}= 2.0	2.05 (0.08)	0.048	0.069 (0.12)	0.028 (0.08)	0.048 (0.08)
β_{13}= 0.25	0.23 (0.02)	-0.017	-0.020 (0.04)	-0.008 (0.03)	-0.016 (0.02)
β_{14}= -0.05	-0.05 (0.00)	0.002	0.002 (0.00)	0.001 (0.00)	0.002 (0.00)
β_{21}= 4.5	4.55 (0.08)	0.047	0.032 (0.11)	0.045 (0.08)	0.047 (0.08)
β_{22}= 2.5	2.48 (0.06)	-0.016	-0.069 (0.09)	-0.016 (0.06)	-0.018 (0.06)
β_{23}= 0.4	0.40 (0.02)	0.000	0.011 (0.03)	0.000 (0.02)	0.001 (0.02)
β_{24}= -0.05	-0.05 (0.00)	0.000	-0.001 (0.00)	0.000 (0.00)	0.000 (0.00)
σ_a^2= 1.0	1.00 (0.02)	0.003	0.109 (0.05)	0.273 (0.06)	0.004 (0.02)
σ_{b1}^2= 1.0	1.02 (0.11)	0.023	0.006 (0.17)	0.311 (0.19)	0.039 (0.12)
σ_{b2}^2= 1.0	1.00 (0.09)	0.000	0.009 (0.14)	0.261 (0.14)	0.001 (0.09)

[a] Fitting a GMM with 2 latent classes using Mplus to the complete data without any missing values.
[b] Multiple imputation via the MICE R-package.
[c] Amelia II R-package using a bootstrap-based EM algorithm to impute missing values.
[d] Robust full information ML (FIML) estimation via Mplus.

9.8 Summary

In this chapter, we considered two types of problems in the framework of mixture modelling commonly seen in medical and health sciences research. These problems concern the modelling of multivariate data with missing values using mixtures of multivariate normal distributions and the clustering of trajectory data with missing observations in longitudinal studies. We outlined a number of various approaches for handling missing data in these situations within the mixture model framework. Comparison of these approaches was performed under different missing-value mechanisms including MCAR and MAR.

Within the mixture model framework, the methods that handle missing data can be divided into two main types: (1) those that ignore the group structure of the data and (2) those that are formulated under the mixture model framework, accounting for the group structure of the data. For type (1), there is an enormous amount of literature on missing data methods. We refer the reader to books such as Hedeker and Gibbons (2006, Chapter 14), Molenberghs and Kenward (2007), Daniels and Hogan (2008), Muthén and Muthén (1998-2015, Chapter 11) and the references therein. Most of the methods for type (2) assume normality of data and focus on the case of ignorable MAR mechanism. The following papers by Ghahramani and Jordan (1994), Hunt and Jorgensen (2003), Di Zio, Guarnera, and Luzi (2007), Graham (2009), Eirola et al. (2014), and Hunt (2017) provide excellent accounts of the relevant topics.

The two examples presented in this chapter were general and common in the various scientific fields concerning the modelling of multivariate data or longitudinal data. The comparison results given in Sections 9.5 and 9.7 indicated generally that missingness that occurs only at certain components of the mixture model is easier to handle, compared to the case where missingness occurs at all components randomly. The methods considered can also handle missing values in covariates, either simultaneously with or separately from the outcome measures. As described in Section 9.1, while these methods may be capable of obtaining valid and reliable parameter estimates in data analysis with missing values, there is always a concern about the potential bias in the interpretation and generalization of findings due to missing data. A sensitivity study should be conducted in order to assess the robustness of the findings obtained from each missing-data handling method.

10

Miscellaneous: Cluster Analysis of "Big Data" Using Mixture Models

10.1 Introduction

Big data in research in the fields of medical and health sciences is now commonplace. Important information about the unknown group structures extracted from these data can contribute to improving the quality, effectiveness and cost-effectiveness of prevention, treatment and care for a sustainable health system. In this chapter, we will discuss issues regarding the use of mixture models for cluster analysis of big data. It is not only the sheer size of the data which imposes difficulty in direct application of conventional mixture modelling techniques; big data often exhibit a multilevel structure with complex correlations among observations and/or a mix of variable types. To this end, this chapter will cover how the EM algorithm can be speeded up for iterative computation of the ML parameter estimates for big data; how random effects modelling can be adopted to cluster big data with complex correlation structures; and how mixtures of multivariate generalized Bernoulli distributions can be used to develop a multitask clustering mechanism for cluster analysis of big data.

Over the past few decades, there has been an ever-increasing use of mixture distributions to provide a model-based approach to the cluster analysis of data arising in a wide variety of scientific fields, including bioinformatics, biomedicine, health, biometrics and biostatistics, among many others; see McLachlan and Peel (2000), McLachlan, Do, and Ambroise (2004), McNicholas (2016), Schlattmann (2009) and the references therein. In particular, multivariate normal mixture models have been used as a powerful tool for model-based clustering of continuous multivariate data; see, for example, McLachlan, Bean, and Ben-Tovim Jones (2006), Ng et al. (2006), and Wang, Ng, and McLachlan (2009), where

$$f(y_j;\Psi) = \sum_{h=1}^{g} \pi_h(x_j;\alpha)\phi(y_j;\mu_h,\Sigma_h) \qquad (j=1,\ldots,n). \qquad (10.1)$$

The mixing proportions $\pi_h(x_j;\alpha)$ depend on a vector of covariates x_j associated with the feature vector y_j, where α contains the unknown parameters in the mixing proportions. The component density function, $\phi(y;\mu,\Sigma)$, denotes the multivariate normal density function with mean vector μ and covariance matrix Σ; see the details of normal mixture models in Chapter 2.

Massive data collection in the fields of medical and health sciences is a fact of twenty-first century science, meaning that we are living in an era of "Big Data", as discussed in Brown, Chui, and Manyika (2011) and Haendel, Chute, and Robinson (2018). For example, due to increasingly information-rich governments and sophisticated data management, data linkage technologies and use of longitudinal designs in national surveys with multiple follow-ups, big data in health sciences research is now commonplace. The extraction of important information about the unknown group structures of the underlying population can contribute to the evidence-base that will provide the impetus for improving the quality, effectiveness and cost effectiveness of disease prevention, treatment and care for a sustainable health system.

As illustrated in Chapter 1, the EM algorithm is a broadly applicable approach to the iterative computation of ML estimates involving mixture models, where the estimation of unknown parameters is formulated as an incomplete-data problem. The EM algorithm has a number of desirable properties including its numerical stability, reliable global convergence, and simplicity of implementation. However, the EM algorithm is not without its limitations. In its basic form, the EM algorithm is sometimes very slow to converge, especially for the analysis of big data. In this chapter, we present developments on an incremental scheme of the EM algorithm for the fitting of normal mixture models to cluster big data. Based on this incremental EM (IEM) algorithm, other variant algorithms to speed up the implementation of the EM algorithm will be discussed. These algorithms not only show an improvement in the rate of convergence, but also preserve the simplicity and stability of the EM algorithm.

For clustering big data, it is not only the sheer size of the data that imposes difficulty in direct application of conventional clustering methods such as mixture models. Big data in medical and health sciences often involve data collected from multiple sources or platforms, possibly on both individual and population levels via data linkage, as illustrated in Ng and McLachlan (2016). Typical data sources in medical research include clinical decision support systems (laboratory, pharmacy, medical imaging, immunological- and genomic-markers expression, and other health administrative data), patient records (demographic, socioeconomic and epidemiological data), hospital emergency department records, medicare, health insurance records, and social media posts; see, for example, Raghupathi and Raghupathi (2014) and Kim, Trimi, and Chung (2014). Big data with complex correlation structures and/or possibly a mix of variable types make them difficult to be analyzed by conventional clustering methods. Moreover, the emergence of big data applications in which observations are acquired in a design setting such that the data may exhibit a hierarchical structure with higher level of units such as hospitals or institutions (Ng and McLachlan, 2007, 2016). The independence assumption about individual observed data y_j $(j = 1,\dots,n)$, where the likelihood function can be expressed as multiplication of individual density functions, may no longer be valid. In this chapter, we present the use of random-effects components for multivariate normal mixture models to cluster big data with a hierarchical structure, and the mixture of multivariate generalized Bernoulli distributions (Ng, 2015) to develop a multitask clustering mechanism for the analysis of big data.

10.2 Speeding up the EM Algorithm for Multivariate Normal Mixtures

Chapter 2 describes the mixture model of (multivariate) normal distributions, which are the most widely used mixture models for model-based cluster analysis of data from various scientific fields; see McLachlan and Peel (2000) and Ng et al. (2006) for more information. As described in Section 2.2, it is computationally advantageous to work in the E-step for mixtures of normal component densities in terms of the sufficient statistics given by

$$\begin{aligned} T_{h1}^{(k)} &= \sum_{j=1}^{n} \tau_h(y_j;\Psi^{(k)}) \\ T_{h2}^{(k)} &= \sum_{j=1}^{n} \tau_h(y_j;\Psi^{(k)})y_j \\ T_{h3}^{(k)} &= \sum_{j=1}^{n} \tau_h(y_j;\Psi^{(k)})y_j y_j^T, \end{aligned} \tag{10.2}$$

where

$$\tau_h(y_j;\Psi^{(k)}) = \frac{\pi_h^{(k)}\phi(y_j;\mu_h^{(k)},\Sigma_h^{(k)})}{\sum_{l=1}^{g}\pi_l^{(k)}\phi(y_j;\mu_l^{(k)},\Sigma_l^{(k)})}$$

is the estimate of the posterior probability that y_j belongs to the hth component, based on the estimate of Ψ after the kth iteration of the EM algorithm $\Psi^{(k)}$ ($h = 1,\ldots,g$; $j = 1,\ldots,n$). In fitting multivariate normal mixtures where the M-step is computationally simple, the rate of convergence of the EM algorithm depends mainly on the computation time of an E-step. From (10.2), it can be seen that each E-step visits each data point y_j on a given iteration. Thus the EM algorithm requires considerable computation time in its application to big data.

On the basis of the sufficient statistics in (10.2), the M-step is simplified as

$$\begin{aligned} \pi_h^{(k+1)} &= T_{h1}^{(k)}/n \\ \mu_h^{(k+1)} &= T_{h2}^{(k)}/T_{h1}^{(k)} \\ \Sigma_h^{(k+1)} &= \{T_{h3}^{(k)} - T_{h1}^{(k)^{-1}} T_{h2}^{(k)} T_{h2}^{(k)^T}\}/T_{h1}^{(k)}, \end{aligned} \tag{10.3}$$

and for homoscedastic normal components, the updated estimate of the common component-covariance matrix Σ is given by

$$\Sigma^{(k+1)} = \sum_{h=1}^{g} T_{h1}^{(k)}\Sigma_h^{(k+1)}/n, \tag{10.4}$$

where $\Sigma_h^{(k+1)}$ is given by (10.3).

An incremental version of the EM algorithm was proposed by Neal and Hinton (1998) to improve the rate of convergence of the EM algorithm. The incremental EM (IEM) algorithm proceeds by dividing the data into B ($B \leq n$) blocks of equal or near-equal size and implementing the "partial" E-step for only a block of data at a time before performing an M-step. A "scan" of the IEM algorithm thus consists of B partial E-steps and B M-steps; see Ng and McLachlan (2003b) for more details. Let $\Psi^{(k+b/B)}$ denote the estimate of Ψ after the bth iteration on the $(k+1)$th scan ($b = 1, \ldots, B$). For the first scan ($k = 0$), a full E-step is performed to avoid the premature component starvation problem. On subsequent scans, the conditional expectations of the sufficient statistics, $T_{hl}^{(k+b/B)}$ ($l = 1,2,3$), are calculated for only a block of data at a time on the E-step. For example, on the $(b+1)$th iteration of the $(k+1)$th scan ($b = 0, \ldots, B$-1), they are obtained for $h = 1, \ldots, g$, using the relationship

$$T_{hl}^{(k+b/B)} = T_{hl}^{(k+(b-1)/B)} - T_{hl,b+1}^{(k-1+b/B)} + T_{hl,b+1}^{(k+b/B)} \tag{10.5}$$

for $l = 1,2,3; b = 0, \ldots, B$-1, where only the last term $T_{hl,b+1}^{(k+b/B)}$ has to be calculated for $l = 1,2,3$ for only a block of y_j that belong to the $(b+1)$th block S_{b+1}. The first and the second terms on the right-hand side of (10.5) are available from the previous iteration and previous scan, respectively, as discussed in Ng and McLachlan (2003b).

Based on the conditional expectations of the sufficient statistics $T_{hl}^{(k+b/B)}$ ($l = 1,2,3$), the estimates of Ψ are updated in the M-step as follows:

$$\begin{aligned}
\pi_h^{(k+(b+1)/B)} &= T_{h1}^{(k+b/B)}/n, \\
\mu_h^{(k+(b+1)/B)} &= T_{h2}^{(k+b/B)}/T_{h1}^{(k+b/B)}, \\
\Sigma_h^{(k+(b+1)/B)} &= \left\{T_{h3}^{(k+b/B)} - T_{h1}^{(k+b/B)^{-1}} T_{h2}^{(k+b/B)} T_{h2}^{(k+b/B)^T}\right\}/T_{h1}^{(k+b/B)}.
\end{aligned} \tag{10.6}$$

The argument for improved rate of convergence is that the IEM algorithm exploits new information more quickly rather than waiting for a complete scan of the data before parameters are updated by an M-step (Neal and Hinton, 1998). Considering the convergence properties of the IEM algorithm, Gunawardana and Byrne (1999) showed that the IEM algorithm will converge to a stationary point in the log-likelihood under slightly stronger conditions than those assumed for obtaining the convergence results for the standard EM algorithm (Wu, 1983). Although the IEM algorithm can possess stable convergence to stationary points in the log-likelihood, the current theoretical results for the IEM algorithm do not quarantine monotonic behaviour of the log-likelihood as the EM algorithm does; see, for example Ng and McLachlan (2003b) and Ng, Krishnan, and McLachlan (2012). However, by reviewing the IEM algorithm as alternating minimization of a joint function in terms of the log-likelihood and the Kullback-Leibler divergence of two distributions over the unobserved variables,

$$D(P, \Psi) = -\log L(\Psi) + KL[P, f(z|y; \Psi)], \tag{10.7}$$

it can be shown that the lower bound of the log-likelihood is monotonic increasing after each iteration, as illustrated in Ng, Krishnan, and McLachlan (2012). In

(10.7), P is any distribution defined over the support of Z, $f(z|y;\Psi)$ is the conditional distribution of Z given the observed data, and the Kullback-Leibler divergence is non-negative. In practice, the IEM algorithm in general converges with fewer scans and hence is faster than the EM algorithm and also increases the likelihood at each scan (McLachlan and Ng, 2009); see the discussion in Ng and McLachlan (2004a).

A sparse version of the IEM algorithm (SPIEM) can be formulated by implementing a "sparse" E-step of Neal and Hinton (1998) within the incremental scheme of the IEM algorithm. In the sparse E-step, small posterior probabilities (such as < 0.005) are held fixed, while only those for the remaining components in the mixture are updated. Thus, the computational time in the sparse E-step is reduced, proportional to the number of components to be updated. Suppose that the sparse E-step is to be implemented on the subsequent B iterations of the $(k+1)$th scan, then on the E-step of the $(b+1)$th iteration for $b = 0,\ldots,B-1$, the posterior probabilities of component membership for all individuals j belong to the $(b+1)$th block S_{b+1}, are updated. Denote A_j be a subset of $\{1,\ldots,g\}$ containing the components which posterior probabilities are close to zero and held fixed. The update is proceeded as

$$\sum_{l \notin A_j} \tau_l(y_j;\Psi^{(k-1+b/B)}) \frac{\tau_h(y_j;\Psi^{(k+b/B)})}{\sum_{l \notin A_j} \tau_l(y_j;\Psi^{(k+b/B)})} \tag{10.8}$$

for those components h which do not belong to A_j; otherwise, leave the posterior probabilities unchanged as $\tau_h(y_j;\Psi^{(k-1+b/B)})$. With the SPIEM algorithm, the M-step can also be performed efficiently by updating only the contribution to the sufficient statistics for those components $h \notin A_j$. For example,

$$T_{h1,b+1}^{(k+b/B)} = \sum_{j \in S_{b+1}} I_{A_j}(h)\tau_h(y_j;\Psi^{(k-1+b/B)}) + \sum_{j \in S_{b+1}} I_{A_j^c}(h)\tau_h(y_j;\Psi^{(k+b/B)}), \tag{10.9}$$

where A_j^c is the complement of A_j and $I_{A_j}(h)$ is the indicator function for the set A_j. The first term on the right-hand side of (10.9) is calculated at the $(b+1)$th iteration of the kth scan and can be saved for use in the subsequent SPIEM iterations on the $(k+1)$th scan. Similar arguments apply to $T_{h2,b+1}^{(k+b/B)}$ and $T_{h3,b+1}^{(k+b/B)}$. The scheme proposed by Ng and McLachlan (2003b) suggest running a standard EM iteration for the first scan and then performing the IEM for five scans before running the sparse version SPIEM. After running the sparse version for five scans, the IEM is performed for a scan, and a new set A_j $(j = 1,\ldots,n)$ is selected. Ng and McLachlan (2003b) studied the relative performances of these algorithms with various numbers of blocks B for the fitting of multivariate normal mixtures. They proposed to choose B to be that factor of n that is the closest to $B^* = \text{round}(n^{2/5})$ for unrestricted component-covariance matrices and $B^* = \text{round}(n^{1/3})$ for diagonal component-covariance matrices, where the "round" function obtains the nearest integer of the argument of the function. With this SPIEM scheme, the likelihood was found to be increased after each scan (Ng and McLachlan, 2003b).

To further speed up the IEM and SPIEM algorithms for the fitting of multivariate normal mixtures, Ng and McLachlan (2004a) proposed to impose a multiresolution kd-tree (*mrk*d-tree) structure in performing the partial E-step of the IEM algorithm or the sparse and partial E-step of the SPIEM algorithm. Here kd stands for k-dimensional where, in our notation, $k = p$, the dimension of an observation y_j. The *mrk*d-tree is a binary tree that recursively splits the whole set of observations into partition; see Moore (1999) for more details. It is constructed top-down, starting from the root node containing all the observations and the splitting procedure continues until the range of observations in the widest dimension of a descendant node is smaller than some threshold γ. This node is then declared to be a leaf node and is left unsplit. The contribution of all the observations in a leaf node to the sufficient statistics is simplified by calculating at the mean of these data points to save time. Let n_L be the total number of leaf nodes and n_m be the number of observations in the leaf node LN_m $(m = 1,\ldots,n_L)$. It is noted that once the *mrk*d-tree is constructed, the number of leaf nodes n_L is unchanged at each scan. The *mrk*d-tree IEM and SPIEM algorithms can be performed by dividing the leaf nodes into B blocks. At each scan, the partial E-step is implemented for only a block of leaf nodes at a time before the next M-step is performed. The conditional expectations of the sufficient statistics for the *mrk*d-tree approach in the sparse step are approximated as

$$T_{hl,b+1}^{(k+b/B)} \approx \sum_{m \in S_{b+1}} T_{hl,m}^{(k+b/B)} \qquad (h = 1,\ldots,g;\ l = 1,2,3), \tag{10.10}$$

where

$$\begin{aligned} T_{h1,m}^{(k+b/B)} &= \tau_h(\bar{y}_m;\Psi^{(k+b/B)})n_m, \\ T_{h2,m}^{(k+b/B)} &= \tau_h(\bar{y}_m;\Psi^{(k+b/B)})n_m\bar{y}_m, \\ T_{h3,m}^{(k+b/B)} &= \tau_h(\bar{y}_m;\Psi^{(k+b/B)}) \sum_{j \in S_{b+1}} y_j y_j^T \end{aligned} \tag{10.11}$$

for $h = 1,\ldots,g$ and S_{b+1} now denote a subset of $\{1,\ldots,n_L\}$ containing the subscripts of those leaf nodes LN_m that belong to the $(b+1)$th block. In (10.11), $\bar{y}_m$ is the mean of the n_m observations belonging to the leaf node LN_m. With the *mrk*d-tree SPIEM algorithm, the sparse E-step (10.9) is now replaced by

$$\begin{aligned} T_{h1,b+1}^{(k+b/B)} &= \sum_{m \in S_{b+1}} I_{A_m}(h)\tau_h(\bar{y}_m;\Psi^{(k-1+b/B)})n_m \\ &\quad + \sum_{m \in S_{b+1}} I_{A_m^c}(h)\tau_h(\bar{y}_m;\Psi^{(k+b/B)})n_m \end{aligned} \tag{10.12}$$

for those LN_m $(m = 1,\ldots,n_L)$ in the $(b+1)$th block S_{b+1}, where $I_{A_m}(h)$ is the indicator function for the set A_m, which contains the components that are held fixed for the leaf node LN_m (Ng and McLachlan, 2004a).

Because of the approximation in (10.10), the *mrk*d-tree approach does not guarantee the desirable reliable convergence properties of the EM algorithm. The leaf nodes have to be very small (i.e., small γ) in order that (10.10) is applicable for approximating the conditional expectations of the sufficient statistics. It has been illustrated empirically that the *mrk*d-tree IEM and SPIEM algorithms showed a monotonic convergence as reliable as the EM algorithm when the sizes of leaf nodes are sufficiently small such as $\gamma < 0.7\%$ of the range in the splitting dimension of the whole data set; see Ng and McLachlan (2004a). Another version of the *mrk*d-tree approach involves "pruning" the tree-nodes by obtaining the upper and lower bounds on the probability density function at each tree node with the use of a new analytical geometry approach that has been proposed by Ng and McLachlan (2004a). With this pruning process, the division of nodes in the partial E-step is performed at some level of the *mrk*d-tree rather than at the leaf nodes. This process improves the rate of convergence of the *mrk*d-tree IEM and SPIEM algorithms when γ is very small (such as $\gamma = 0.3\%$); however, the empirical monotonic convergence of the *mrk*d-tree IEM and SPIEM algorithms is no longer achieved.

As described in Chapter 1, the EM algorithm has to be implemented from a variety of starting values in an attempt to locate all local maxima; see also Ng (2013). In cluster analysis of big data, the computational burden for each starting value is large and hence the scheme of random selection of starting values is not always feasible. The Generalized Learning Vector Quantization (GLVQ) of Pal, Bezdek, and Tsao (1993) has been adopted to obtain starting values for the EM algorithm for automatic segmentation of large imaging data, as illustrated in Ng and Lam (2013). The GLVQ was explicitly designed as a clustering algorithm for segmentation that updates all nodes for given input centroids, where the GLVQ-generated centroids are relatively invariant to the number of iterations, learning coefficient, and the choice of initial centroids in the GLVQ procedure. A simulation study has been conducted in the context of image segmentation to show that the likelihood value obtained using the GLVQ starting values is very close to the largest of the likelihood values obtained from fifty random starting values (Ng and Lam, 2013). A potential drawback is that the GLVQ procedure may converge to a k-means solution or numerical instability of the algorithm may occur when the number of components g or the dimension ranges of the data are large, as discussed in Gonzalez, Graña, and D'Anjou (1995). This situation corresponds to the initialization of GLVQ with centroids outside of the convex hull of the sample space. In the context of image segmentation, the following scheme for choosing the initial centroids may be used to avoid the degradation and instability of GLVQ. The initialization procedure starts with scaling the image intensities to the range of [0.0, 20.0] and then divides the intensity values uniformly into g intervals for each dimension of the data. Finally, the image intensities of the initial centroids are taken to be the midpoints of these intervals. It has been shown in Ng and Lam (2013) that this initialization procedure preserves the stability and the invariant advantage of GLVQ.

10.2.1 Example 10.1: Segmentation of Magnetic Resonance Images of the Human Brain

We consider the segmentation of three-dimensional (3D) magnetic resonance (MR) images of the human brain, where a continuous MR image is partitioned into a set of disjoint voxels labelled 1 to n. Typical MR imaging permits the acquisition of images with three tissue-dependent parameters: the spin-lattice relaxation time T_1-weighted; the spin-spin relaxation time T_2-weighted; and the proton density ρ_D-weighted MR images. Let y_j denote the three-dimensional feature vector containing the values of the variables T_1, T_2, and ρ_D for the jth voxel $(j = 1, \ldots, n)$. Within the context of image segmentation, the problem is to infer the unknown vector of component indicator z from the observed data as well as the tissue-specific intensity values for T_1, T_2, and ρ_D parameters. Figure 10.1 displays the T_1-, T_2-, and ρ_D-weighted images of a slice of a 3D MR image of the human brain acquired by a two-Tesla Bruker Medspac whole body scanner. It can be seen that the contrast of the characteristics of a specific tissue is enhanced in various types of MR images in Figure 10.1. For instance, the contrast between brain parenchyma and cerebral spinal flow (CSF) is emphasized in the T_2-weighted image, while the contrast between white and gray matters is enhanced in the ρ_D-weighted image. The consideration of multispectral characteristics of the MR images in the cluster analysis suggests that any tissue type, which has distinct image intensity distribution within any of the three images, can be distinguishable. Tissue-segmentation of MR images of the human brain has a large potential to facilitate an imaging-based medical diagnosis, providing an aid to surgery and treatment planning (Freund et al., 2001). They can also be used to study the effect of the locality of abnormal tissues in neurologic disease (Wolf et al., 2001) as well as the human brain activation effects to stimuli; see, for example, Jancke et al. (2001) and Ng and McLachlan (2004a).

With a mixture model-based approach to image segmentation, the first step is to fit a multivariate normal mixture model to voxel intensities, y_j $(j = 1, \ldots, n)$, by ignoring the spatial correlation among neighbouring voxels. This "noncontextual" process provides soft (fuzzy) classification of tissue type in terms of the posterior probability τ_{hj} for each voxel and the estimates of unknown parameter Ψ. By assuming y_j is independent and identically distributed, the estimation of Ψ corresponds to the maximum likelihood estimation from incomplete data via the EM algorithm (Ng and McLachlan, 2004a). Thus, the process can be speeded up by the variants of the EM algorithm described in Section 10.2. The second step of the image segmentation adopts a "contextual" process, where the spatial correlation in image intensity between voxels and their neighbours is captured by a Markov random field prior in which

$$\log \pi_{hj}^{(k+1)} \propto \xi(\sum_{\delta j} \tau_{h\delta}^{(k)} + 1/\sqrt{2}\sum_{\delta j} \tau_{h\delta}^{(k)}) + 1/\sqrt{3}\sum_{\delta j} \tau_{h\delta}^{(k)}), \tag{10.13}$$

where

$$\pi_{hj}^{(k+1)} = \text{pr}\{Z_{hj} = 1 |\, z_{\delta j} = \hat{z}_{\delta j}^{(k)} = \tau_{\delta j}^{(k)}\} \tag{10.14}$$

is the probability that the jth voxel belongs to the hth tissue type given the component-membership of its specified neighbours δj as implied by $\tau_{\delta j}^{(k)}$ on the kth iteration. The prior parameter ξ in (10.13) governs the influence of neighbouring voxels in forming the estimates of the prior probabilities of component-membership of each voxel. The use of $\tau_{\delta j}$ for $\hat{z}_{\delta j}$ in (10.14) avoids the discretization in counting the neighbours of the jth voxel; see Ng and McLachlan (2004a) and Ng and Lam (2013) for more information. This soft decision approach is more effective in estimation of the tissue parameters than a hard decision approach. On the right-hand side

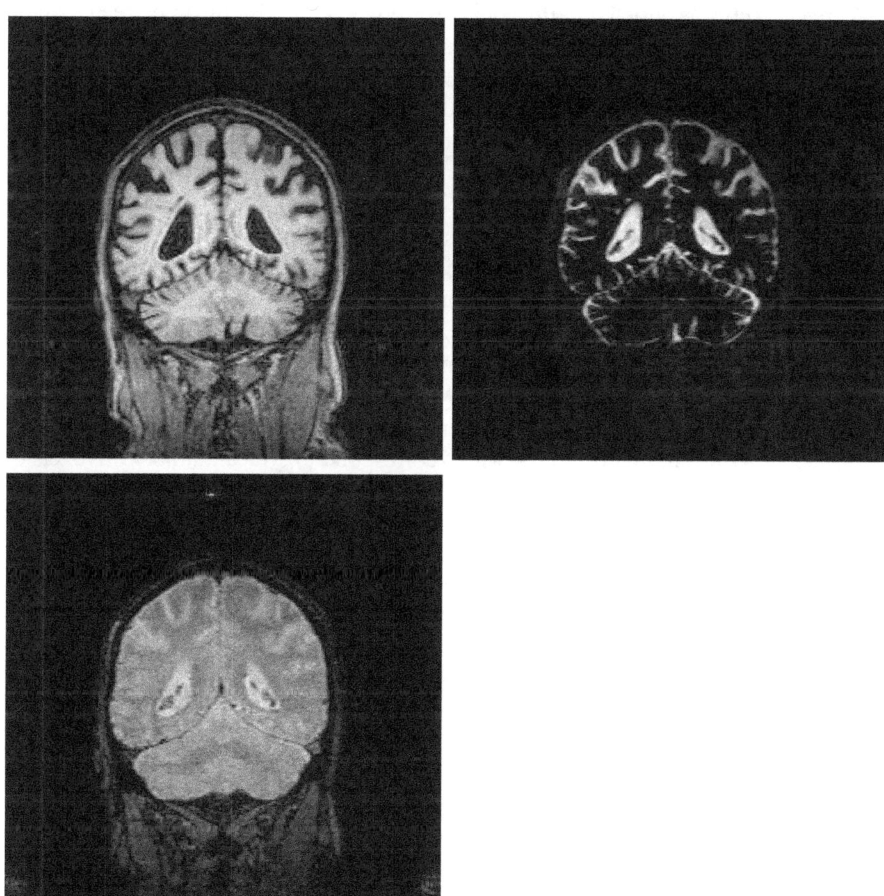

FIGURE 10.1
Real MR data set: T_1-weighted image (top left); T_2-weighted image (top right); ρ_D-weighted image (bottom). Adapted from Ng and McLachlan (2004a).

of (10.13), the summations are over the prescribed first-, second-, and third-order neighbours, respectively, of the jth voxel. Smaller coefficients $1/\sqrt{2}$ and $1/\sqrt{3}$ are adopted to reflect the spatial relatedness between a central pixel and its second- and third-order neighbouring voxels, respectively.

Figure 10.2 displays a slice of the final segmented images of the three main tissue types, white matter, gray matter, and CSF using various EM-based algorithms, where $\xi = 0.6$ was adopted in the contextual process (10.13). It can be seen that the details of the three main tissue types are all well classified by the algorithms. As expected, the SPIEM *mrk*d-tree algorithm with a smaller γ (0.3% versus 0.7%) obtains a segmentation result close to that using the standard EM algorithm.

The following presents a simulation study that evaluates the relative performance of different variants of the EM algorithm for speeding up the noncontextual segmentation of big MR image data (Ng and McLachlan, 2004a). In the simulation study, a random sample of size n observations was generated from a seven-component trivariate normal mixture ($g = 7,\ p = 3$), corresponding to seven tissue types (outer table of the skull and skin; inner table of the skull; temporalis muscle and internal occipital protuberance; cerebral spinal flow space; gray matter; subcutaneous fat and diploic space; white matter) in the segmentation of a 2D MR image of the human brain; see Liang, MacFall, and Harrington (1994). The parameter values were provided in Table II of their paper. Two different sample sizes of $n = 128^3$ and $n = 256^3$ were considered in the simulation study, which correspond to typical number of voxels of a 3D MR image. All the algorithms used the same initial GLVQ estimates as starting values and were terminated when the absolute values of the relative changes in the estimates of the means μ_h all fell below 0.0001; see Ng and McLachlan (2004a) for more information. Table 10.1 reports the results of the simulation study for $n = 128^3$ and $n = 256^3$. Key conclusions are: (a) both the IEM and SPIEM algorithms speed up the EM algorithm and have reliable convergence as that of the standard EM algorithm. The speed-up factor, however, remains similar when the sample size of the data increases; (b) the *mrk*d-tree SPIEM algorithm improves the rate of convergence of the EM algorithm. The speed-up factor increases when the sample size increases, as the number of leaf nodes n_L does not linearly increase with the sample size. The algorithm, however, is inexact for the approximation in calculating the conditional expectations of sufficient statistics in (10.10), but it has shown reliable monotonic convergence when the leaf nodes are of a small size (such as $\gamma \leq 0.7\%$); and (c) with this simulation study, some variants of the EM algorithm provided slightly better estimates (and likelihood values) compared to the standard EM algorithm.

10.2.2 Example 10.2: Segmentation of Molecular Pathology Images of Cancer Patients

The segmentation of molecular pathology images into different tissue components is important for the assessment of clinical behaviour of disease conditions. Estimation of tissue parameters helps to quantify the size of various tissue components and can be used to assess progression of disease or to evaluate the effect of drug therapy in order to improve diagnostic and prognostic capabilities; see, for example,

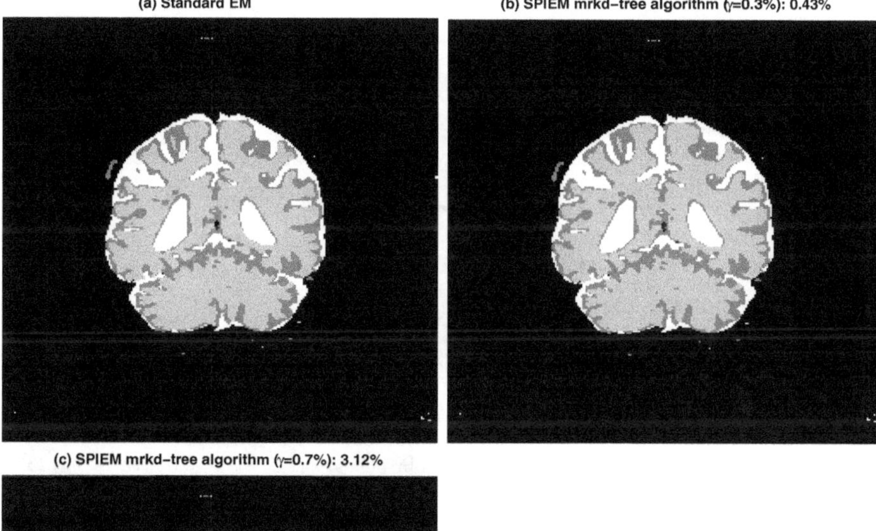

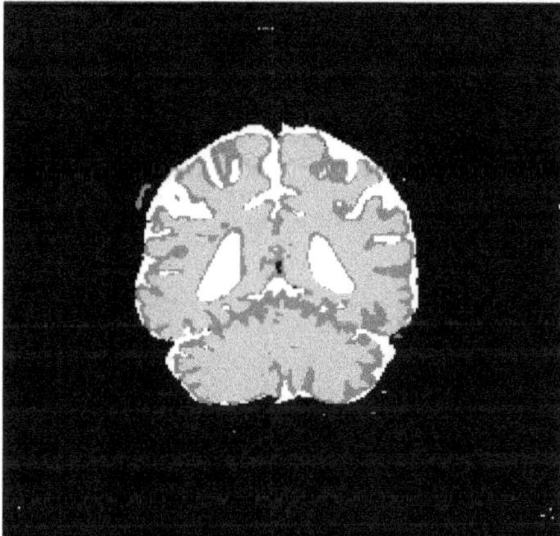

FIGURE 10.2
Final segmented images using various algorithms: (a) Standard EM algorithm (top left); (b) SPIEM *mrk*d-tree algorithm with $\gamma = 0.3\%$ (top right); (c) SPIEM *mrk*d-tree algorithm with $\gamma = 0.7\%$ (bottom). The percentages of voxels allocated to different tissue types compared to the segmentation obtained by the standard EM algorithm are given. Adapted from Ng and McLachlan (2004a).

TABLE 10.1
Speeding up the EM algorithm (Summary of the simulation study). Adapted from Ng and McLachlan (2004a).

Algorithm[a]	**nscan**[b]	**Log-likelihood**	**Speed-up**[c]	**RMSE**[d] **in** π_h	**RMSE in** μ_h
(a) Simulation 1 ($n = 128^3$)					
Standard EM	95	-11,749,600	1.0	0.00069	0.01458
IEM (*B*=256)	54	-11,749,599	1.5	0.00062	0.01313
SPIEM (*B*=256)	60	-11,749,599	2.4	0.00063	0.01320
SPIEM *mrk*d-tree					
$\gamma = 0.3\%$	61	-11,749,600	7.5	0.00078	0.01235
$\gamma = 0.7\%$	61	-11,749,649	23.5	0.00169	0.01228
(b) Simulation 2 ($n = 256^3$)					
Standard EM	95	-94,015,922	1.0	0.00026	0.00401
IEM (*B*=1,024)	54	-94,015,918	1.5	0.00016	0.00234
SPIEM (*B*=1,024)	62	-94,015,918	2.5	0.00017	0.00258
SPIEM *mrk*d-tree					
$\gamma = 0.3\%$	61	-94,015,937	20.3	0.00031	0.00580
$\gamma = 0.7\%$	62	-94,016,387	52.4	0.00140	0.01298

[a] All algorithms have the log likelihood value at each scan monotonic increasing.
[b] nscan indicates the number of scans to convergence.
[c] Speed-up is the ratio of the CPU time compared to that of the standard EM algorithm (for *mrk*d-tree-based algorithms, the CPU time includes the time to construct the *mrk*d-tree). The algorithms were written in FORTRAN and the simulations were all run on a Sun unix workstation.
[d] RMSE presents the root mean square error of estimates in the mixing proportions π_h and means μ_h relative to the true parameter values.

Kong et al. (2011) and Madabhushi et al. (2011). Digital microscopy or technology to transform microscopic glass slides into digital images is becoming increasingly popular in pathology, allowing the counting and separation of different components on a histological section to adopt a more objective and less labour-intensive way (Ng and Lam, 2013). Automatic segmentation of digital pathology images thus has a large potential to facilitate the assessment of clinical behaviour of many diseases including cancer, by providing a new tool to quantify and observe the molecular variation within tissue samples, as discussed in Cooper et al. (2012) and Mulrane et al. (2008).

In the segmentation of 2D digital pathology images, it is assumed that y_j represents a p-dimensional vector of image intensities for the jth pixel $(j = 1, \dots, n)$ and that each pixel has a true colour lying in a prescribed set. In this example, the colour is unordered, representing only the category of the pixel; see Ng and Lam (2013). Here we consider a 2D pathology image obtained from a four-micrometer formalin fixed paraffin embedded section of a patient with adenocarcinoma of the transverse colon. The section was stained with Hematoxylin and Eosin (H&E) for the analysis of morphological features. The stained section was then scanned at 40x magnification using an Aperio ScanScope CS slide scanner (Aperio Technologies, Vista, CA, USA) to capture the digital image of tissue microarrays. In this digital pathology image, the acquisition matrix was 580×580. The standard EM, IEM, SPIEM, and the SPIEM *mrk*d-tree algorithms were used separately to obtain noncontextual segmentation of the digital pathology image, using the same initial GLVQ estimates as starting values. Based on the BIC model selection method, the number of components was taken to be seven $(g = 7)$. The noncontextual segmentation process was completed when the absolute values of the relative changes in the estimates of the means μ_h all fell below 0.0001. Contextual segmentation was then carried out with the prior parameter $\xi = 0.6$. Figure 10.3 displays the contextual segmentation of the 2D pathology image on the basis of estimates obtained from various EM-based algorithms. With this high spatial resolution image, the details of the carcinomatous colon tissues including the main features such as lesion, muscle, and nuclei were well identified. Comparison between variants of the EM algorithm is presented in Table 10.2. Most comparison conclusions remain regarding the relative performance of these EM-based algorithms given in the simulation study for segmenting MR images (Example 10.1). A key deviation is that the number of scans in the IEM and the SPIEM-based algorithms in this real data set was not much smaller than that of the EM algorithm, implying a limited gain in speeding up the EM algorithm. For the *mrk*d-tree SPIEM algorithm, reliable monotonic convergence is obtained only when the size of leaf nodes is small (such as $\gamma = 0.3\%$). As shown in the simulation study in Section 10.2.1, some variants of the EM algorithm gave slightly better likelihood values compared to the standard EM algorithm. While the noncontextual results were somewhat different between these algorithms, the final contextual segmentation of this 2D pathology image after adjusting for spatial correlation was similar, in terms of the percentages of pixels allocated to different tissue types indicated in Figure 10.3.

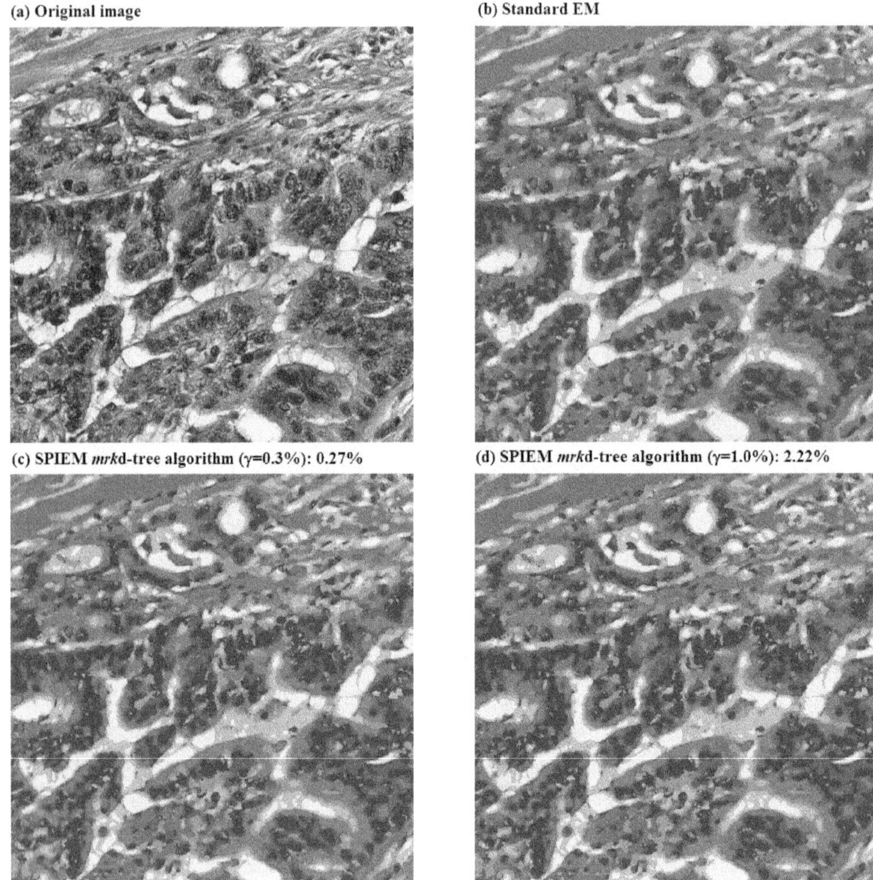

FIGURE 10.3
Segmentation of Pathology Image (Example 10.2): (a) Original image; and contextual segmentation using estimates from (b) Standard EM algorithm; (c) SPIEM *mrk*d-tree algorithm with $\gamma = 0.3\%$; (d) SPIEM *mrk*d-tree algorithm with $\gamma = 1.0\%$. The percentages of pixels allocated to different tissue types compared to the segmentation obtained by the standard EM algorithm are provided. Adapted from Ng and Lam (2013).

TABLE 10.2
Speeding up the EM algorithm (Segmentation of a pathology image in Example 10.2).

Algorithm[a]	**nscan**[b]	**Log-likelihood**	**Speed-up**[c]	**RMSD**[d] **in** π_h	**RMSD in** μ_h
Standard EM	291	-1,951,116	1.0	Reference	Reference
IEM (B=145)	259	-1,951,113	1.03	0.00400	0.03939
SPIEM (B=145)	286	-1,951,114	1.60	0.00313	0.03078
SPIEM *mrk*d-tree					
$\gamma = 0.3\%$	280	-1,951,114	3.37	0.00290	0.02838
$\gamma = 1.0\%$	364	-1,951,135	6.37	0.01751	0.18335

[a] All algorithms (except SPIEM *mrk*d-tree with $\gamma = 1.0\%$) have the log-likelihood value at each scan monotonic increasing.
[b] nscan indicates the number of scans to convergence.
[c] Speed-up is the ratio of the CPU time compared to that of the standard EM algorithm (for *mrk*d-tree-based algorithms, the CPU time includes the time to construct the *mrk*d-tree). The algorithms were written in FORTRAN and the computations were all run on a High Performance Computing Cluster with Intel Xeon CPU X5650 processor at 2.67GHz.
[d] RMSD presents the root mean square deviation of estimates in the mixing proportions π_h and means μ_h relative to those obtained by the standard EM algorithm.

10.3 Mixtures of Linear Mixed Models for Clustering Big Data with a Hierarchical Structure

For many applied problems in medical and health science research, modern study designs often involve the acquisition of observations of outcome measures from various sampling units such as hospitals, schools, or any natural hierarchical units in clustered sampling. The observed data collected could exhibit a multilevel structure. Let m denote the number of higher-level units and there are n_i individuals in the ith unit $(i = 1,\dots,m)$. The total number of individuals is thus $n = \sum_{i=1}^{m} n_i$. Denote the dimension of each individual feature vector by p and that these p features are obtained from q different data sources as described in Section 10.1. Thus each feature vector is presented by $y_{ij} = (y_{1ij},\dots,y_{pij})^T$ for the jth individual in the ith unit $(i = 1,\dots,m;\ j = 1,\dots,n_i)$, where $p = \sum_{l=1}^{q} q_l$ and q_l is the number of features in the lth data source $(l = 1,\dots,q)$. It is assumed that a vector of risk factors is associated with Y_{ij}. These risk factors, denoted by x_{ij}, are incorporated into the normal mixture model via a logistic function on the mixing proportions $\pi_h(x_{ij};\alpha)$ in (10.1):

$$\pi_h(x_{ij};\alpha) = \frac{\exp(v_h^T x_{ij})}{1 + \sum_{l=1}^{g-1} \exp(v_l^T x_{ij})} \tag{10.15}$$

for $h = 1,\dots,g-1$ and $\pi_g(x_{ij}) = 1 - \sum_{l=1}^{g-1} \pi_l(x_{ij})$, where the first element of x_{ij} is assumed to be one to account for an intercept term. In (10.15), α contains the elements in v_h, which are fixed effects (unknown constants) representing the log odds ratio of cluster membership for each risk factor in x_{ij}; see, for example, Ng and McLachlan (2016).

Mixtures of linear mixed models (LMMs) with two sets of random effects described in Chapter 8 can be extended to capture correlation within individuals across the q data sources and between individuals from the same hierarchical unit, as illustrated in Ng and McLachlan (2014b). Precisely, the two sets of random effects are linked to the mean of y_{ij} as

$$E(y_{ij}|b_{hij}, c_{hi}) = W\beta_h + Ub_{hij} + Vc_{hi}, \quad (10.16)$$

given that y_{ij} belongs to the hth cluster $(h = 1, \ldots, g)$, where β_h is a p-dimensional vector of fixed effects modelling the conditional mean of y_{ij} in the hth component $(h = 1, \ldots, g)$ and W, U, and V denote the known design matrices corresponding to the fixed and two sets of random effects, respectively. In (10.16), the individual-specific random-effects vector $b_{hij} = (b_{1hij}, \ldots, b_{qhij})^T$ captures individual variation in observations that vary around the mean vector within a unit across the q data sources. The vector $c_{hi} = (c_{1hi}, \ldots, c_{mhi})^T$ contains the unobservable unit-specific random effects, capturing the effects imposed by the m hierarchical units and shared among observations collected from the same unit. The measurement error vector in (10.16) is taken to be multivariate normal $N_p(0, A_h)$, where $A_h = \theta_{ah}I$ is a diagonal matrix $(h = 1, \ldots, g)$. Similar to the mixture of LMMs described in Chapter 8, random effects b_{hij} and c_{hi} are, respectively, taken to be multivariate normal $N_q(0, B_h)$ and $N_m(0, C_h)$, where the variance components $B_h = \theta_{bh}I$ and $C_h = \theta_{ch}I$ are assumed to be diagonal $(h = 1, \ldots, g)$. Maximum likelihood estimation of unknown parameters Ψ can be undertaken by proceeding conditionally on the unit-specific random effects c_{hi} within the framework of the EM algorithm; see Chapter 8 and the references therein.

10.3.1 Clustering of Multilevel Data from Multiple Sources

We report here a simulation study based on a multicentre data structure, where there were $m = 20$ clinical centres and within each centre there were 1,000 participants. In this simulation study, $p = 9$ continuous outcome measures collected from $q = 3$ data sources were considered (a low-dimensional setting), where q_l were 2, 3, and 4 for $l = 1, 2, 3$, respectively, and hence $\sum_{l=1}^{q} q_l = p = 9$. We took $g = 2$ clusters, with conditional means β_1 and β_2, respectively, and assumed a common θ_a for θ_{a1} and θ_{a2}. The measurement error vectors were generated from $N(0, \theta_a I_9)$, whereas the random effects b_{hij} and c_{hi} were generated from $N(0, \theta_{bh} I_3)$ and $N(0, \theta_{ch} I_9)$, respectively, for $h = 1, 2$.

The vector of risk factors x_{ij} was taken to be a three-dimensional vector with the first element equal to one and the last two elements generated independently from $N(0, I_2)$. Realizations of cluster membership were then simulated in which an individual has a probability of $\pi_1(x_{ij}; \alpha)$ to be in the first cluster G_1 and has a probability of $(1 - \pi_1(x_{ij}; \alpha))$ to be in the second cluster G_2. Here $\alpha = ((v_1)_1, (v_1)_2, (v_1)_3)^T$ containing the elements of v_1; see (10.15). Simulated data with 100 replications were generated.

The mixture of LMMs (10.16) was fitted to the simulated multilevel data with $g = 2$ clusters and a common θ_a for both clusters. Table 10.3 presents the results of the simulation study along with the true parameter values. The performances of the estimators are assessed in terms of their biases and sample standard errors of the estimates over the 100 replications. It can be seen from Table 10.3 that, with $n = 20 \times 1{,}000 = 20{,}000$, no appreciable bias is observed in the estimation of all the parameters in mixing proportions, conditional means, and random-effects terms. The result for a small sample situation ($n = 20 \times 50 = 1{,}000$) was included in Table 10.3 for comparison. It was observed that the average biases and sample standard errors generally became larger when the sample size was smaller, especially for the parameter β_h and the variance component θ_{bh} for the unobservable individual-specific random effects ($h = 1,2$); see Ng and McLachlan (2016) for more simulation results about the applicability of the mixture of LMMs in small sample situations. In the analysis of big data, the number of data sources q and/or the number of features in each source q_l ($l = 1,\ldots,q$) may be larger. It is anticipated that dimensionality reduction methods are required to reduce the dimensionality of the data. The simulation study presented here assumes that only key features are considered for cluster analysis of multilevel data from multiple sources.

10.4 Mixtures of Multivariate Generalized Bernoulli Distributions for Consensus Clustering

Many real problems in cluster analysis of big data involve multivariate data with a mix of variable types as a result of experimental designs in which measurements from different data sources are collected. For example in medical and health sciences, feature variables are measurements on individual patients from complementary data sources such as pathology, radiology images, immunological- and genomic-markers expressions, epidemiological and clinical data, medicare and healthcare service use data via data linkage. Mixtures of LMMs presented in Chapter 8 and the previous section are useful for clustering continuous multivariate data. For multisource data with a mix of variable types, a consensus clustering mechanism (Topchy, Jain, and Punch, 2005) can be adopted where in the first stage mixture models with different distributional families are used to cluster patients, separately, on the basis of a subset of feature variables collected from a single data source. In the second stage, cluster results obtained from each data source are then aggregated to obtain a final clustering of patients using the mixture of multivariate generalized Bernoulli distributions.

Let p denote the number of data sources. The data matrix in the second stage of a consensus clustering process is an $n \times p$ matrix, where each column represents the partition of patients obtained from each of the p data sources in the first stage. Let $y_j = (y_{1j},\ldots,y_{pj})^T$ contain the p categorical indicator variables of cluster membership for the jth patient ($j = 1,\ldots,n$), where y_{lj} is the indicator variable for the partition corresponding to the lth data source, taking on d_l distinct cluster labels

TABLE 10.3
Estimated biases and sample standard errors of estimators for the mixture of LMMs (Multilevel data from multiple sources). Adapted from Ng and McLachlan (2016).

Parameter	True value	Average bias (Sample standard error)	
		Set 1 ($n = 20 \times 1000$)	Set 2 ($n = 20 \times 50$)
$(v_1)_1$	0.25	-0.0026 (0.014)	0.0060 (0.067)
$(v_1)_2$	-0.5	0.0042 (0.015)	-0.0002 (0.067)
$(v_1)_3$	-0.25	0.0024 (0.015)	-0.0076 (0.066)
$(\beta_1)_1$	10.0	-0.0016 (0.007)	0.0196 (0.091)
$(\beta_1)_2$	8.0	-0.0001 (0.008)	0.0026 (0.092)
$(\beta_1)_3$	4.0	-0.0002 (0.007)	0.0074 (0.093)
$(\beta_1)_4$	12.0	0.0004 (0.007)	-0.0118 (0.095)
$(\beta_1)_5$	-10.5	0.0006 (0.007)	-0.0040 (0.083)
$(\beta_1)_6$	9.5	0.0002 (0.008)	-0.0002 (0.083)
$(\beta_1)_7$	-10.5	0.0009 (0.007)	-0.0032 (0.095)
$(\beta_1)_8$	8.0	0.0000 (0.008)	-0.0171 (0.095)
$(\beta_1)_9$	-6.5	-0.0004 (0.008)	-0.0088 (0.084)
$(\beta_2)_1$	15.0	-0.0002 (0.008)	-0.0027 (0.120)
$(\beta_2)_2$	15.0	-0.0013 (0.010)	-0.0046 (0.109)
$(\beta_2)_3$	9.0	-0.0001 (0.008)	-0.0033 (0.112)
$(\beta_2)_4$	5.0	0.0004 (0.009)	0.0020 (0.110)
$(\beta_2)_5$	-4.5	0.0005 (0.009)	0.0201 (0.117)
$(\beta_2)_6$	5.0	-0.0004 (0.009)	0.0117 (0.094)
$(\beta_2)_7$	5.0	0.0006 (0.008)	-0.0059 (0.126)
$(\beta_2)_8$	12.5	-0.0008 (0.009)	-0.0071 (0.103)
$(\beta_2)_9$	-9.0	0.0007 (0.008)	-0.0181 (0.113)
θ_a	1.0	0.0046 (0.007)	0.0098 (0.019)
θ_{b1}	1.0	0.0145 (0.116)	0.0997 (0.151)
θ_{b2}	1.0	0.0117 (0.102)	0.1317 (0.171)
θ_{c1}	1.0	-0.0001 (0.010)	-0.0081 (0.046)
θ_{c2}	1.0	-0.0001 (0.012)	-0.0007 (0.062)

$(l = 1,\ldots,p)$. The clustering of the y_j $(j = 1,\ldots,n)$ is implemented using a mixture model, with component-density $f_h(y_j;\theta_h)$ given by a multivariate generalized Bernoulli distribution consisting of one draw on d_l cluster labels with probabilities $\theta_{hl1},\ldots,\theta_{hld_l}$ for each cluster grouping $l = 1,\ldots,p$, and where $\theta_{hld_l} = 1 - \sum_{r=1}^{d_l-1}\theta_{hlr}$; see Section 4.6 for more details. Assuming that the categorical variables $y_{1j},\ldots,y_{pj}$ are independent of each other, the mixture of multivariate generalized Bernoulli distributions is given by

$$f(y_j;\Psi) = \sum_{h=1}^{g}\pi_h\prod_{l=1}^{p}\prod_{r=1}^{d_l}\theta_{hlr}^{I(y_{lj},r)} \qquad (j = 1,\ldots,n), \tag{10.17}$$

where $\pi_1,\ldots,\pi_g$ are the mixing proportions that sum to one and $I(y_{lj},r)$ is an indicator function which is equal to one if $y_{lj} = r$ and is zero otherwise $(r = 1,\ldots,d_l)$. With (10.17), the vector of unknown parameters Ψ contains the mixing proportions $\pi_1,\ldots,\pi_{g-1}$, and parameters $\theta_1,\ldots,\theta_g$ for component densities. The unknown parameter vector Ψ can be estimated by the ML method via the EM algorithm, where the E and M steps are in closed form; see Section 4.6 for more information on the E- and M-steps as well as the initialization procedure.

As described in Section 4.6, mixtures of multivariate generalized Bernoulli distributions are identifiable under the condition that $p \geq 2\log_d g + 1$, where d is the minimum value of d_l $(l = 1,\ldots,p)$. That is, the number of data sources p is sufficiently large, which is usually the case for cluster analysis of big data. Another condition is $g\sum_{l=1}^{p}(d_l - 1) + g - 1 \leq S - 1$, where S is the number of distinct observed patterns of y_j, which has an upper bound of $\prod_{l=1}^{p} d_l$ corresponding to the dimension of the distribution space. This condition corresponds to the situation where the dimension of the parameter space for Ψ is smaller than that of the observed sample space.

10.4.1 Consensus Clustering of Data from Multiple Sources

For illustrative purposes, we consider again the National Health and Nutrition Examination Survey (NHANES) 2013-2014 data sets described in Chapter 3, where the Public Use Files (PUFs) "SMQ_H.XPT" (updated version of September 2016), "HUQ_H.XPT" (released October 2015), and "DR1TOT_H.XPT" (released September 2016) were retrieved to provide respondent's cigarette-smoking use information, hospital utilization data, and record of daily nutrient intakes from foods and beverages, respectively; see Ng and McLachlan (2016) for more details. In this example, the mixture model (10.17) was applied to aggregate partitions of respondents based on seven health and nutrient intake variables from these three PUFs and obtain a final consensus clustering of respondents. It is noted that some of these seven variables are categorical, thus providing a direct way to allocate each respondent into clusters on the basis of the category in which the respondent belongs to. Specifically, partition of respondents has been obtained separately using: (i) smoking status (A: non-smoker; B: previous smoker; C: current smoker); (ii) number of general practitioner (GP) visits over past year (A: 0-4; B: 5 or more, obtained by a two-component mixture

TABLE 10.4
Model selection for the mixture of multivariate generalized Bernoulli distributions (NHANES data).

No. of components (g)	Log-likelihood	AIC	BIC
1	-26052.5	52127	52199
2	-24118.3	48279	48417
3	-23913.2	47890	48101
4	-23738.7	47557	47821
5	-23667.3	47433	47755*
6	-23638.4	47399*	47800

* Number of components selected based on the criterion.

of gamma distributions (see Chapter 3)); (iii) overnight hospitalization (A: No; B: Yes); and daily intakes of (iv) energy; (v) protein; (vi) saturated fatty acids, where three clusters of intakes, namely (A) low; (B) medium; (C) high, were obtained using 3-component mixtures of normal distributions (see Chapter 2), separately, and (vii) fish eaten in past 30 days (A: No; B: Yes).

We fitted the data matrix (n=5,335, p=7) with the mixture model of multivariate generalized Bernoulli distributions (10.17). The BIC indicates there are $g = 5$ clusters of respondents (Table 10.4). The clustering results are provided in Table 10.5 where the standard errors of the estimates of Ψ are obtained by the bootstrap resampling method with replacement described in Chapter 1. The number of bootstrap replications is taken to be 100. The probabilities of the membership in partitions in the seven partition variables are presented in Figure 10.4. Cluster 1 (38.3%) is the largest group that consists of respondents with a higher probability of non-smoking, very low health service use, and low nutrient intakes. Respondents in Cluster 2 (28.6%) also have low nutrient intakes, but they have higher service use and smoking level. Clusters 3 (20.9%) and 4 (9.6%) consist of respondents with higher nutrient intakes, but are distinguished by different levels of health service utilization. The smallest group is Cluster 5 (2.6%) which comprises respondents with very high nutrient intakes, especially saturated fat, and being a current smoker.

TABLE 10.5
Results of fitting a five-component mixture of multivariate generalized Bernoulli distributions to the NHANES data.

Parameter	Cluster 1	Cluster 2	Cluster 3	Cluster 4	Cluster 5
π_h	0.383 (0.045)	0.286 (0.027)	0.209 (0.033)	0.096 (0.001)	0.026
θ_{h11}	0.656 (0.045)	0.531 (0.051)	0.568 (0.065)	0.521 (0.032)	0.398 (0.012)
θ_{h12}	0.183 (0.026)	0.272 (0.029)	0.191 (0.048)	0.288 (0.044)	0.236 (0.022)
θ_{h13}	0.161	0.198	0.241	0.192	0.366
θ_{h21}	0.989 (0.003)	0.297 (0.023)	0.922 (0.009)	0.278 (0.033)	0.757 (0.032)
θ_{h22}	0.012	0.703	0.078	0.722	0.243
θ_{h31}	0.000 (0.020)	0.280 (0.034)	0.000 (0.032)	0.300 (0.055)	0.145 (0.033)
θ_{h32}	1.000	0.721	1.000	0.700	0.855
θ_{h41}	0.969 (0.016)	0.992 (0.095)	0.132 (0.064)	0.232 (0.038)	0.000 (0.011)
θ_{h42}	0.031 (0.015)	0.008 (0.064)	0.858 (0.053)	0.764 (0.053)	0.332 (0.033)
θ_{h43}	0.000	0.000	0.010	0.004	0.669
θ_{h51}	0.828 (0.003)	0.888 (0.022)	0.211 (0.012)	0.232 (0.055)	0.000 (0.011)
θ_{h52}	0.170 (0.026)	0.112 (0.029)	0.735 (0.048)	0.743 (0.044)	0.444 (0.022)
θ_{h53}	0.002	0.000	0.054	0.025	0.556
θ_{h61}	0.649 (0.022)	0.668 (0.056)	0.046 (0.038)	0.040 (0.081)	0.000 (0.011)
θ_{h62}	0.351 (0.013)	0.332 (0.025)	0.772 (0.022)	0.828 (0.038)	0.126 (0.023)
θ_{h63}	0.000	0.000	0.182	0.132	0.874
θ_{h71}	0.695 (0.004)	0.699 (0.022)	0.680 (0.014)	0.714 (0.034)	0.639 (0.022)
θ_{h72}	0.305	0.301	0.320	0.286	0.361

Note: Standard error of parameter estimates given in parentheses.

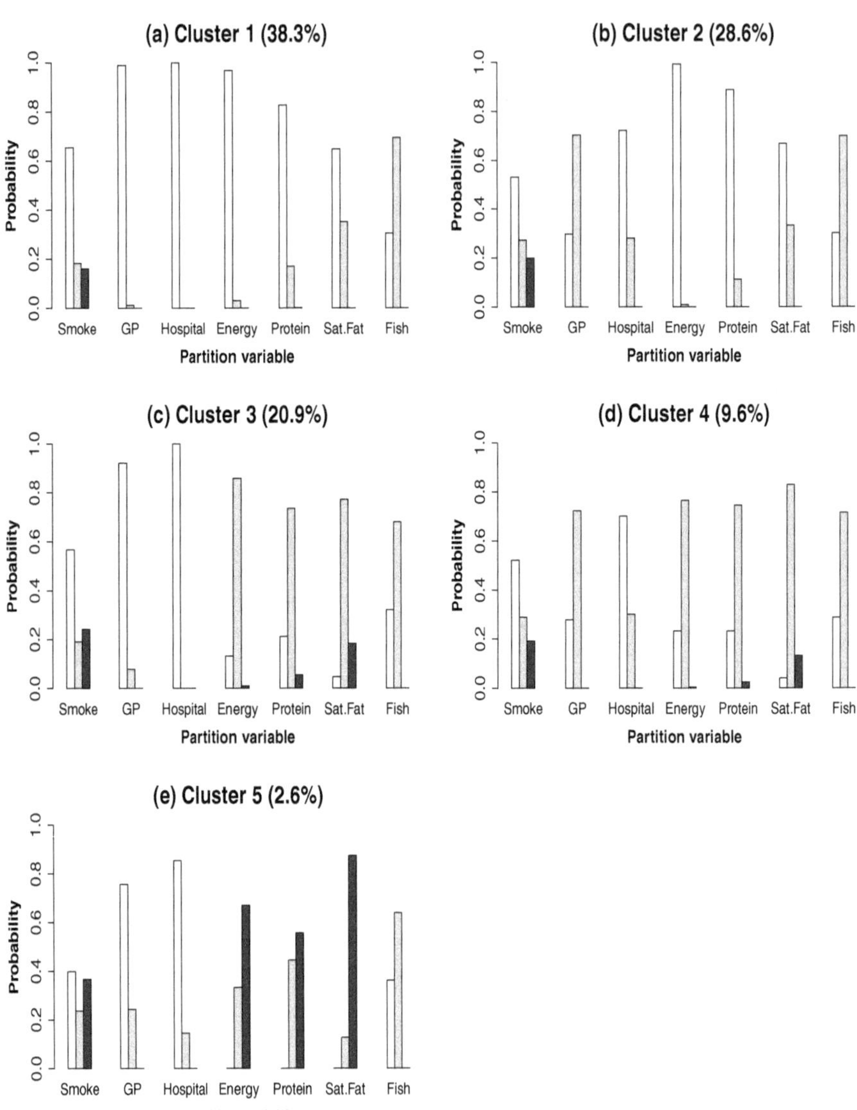

FIGURE 10.4
Five clusters with different probabilities of membership in Partitions A (white bar), B (light grey) and C (dark grey) in seven partition variables. Adapted from Ng and McLachlan (2016).

Bibliography

Aalen, O.O., and S. Johansen. 1978. "An empirical transition matrix for non-homogenous Markov chains based on censored observations." *Scandinavian Journal of Statistics* 5:141–150.

Abramowitz, M., and I.A. Stegun (eds.). 1964. *Handbook of Mathematical Functions with Formulas, Graphs, and Mathematical Tables*, NBS Applied Mathematics Series 55. Washington, D.C.: National Bureau of Standards.

Achenbach, R.M., and C.S. Edlebrock. 1981. "Behavioral problems and competencies reported by parents of normal and disturbed children aged four through sixteen." *Monograph of the Society for Research in Child Development* 46:188.

Agency for Healthcare Research and Quality. 2015. *Medical Expenditure Panel Survey Household Component.* Rockville: Agency for Healthcare Research / Quality.

Aitkin, M. 1997. "Contribution to the discussion of paper by S. Richardson and P.J. Green." *Journal of the Royal Statistical Society B* 59:766.

———. 1999. "Meta-analysis by random effect modelling in generalized linear models." *Statistics in Medicine* 18:2343–2351.

Akram, M.U., A. Tariq, S.A. Khan, and M.Y. Javed. 2014. "Automated detection of exudates and macula for grading of diabetic macular edema." *Computer Methods and Programs in Biomedicine* 114:141–152.

Alava, M.H., A. Wailoo, E. Wolfe, and K. Michaud. 2013. "The relationship between EQ-5D, HAQ and pain in patients with rheumatoid arthritis." *Rheumatology* 52:944–950.

Alfo, M., and A. Maruotti. 2010. "Two-part regression models for longitudinal zero-inflated count data." *Canadian Journal of Statistics* 38:197–216.

Allman, E., C. Matias, and J. Rhodes. 2009. "Identifiability of parameters in latent structure models with many observed variables." *Annals of Statistics* 37:3099–3132.

Ambroise, C., and C. Matias. 2012. "New consistent and asymptotically normal parameter estimates for random-graph mixture models." *Journal of the Royal Statistical Society B* 74:3–35.

Amugsi, D.A., Z.T. Dimbuene, P. Bakibinga, E.W. Kimani-Murage, T.N. Haregu, and B. Mberu. 2016. "Dietary diversity, socioeconomic status and maternal body mass index (BMI): quantile regression analysis of nationally representative data from Ghana, Namibia and Sao Tome and Principe." *BMJ Open* 6:e012615.

Andrews, D.F., and A.M. Herzberg. 1985. *Data: a Collection of Problems from Many Fields for the Student and Research Worker.* New York: Springer, pp. 261–274.

Andrews, J.L., and P.D. McNicholas. 2012. "Model-based clustering, classification, and discriminant analysis via mixtures of multivariate t-distributions." *Statistics and Computing* 22:1021–1029.

Arbuckle, J.L. 2013. *IBM SPSS Amos 22 User's Guide.* Chicago: IBM, Chapters 35 & 36.

Arnold, B.C., and R.J. Beaver. 2002. "Skewed multivariate models related to hidden truncation and/or selective reporting." *Test* 11:7–54.

Atherton, K., E. Fuller, P. Shepherd, D.P. Strachan, and C. Power. 2008. "Loss and representativeness in a biomedical survey at age 45 years: 1958 British birth cohort." *Journal of Epidemiology and Community Health* 62:216–223.

Australian Bureau of Statistics. 2009. *National Health Survey 2007-08: User's guide, Cat. No. 4363.0.55.001.* Canberra: Australian Bureau of Statistics. Accessed November 3, 2014. `http://www.abs.gov.au`.

Axon, R.N., M. Gebregziabher, C.J. Everett, P. Heidenreich, and K.J. Hunt. 2016. "Dual health care system use is associated with higher rates of hospitalization and hospital readmission among veterans with heart failure." *American Heart Journal* 174:157–163.

Bache, K., and M. Lichman. 2013. *UCI Machine Learning Repository.* Irvine, CA: University of California, School of Information and Computer Science. Accessed November 3, 2014. `http://archive.ics.uci.edu/ml`.

Baek, J., G.J. McLachlan, and L.K. Flack. 2010. "Mixtures of factor analyzers with common factor loadings: applications to the clustering and visualization of high-dimensional data." *IEEE Transactions on Pattern Analysis and Machine Intelligence* 32:1298–1309.

Bambs, C.E., K.E. Kip, S.R. Mulukutla, A.N. Aiyer, C. Johnson, L.A. McDowell, K. Matthews, and S.E. Reis. 2013. "Sociodemographic, clinical, and psychological factors associated with attrition in a prospective study of cardiovascular prevention: the Heart Strategies Concentrating on Risk Evaluation study." *Annals of Epidemiology* 23:328–333.

Bang, H., and J.M. Robins. 2005. "Doubly robust estimation in missing data and causal inference models." *Biometrics* 61:962–972.

Barnett, K., S.W. Mercer, M. Norbury, G. Watt, S. Wyke, and B. Guthrie. 2012. "Epidemiology of multimorbidity and implications for health care, research, and medical education: a cross-sectional study." *Lancet* 380:37–43.

Batstra, L., E.H. Bos, and J. Neeleman. 2002. "Quantifying psychiatric comorbidity – lessions from chronic disease epidemiology." *Social Psychiatry and Psychiatric Epidemiology* 37:105–111.

Bedrick, E.J., R. Christensen, and W. Johnson. 1996. "A new perspective on priors for generalized linear models." *Journal of the American Statistical Association* 91:1450–1460.

Benjamini, Y., and Y. Hochberg. 1995. "Controlling the false discovery rate: a practical and powerful approach to multiple testing." *Journal of the Royal Statistical Society B* 57:259–300.

Bennett, D.A. 2001. "How can I deal with missing data in my study?" *Australian and New Zealand Journal of Public Health* 25:464–469.

Ben-Tovim Jones, L., S.K. Ng, C. Ambroise, K. Monico, N. Khan, and G.J. McLachlan. 2005. "Use of microarray data via model-based classification in the study and prediction of survival from lung cancer." In *Methods of Microarray Data Analysis IV,* edited by J.S. Shoemaker and S.M. Lin, 163–173. New York: Springer.

Berg'e, L., C. Bouveyron, and S. Girard. 2012. "HDclassif: An R package for model-based clustering and discriminant analysis of high-dimensional data." *Journal of Statistical Software* 46:Issue 6.

Berlinet, A.F., and C. Roland. 2012. "Acceleration of the EM algorithm: P-EM versus epsilon algorithm." *Computational Statistics and Data Analysis* 56:4122–4137.

Biernacki, C., G. Celeux, and G. Govaert. 1998. "Assessing a mixture model for clustering with the integrated classification likelihood." *IEEE Transactions on Pattern Analysis and Machine Intelligence* 22:719–725.

———. 2003. "Choosing starting values for the EM algorithm for getting the highest likelihood in multivariate Gaussian mixture models." *Computational Statistics and Data Analysis* 41:561–575.

Blackstone, E.H., D.C. Naftel, and M.E. Turner. 1986. "The decomposition of time-varying hazard into phases, each incorporating a separate stream of concomitant information." *Journal of the American Statistical Association* 81:615–624.

Böhning, D., and P. Schlattmann. 1992. "Computer-Assisted Analysis of Mixtures (C.A.MAN): Statistical algorithms." *Biometrics* 48:283–303.

Böhning, D., P. Schlattmann, and B. Lindsay. 1998. "Recent developments in computer-assisted analysis of mixtures." *Biometrics* 54:525–536.

Boldea, O., and J. Magnus. 2009. "Maximum likelihood estimation of the multivariate noraml mixture model." *Journal of the American Statistical Association* 104:1539–1549.

Booth, J.G., G. Casella, J.E.K. Cooke, and J.M. Davis. 2004. *Statistical approaches to analysing microarray data representing periodic biological processes: a case study using the yeast cell cycle.* Technical report, Department of Biological Statistics & Computational Biology, Cornell University.

Borah, B.J., and A. Basu. 2013. "Highlighting differences between conditional and unconditional quantile regression approaches through an application to assess medication adherence." *Health Economics* 22:1052–1070.

Borgatti, S.P., M.G. Everett, and L.C. Freeman. 2002. *Ucinet for Windows: Software for Social Network Analysis.* Analytic Technologies, Harvard, MA. `http://www.analytictech.com/`.

Breslow, N.E. 1974. "Covariance analysis of censored survival data." *Biometrics* 30:89–99.

Breslow, N.E., and D.G. Clayton. 1993. "Approximate inference in generalized linear mixed models." *Journal of the American Statistical Association* 88:9–25.

Brown, B., M. Chui, and J. Manyika. 2011. "Are you ready for the era of 'big data'?" *McKinsey Quarterly* 4:24–35.

Button, K.S., J.P.A. Ioannidis, C. Mokrysz, B.A. Nosek, J. Flint, S.J. Robinson, and M.R. Munafo. 2013. "Power failure: why small sample size undermines the reliability of neuroscience." *Nature Reviews Neuroscience* 14:365–376.

Cai, C., S. Wang, W. Lu, and J. Zhang. 2014. "NPHMC: An R-package for estimating sample size of proportional hazards mixture cure model." *Computer Methods and Programs in Biomedicine* 113:290–300.

Cai, C., Y. Zou, Y. Peng, and J. Zhang. 2012. "smcure: An R package for estimating semiparametric mixture cure models." *Computer Methods and Programs in Biomedicine* 108:1255–1260.

Cameron, C.M., R. Shibl, R.J. McClure, S.K. Ng, and A.P. Hills. 2014. "Maternal pregravid body mass index and child hospital admissions in the first 5 years of life: results from an Australian birth cohort." *International Journal of Obesity* 38:1268–1274.

Cantor, A.B., and J.J. Shuster. 1992. "Parametric versus non-parametric methods for estimating cure rates based on censored survival data." *Statistics in Medicine* 11:931–937.

Carlson, M.J., and M.E. Corcoran. 2001. "Family structure and children's behavioral and cognitive outcomes." *Journal of Marriage and Family* 63:779–792.

Carpenter, J.R., M.G. Kenward, and S. Vansteelandt. 2006. "A comparison of multiple imputation and doubly robust estimation for analyses with missing data." *Journal of the Royal Statistical Society A* 169:571–584.

Caughey, G.E., and E.E. Roughead. 2011. "Multimorbidity research challenges: where to go from here?" *Journal of Comorbidity* 1:8–10.

Celeux, G., O. Martin, and C. Lavergne. 2005. "Mixture of linear mixed models for clustering gene expression profiles from repeated microarray experiments." *Statistical Modelling* 5:243–267.

Center for Human Resource Research. 2002. *NLSY79 Child and Young Adult Data Users Guide: A guide to the 1986-2000 child data, 1994-2000 young adult data.* Columbus: Ohio State University.

Chambers, S.K., S.K. Ng, P. Baade, J. Aitken, M.K. Hyde, G. Wittert, M. Frydenberg, and J. Dunn. 2017. "Trajectories of quality of life, life satisfaction and psychological adjustment after prostate cancer." *Psycho-Oncology* 26:1576–1585.

Chang, J.J., C.T. Halpern, and J.S. Kaufman. 2007. "Maternal depression symptoms, father's involvement, and the trajectories of child problem behaviors in a US national sample." *Archives of Pediatrics and Adolescent Medicine* 161:697–703.

Charan, J., and T. Biswas. 2013. "How to calculate sample size for different study designs in medical research." *Indian Journal of Psychological Medicine* 35:121–126.

Chaudhuri, P., and P.A. Mykland. 1993. "Nonlinear experiments: optimal design and inference based on likelihood." *Journal of the American Statistical Association* 88:538–546.

Chen, T., V. Filkov, and S.S. Skiena. 2001. "Identifying gene regulatory networks from experimental data." *Parallel Computing* 27:141–162.

Cho, R.J., M.J. Campbell, E.A. Winzeler, L. Steinmetz, A. Conway, L. Wodicka, T.G. Wolfsberg, et al. 1998. "A genome-wide transcriptional analysis of the mitotic cell cycle." *Molecular Cell* 2:65–73.

Concordet, D., A. Geffr'e, J.P. Braun, and C. Trumel. 2009. "A new approach for the determination of reference intervals from hospital-based data." *Clinica Chimica Acta* 405:43–48.

Cook, R.D. 1986. "Assessment of local influence (with discussion)." *Journal of the Royal Statistical Society B* 48:133–169.

Cooper, L.A.D., K. Jun, D.A. Gutman, F. Wang, J. Gao, C. Appin, S. Cholleti, et al. 2012. "Integrated morphologic analysis for the identification and characterization of disease subtypes." *Journal of the American Medical Informatics Association* 19:317–323.

Corbière, F., and P. Joly. 2007. "A SAS macro for parametric and semiparametric mixture cure models." *Computer Methods and Programs in Biomedicine* 85:173–180.

Cox, D.R. 1972. "Regression models and life tables (with discussion)." *Journal of the Royal Statistical Society B* 34:187–220.

———. 1975. "Partial likelihood." *Biometrika* 62:269–276.

Crawford, S.L. 1994. "An application of the Laplace method to finite mixture distributions." *Journal of the American Statistical Association* 89:259–267.

Dahl, D.B., and M.A. Newton. 2007. "Multiple hypothesis testing by clustering treatment effects." *Journal of the American Statistical Association* 102:517–526.

Dal Maso, L., S. Guzzinati, C. Buzzoni, R. Capocaccia, D. Serraino, A. Caldarella, A.P. Dei Tos, et al. 2014. "Long-term survival, prevalence, and cure of cancer: a population-based estimation for 818,902 Italian patients and 26 cancer types." *Annals of Oncology* 25:2251–2260.

Daniels, M.J., and J.W. Hogan. 2008. *Missing Data in Longitudinal Studies: Strategies for Bayesian Modeling and Sensitivity Analysis.* London: Chapman & Hall.

Deb, P., and P.K. Trivedi. 1997. "Demand for medical care by the elderly: A finite mixture approach." *Journal of Applied Econometrics* 12:313–336.

Dempster, A.P., N.M. Laird, and D.B. Rubin. 1977. "Maximum likelihood from incomplete data via the EM algorithm." *Journal of the Royal Statistical Society B* 39:1–38.

Dennis, G.Jr., B.T. Sherman, D.A. Hosack, J. Yang, W. Gao, H.C. Lane, and R.A. Lempicki. 2003. "DAVID: Database for Annotation, Visualization, and Integrated Discovery." *Genome Biology* 4:P3.

DeRisi, J.L., V.R. Iyer, and P.O. Brown. 1997. "Exploring the metabolic and genetic control of gene expression on a genomic scale." *Science* 278:680–686.

DeRoon-Cassini, T.A., A.D. Mancini, M.D. Rusch, and G.A. Bonanno. 2010. "Psychopathology and resilience following traumatic injury: A latent growth mixture model analysis." *Rehabilitation Psychology* 55:1–11.

Destrempes, F., J. Meunier, M.-F. Giroux, G. Soulez, and G. Cloutier. 2009. "Segmentation in ultrasonic B-mode images of healthy carotid arteries using mixtures of Nakagami distributions and stochastic optimization." *IEEE Transactions on Medical Imaging* 28:215–229.

D'haeseleer, P., S. Liang, and R. Somogyi. 2000. "Genetic network inference: from co-expression clustering to reverse engineering." *Bioinformatics* 16:707–726.

Di Zio, M., U. Guarnera, and O. Luzi. 2007. "Imputation through finite Gaussian mixture models." *Computational Statistics and Data Analysis* 51:5305–5316.

Dietz, E., and D. Böhning. 2000. "On estimation of the Poisson parameter in zero-modified Poisson models." *Computational Statistics and Data Analysis* 34:441–459.

Diggle, P., and M.G. Kenward. 1994. "Informative drop-out in longitudinal data analysis (with discussion)." *Applied Statistics* 43:49–93.

Dogru, F.Z., Y.M. Bulut, and O. Arslan. 2016. "Finite mixtures of matrix variate *t* distributions." *Journal of Science* 25:335–341.

Dunn, J., S.K. Ng, W. Breitbart, J. Aitken, P. Youl, P.D. Baade, and S.K. Chambers. 2013. "Health-related quality of life and life satisfaction in colorectal cancer survivors: trajectories of adjustment." *Health and Quality of Life Outcomes* 11:46.

Dunn, J., S.K. Ng, J. Holland, J. Aitken, P. Youl, P.D. Baade, and S.K. Chambers. 2013. "Trajectories of psychological distress after colorectal cancer." *Psycho-Oncology* 22:1759–1765.

Efron, B. 1977. "The efficiency of Cox's likelihood function for censored data." *Journal of the American statistical Association* 72:557–565.

Efron, B., and R.J. Tibshirani. 1993. *An Introduction to the Bootstrap.* New York: Chapman & Hall.

Eirola, E., A. Lendasse, V. Vandewalle, and C. Biernacki. 2014. "Mixture of Gaussians for distance estimation with missing data." *Neurocomputing* 131:32–42.

Eisen, M.B., P.T. Spellman, P.O. Brown, and D. Botstein. 1998. "Cluster analysis and display of genome-wide expression patterns." *Proceedings of the National Academy of Sciences of the United States of America* 95:14863–14868.

Evans, J.D. 1996. *Straightforward Statistics for the Behavioral Sciences.* Pacific Grove, CA: Brooks/Cole Publishing.

Everitt, B.S., and D.J. Hand. 1981. *Finite Mixture Distributions.* London: Chapman & Hall.

Fahey, M.T., P. Ferrari, N. Slimani, J.K. Vermunt, I.R. White, K. Hoffmann, E. Wirfalt, et al. 2012. "Identifying dietary patterns using a normal mixture model: application to the EPIC study." *Journal of Epidemiology and Community Health* 66:89–94.

Farewell, V.T. 1982. "The use of mixture models for the analysis of survival data with long-term survivors." *Biometrics* 38:1041–1046.

———. 1986. "Mixture models in survival analysis: Are they worth the risk?" *Canadian Journal of Statistics* 14:257–262.

Feinstein, A.R. 1970. "The pretherapeutic classification of comorbidity in chronic disease." *Journal of Chronic Disease* 23:455–468.

Fellner, W.H. 1986. "Robust estimation of variance components." *Technometrics* 28:51–60.

———. 1987. "Sparse matrices, and the estimation of variance components by likelihood methods." *Communications in Statistics-Simulation and Computation* 16:439–463.

Fine, J.P. 1999. "Analysing competing risks data with transformation models." *Journal of the Royal Statistical Society B* 61:817–830.

Firpo, S., N.M. Fortin, and T. Lemieux. 2009. "Unconditional quantile regressions." *Econometrica* 77:953–973.

Fisher, R.A. 1936. "The use of multiple measurements in taxonomic problems." *Annals of Eugenics* 7:179–188.

Fleming, T.R., and D.P. Harrington. 1991. *Counting Processes and Survival Analysis*. New York: Wiley.

Foguet-Boreu, Q., C. Violán, T. Rodriguez-Blanco, A. Roso-Llorach, M. Pons-Vigués, E. Pujol-Ribera, Y. Cossio Gil, and J.M. Valderas. 2015. "Multimorbidity patterns in elderly primary health care patients in a South Mediterranean European region: a cluster analysis." *PLoS One* 10:e0141155.

Fraley, C., and A.E. Raftery. 2002. "Model-based clustering, discriminant analysis and density estimation." *Journal of the American Statistical Association* 97:611–631.

———. 2006. *MCLUST: version 3 for R: Normal mixture modeling and model-based clustering*. Technical Report 504, Department of Statistics, University of Washington.

Fraley, C., A.E. Raftery, T.B. Murphy, and L. Scrucca. 2012. *MCLUST version 4 for R: Normal mixture modeling for model-based clustering, classification, and density estimation*. Technical Report Vol. 597, Department of Statistics, University of Washington.

Freund, M., S. Hahnel, M. Thomsen, and K. Sartor. 2001. "Treatment planning in severe scoliosis: the role of MRI." *Neuroradiology* 43:481–484.

Frühwirth-Schnatter, S. 2006. *Finite Mixture and Markov Switching Models*. New York: Springer.

Fujii, Y., S. Kaneko, S.M. Nzou, M. Mwau, S.M. Njenga, C. Tanigawa, J. Kimotho, A.W. Mwangi, I. Kiche, and S. et al. Matsumoto. 2014. "Serological surveillance development for tropical infectious diseases using simultaneous microsphere-based multiplex assays and finite mixture models." *PLoS Neglected Tropical Diseases* 8:e3040.

García-Laencina, P.J., J.L. Sancho-Gómes, and A.R. Figueiras-Vidal. 2010. "Pattern classification with missing data: A review." *Neural Computation* 19:263–282.

Garin, N., A. Koyanagi, S. Chatterji, S. Tyrovolas, B. Olaya, M. Leonardi, E. Lara, et al. 2016. "Global multimorbidity patterns: a cross-sectional, population-based, multi-country study." *Journal of Gerontology Series A: Biological Sciences and Medical Sciences* 71:205–214.

Garrison, C.Z., C.L. Addy, K.L. Jackson, R.E. KcKeown, and J.L. Waller. 1992. "Major depressive disorder and dysthymia in young adolescents." *American Journal of Epidemiology* 135:792–802.

Gelfand, A.E., S.K. Ghosh, C. Christiansen, S.B. Soumeral, and T.J. McLaughlin. 2000. "Proportional hazards models: a latent competing risk approach." *Applied Statistics* 49:385–397.

Ghahramani, Z., and M.I. Jordan. 1994. "Supervised learning from incomplete data via an EM approach." In *Advances in Neural Information Processing Systems 6 (NIPS 93),* edited by J.D. Cowan, G. Tesauro, and J. Alspector, 120–127. San Francisco, CA: Morgan Kaufmann.

Ghitany, M.E., and R.A. Maller. 1992. "Asymptotic results for exponential mixture models with long-term survivors." *Statistics* 23:321–336.

Ghosh, D., and A.M. Chinnaiyan. 2002. "Mixture modelling of gene expression data from microarray experiments." *Bioinformatics* 18:275–286.

Gill, T.M., E.A. Gahbauer, L. Han, and H.G. Allore. 2015. "The role of intervening hospital admissions on trajectories of disability in the last year of life: prospective cohort study of older people." *British Medical Journal* 350:h2361.

Glasser, M. 1980. "Bathtub and related failure rate characterizations." *Journal of the American Statistical Association* 75:667–672.

Golbeck, A.L. 1992. "Bootstrapping current life table estimators." In *Bootstrapping and Related Techniques,* edited by K.H. Jöckel, G. Rothe, and W. Sendler, 197–201. Heidelberg: Springer Verlag.

Gonzalez, A.I., M. Graña, and A. D'Anjou. 1995. "An analysis of the GLVQ algorithm." *IEEE Transactions on Neural Networks* 6:1012–1016.

Gordon, N.H. 1990a. "Application of the theory of finite mixtures for the estimation of 'cure' rates of treated cancer patients." *Statistics in Medicine* 9:397–407.

———. 1990b. "Maximum likelihood estimation for mixtures of two Gompertz distributions when censoring occurs." *Communications in Statistics – Simulation and Computation* 19:737–747.

Gorman, E., A.H. Leyland, G. McCartney, S.V. Katikireddi, L. Rutherford, L. Graham, M. Robinson, and L. Gray. 2017. "Adjustment for survey non-representativeness using record-linkage: refined estimates of alcohol consumption by deprivation in Scotland." *Addiction* 112:1270–1280.

Graham, J.W. 2009. "Missing data analysis: Making it work in the real world." *Annual Review of Psychology* 60:549–576.

Graham, M.W., and D.J. Miller. 2006. "Unsupervised learning of parsimonious mixtures on large spaces with integrated feature and component selection." *IEEE Transactions on Signal Processing* 54:1289–1303.

Gray, L., G. McCartney, I.R. White, S.V. Katikireddi, L. Rutherford, E. Gorman, and A.H. Leyland. 2013. "Use of record-linkage to handle non-response and improve alcohol consumption estimates in health survey data: a study protocol." *BMJ Open* 3:e002647.

Grün, B., T. Scharl, and F. Leisch. 2012. "Modelling time-course gene expression data with finite mixtures of linear additive models." *Bioinformatics* 28:222–228.

Gunawardana, A., and W. Byrne. 1999. *Convergence of EM Variants.* Technical report No. 32, CLSP, Johns Hopkins University.

Haas, J.S., K.A. Phillips, E.P. Gerstenberger, and A.C. Seger. 2005. "Potential savings from substituting generic drugs for brand-name drugs: Medical Expenditure Panel Survey, 1997-2000." *Annals of Internal Medicine* 142:891–897.

Haendel, M.A., C.G. Chute, and P.N. Robinson. 2018. "Classification, ontology, and precision medicine." *New England Journal of Medicine* 379:1452–1462.

Hafemeister, C., I.G. Costa, A. Schönhuth, and A. Schliep. 2011. "Classifying short gene expression time-courses with Bayesian estimation of piecewise constant functions." *Bioinformatics* 27:946–952.

Hahn, C., M.D. Johnson, A. Herrmann, and F. Huber. 2002. "Capturing customer heterogeneity using a finite mixture PLS approach." *Schmalenbach Business Review* 54:243–269.

Han, P. 2014. "Multiply robust estimation in regression analysis with missing data." *Journal of the American Statistical Association* 109:1159–1173.

Hao, J.G., B.P. Koester, T.A. McKay, E.S. Rykoff, E. Rozo, A. Evrard, J. Annis, et al. 2009. "Precision measurements of the cluster red sequence using an error-corrected Gaussian mixture model." *Astrophysical Journal* 702:745–758.

Harkanen, T., R. Kaikkonen, E. Virtala, and S. Koskinen. 2014. "Inverse probability weighting and doubly robust methods in correcting the effects of non-response in the reimbursed medication and self-reported turnout estimates in the ATH survey." *BMC Public Health* 14:1150.

Harville, D.A. 1977. "Maximum likelihood approaches to variance component estimation and to related problems." *Journal of the American Statistical Association* 72:320–338.

Hawkins, D.M. 1981. "A new test for multivariate normality and homoscedasticity." *Technometrics* 23:105–110.

Hawkins, D.M., M.W. Muller, and J.A. ten Krooden. 1982. "Cluster analysis." In *Topics in Applied Multivariate Analysis,* edited by D.M. Hawkins, 303–356. Cambridge: Cambridge University Press.

Haybittle, J.L. 1983. "Is breast cancer ever cured?" *Reviews on Endocrine-Related Cancer* 14:13–18.

Haynie, D.L., T. Farhat, A. Brooks-Russell, J. Wang, B. Barbieri, and R.J. Iannotti. 2013. "Dating violence perpetration and victimization among U.S. adolescents: Prevalence, patterns, and associations with health complaints and substance use." *Journal of Adolescent Health* 53:194–201.

He, Y., and C. Liu. 2012. "The dynamic 'expectation-conditional maximization either' algorithm." *Journal of the Royal Statistical Society B* 74:313–336.

He, Y., W. Pan, and J. Lin. 2006. "Cluster analysis using multivariate normal mixture models to detect differential gene expression with microarray data." *Computational Statistics and Data Analysis* 51:641–658.

Hedeker, D., and R.D. Gibbons. 2006. *Longitudinal Data Analysis.* Hoboken, New Jersey: Wiley.

Henderson, C.R. 1975. "Best linear unbiased estimation and prediction under a selection model." *Biometrics* 31:423–447.

Henderson, C.R., O. Kempthorne, S.R. Searle, and C.M. Von Krosigk. 1959. "The estimation of environmental and genetic trends from records subject to culling." *Biometrics* 15:192–218.

Hennig, C. 2010. "Methods for merging Gaussian mixture components." *Advances in Data Analysis and Classification* 4:3–34.

Hofler, M., H. Pfister, R. Lieb, and H.U. Wittchen. 2005. "The use of weights to account for non-response and drop-out." *Social Psychiatry and Psychiatric Epidemiology* 40:291–299.

Hogan, J.W., J. Roy, and C. Korkontzelou. 2004. "Handling drop-out in longitudinal studies." *Statistics in Medicine* 23:1455–1497.

Holden, L., P.A. Scuffham, M.F. Hilton, A. Muspratt, S.K. Ng, and H.A. Whiteford. 2011. "Patterns of multimorbidity in working Australians." *Population Health Metrics* 9:15.

Honaker, J., and G. King. 2010. "What to do about missing values in time-series cross-section data." *American Journal of Political Science* 54:561–581.

Honaker, J., G. King, and M. Blackwell. 2015. *AMELIA II: A Program for Missing Data.* Analytic Technologies, Harvard, MA. `http://gking.harvard.edu/amelia`.

Hornik, K., and B. Grün. 2014. *movMF: An R package for fitting mixtures of von Mises-Fisher Distributions.* Vol. 58, Issue 10.

Horton, N.J., and S.R. Lipsitz. 2001. "Multiple imputation in practice: Comparison of software packages for regression models with missing variables." *American Statistician* 55:244–254.

Hu, H., Y. Wu, and W. Yao. 2016. "Maximum likelihood estimation of the mixture of log-concave densities." *Computational Statistics and Data Analysis* 101:137–147.

Hunt, L.A. 2017. "Missing data imputation and its effect on the accuracy of classification." In *Studies in Classification, Data Analysis, and Knowledge Organization: Data Science – Innovative Developments in Data Analysis and Clustering,* edited by F. Palumbo, A. Montanari, and M. Vichi, 3–14. Switzerland: Springer.

Hunt, L.A., and M.A. Jorgensen. 1999. "Mixture model clustering using the Multimix program." *Australian and New Zealand Journal of Statistics* 41:153–171.

———. 2003. "Mixture model clustering for mixed data with missing information." *Computational Statistics and Data Analysis* 41:429–440.

Ibrahim, J.G., M.-H. Chen, and H. Chu. 2012. "Bayesian methods in clinical trials: a Bayesian analysis of ECOG trials E1684 and E1690." *BMC Medical Research Methodology* 12:183.

Ibrahim, J.G., M.-H. Chen, and D. Sinha. 2001. "Bayesian semiparametric models for survival data with a cure fraction." *Biometrics* 57:383–388.

Ibrahim, J.G., and G. Molenberghs. 2009. "Missing data methods in longitudinal studies: a review." *TEST* 18:1–43.

Jackson, C.A., M. Jones, L. Tooth, G.D. Mishra, J. Byles, and A. Dobson. 2015. "Multimorbidity patterns are differentially associated with functional ability and decline in a longitudinal cohort of older women." *Age Ageing* 44:810–816.

Jácome, A.A.A., D.R. Wohnrath, C. Scapulatempo Neto, E.C. Carneseca, S.V. Serrano, L.S. Viana, J.S. Nunes, E.Z. Martinez, and J.S. Santos. 2014. "Prognostic value of epidermal growth factor receptors in gastric cancer: a survival analysis by Weibull model incorporating long-term survivors." *Gastric Cancer* 17:76–86.

James, G., and C. Sugar. 2003. "Clustering for sparsely sampled functional data." *Journal of the American Statistical Association* 98:397–408.

Jamshidian, M., and S. Jalal. 2010. "Tests of homoscedasticity, normality, and missing completely at random for incomplete multivariate data." *Psychometrika* 75:649–674.

Jamshidian, M., S. Jalal, and C. Jansen. 2014. "MissMech: An R package for testing homoscedasticity, multivariate normality, and missing completely at random (MCAR)." *Journal of Statistical Software* 56:Issue 6.

Jancke, L., T.W. Buchanan, K. Lutz, and N.J. Shah. 2001. "Focused and nonfocused attention in verbal and emotional dichotic listening: an FMRI study." *Brain and Language* 78:349–363.

Jank, W. 2006. "Implementing and diagnosing the stochastic approximation EM algorithm." *Journal of Computational and Graphical Statistics* 15:803–829.

Jansen, R.C. 1993. "Maximum likelihood in a generalized linear finite mixture model by using the EM algorithm." *Biometrics* 49:227–231.

Jeon, Y., and J.H.T. Kim. 2013. "A gamma kernel density estimation for insurance loss data." *Insurance: Mathematics and Economics* 53:569–579.

Ji, Z.X., Y. Xia, Q.S. Sun, Q. Chen, D.S. Xia, and D.D. Feng. 2012. "Fuzzy local Gaussian mixture model for brain MR image segmentation." *IEEE Transactions on Information Technology in Biomedicine* 16:339–347.

Jones, A.M., J. Lomas, P.T. Moore, and N. Rice. 2016. "A quasi-Monte-Carlo comparison of parametric and semiparametric regression methods for heavy-tailed and non-normal data: an application to healthcare costs." *Journal of the Royal Statistical Society A* 179:951–974.

Jones, B.L., and D.S. Nagin. 2013. "A note on a Stata plugin for estimating group-based trajectory models." *Sociological Methods & Research* 42:608–613.

Jones, B.L., D.S. Nagin, and K. Roeder. 2001. "A SAS procedure based on mixture models for estimating developmental trajectories." *Sociological Methods & Research* 29:374–393.

Jones, P.N., and G.J. McLachlan. 1990. "Algorithm AS 254. Maximum likelihood estimation from grouped and truncated data with finite normal mixture models." *Journal of the Royal Statistical Society C* 39:273–282.

Jorstad, T.S., H. Midelfart, and A.M. Bones. 2008. "A mixture model approach to sample size estimation in two-sample comparative microarray experiments." *BMC Bioinformatics* 9:117.

Jowsey, T., Y.H. Jeon, P. Dugdale, N.J. Glasgow, M. Kljakovic, and T. Usherwood. 2009. "Challenges for co-morbid chronic illness care and policy in Australia: a qualitative study." *Australian and New Zealand Health Policy* 6:22.

Kalbfleisch, J.D., and R.L. Prentice. 2011. *The Statistical Analysis of Failure Time Data (Second Edition).* New Jersey: Wiley.

Kay, R. 1986. "Treatment effects in competing-risks analysis of prostate cancer data." *Biometrics* 42:203–211.

Kenward, M.G. 1998. "Selection models for repeated measurements with non-random dropout: an illustration of sensitivity." *Statistics in Medicine* 17:2723–2732.

Kenward, M.G., and G. Rosenkranz. 2011. "Joint modelling of outcome, observation time and missingness." *Journal of Biopharmaceutical Statistics* 21:252–262.

Keribin, C. 2000. "Consistent estimation of the order of mixture models." *Sankhyā: The Indian Journal of Statistics* 62:49–66.

Kim, G.-H., S. Trimi, and J.-H. Chung. 2014. "Big-data applications in the government sector." *Communications of the ACM* 57:78–85.

Kim, W., D. Gordon, J. Sebat, K.Q. Ye, and S.J. Finch. 2008. "Computing power and sample size for case-control association studies with copy number polymorphism: Application of mixture-based likelihood ratio test." *PLoS One* 3:e3475.

Kirkwood, J.M., J.G. Ibrahim, V.K. Sondak, J. Richards, L.E. Flaherty, M.S. Ernstoff, T.J. Smith, U. Rao, M. Steele, and R.H. Blum. 2000. "High- and low-dose interferon alfa-2b in high-risk melanoma: first analysis of intergroup trial E1690/S9111/C9190." *Journal of Clinical Oncology* 18:2444–2458.

Koenker, R., and K.F. Hallock. 2001. "Quantile regression." *Journal of Economic Perspectives* 15:143–156.

Komárek, A. 2009. "A new R package for Bayesian estimation of multivariate normal mixtures allowing for selection of the number of components and interval-censored data." *Computational Statistics and Data Analysis* 53:3932–3947.

Kong, J., L.A.D. Cooper, F. Wang, D.A. Gutman, J. Gao, C. Chisolm, A. Sharma, et al. 2011. "Integrative, multimodal analysis of glioblastoma using TCGA molecular data, pathology images, and clinical outcomes." *IEEE Transactions on Biomedical Engineering* 58:3469–3474.

Kuk, A.Y.C. 1992. "A semiparametric mixture model for the analysis of competing risks data." *Australian Journal of Statistics* 34:169–180.

———. 1995. "Asymptotically unbiased estimation in generalized linear models with random effects." *Journal of the Royal Statistical Society Series B* 57:395–407.

Kuk, A.Y.C., and C.H. Chen. 1992. "A mixture model combining logistic regression with proportional hazards regression." *Biometrika* 79:531–541.

Lai, X., and K.K.W. Yau. 2008. "Long-term survivor model with bivariate random effects: Applications to bone marrow transplant and carcinoma study data." *Statistics in Medicine* 27:5692–5708.

———. 2009. "Multilevel mixture cure models with random effects." *Biometrical Journal* 51:456–466.

Lambert, D. 1992. "Zero-inflated Poisson regression, with an application to defects in manufacturing." *Technometrics* 34:1–14.

Lambert, P.C., P.W. Dickman, C.L. Weston, and J.R. Thompson. 2010. "Estimating the cure fraction in population based cancer studies by using finite mixture models." *Journal of the Royal Statistical Society A* 59:35–55.

Lange, N., L. Ryan, L. Billard, D. Brillinger, L. Conquest, and J. Greenhouse. 1994. *Case Studies in Biometry*. New York: Wiley.

Larson, M.G., and G.E. Dinse. 1985. "A mixture model for the regression analysis of competing risks data." *Applied Statistics* 34:201–211.

Laska, E.M., and M.J. Meisner. 1992. "Nonparametric estimation and testing in a cure model." *Biometrics* 48:1223–1234.

Lawless, J.F. 2011. *Statistical Models and Methods for Lifetime Data (Second Edition)*. New Jersey: Wiley.

Lee, A.H., S.K. Ng, and K.K.W. Yau. 2001. "Determinants of maternity length of stay: A gamma mixture risk-adjusted model." *Health Care Management Science* 4:249–255.

Lee, A.H., K. Wang, J.A. Scott, K.K.W. Yau, and G.J. McLachlan. 2006. "Multilevel zero-inflated Poisson regression modelling of correlated count data with excess zeros." *Statistical Methods in Medical Research* 15:47–61.

Lee, A.H., K. Wang, and K.K.W. Yau. 2001. "Analysis of zero-inflated Poisson data incorporating extent of exposure." *Biometrical Journal* 43:963–975.

Lee, A.H., K. Wang, K.K.W. Yau, G.J. McLachlan, and S.K. Ng. 2007. "Maternity length of stay modelling by gamma mixture regression with random effects." *Biometrical Journal* 49:750–764.

Lee, A.H., L. Xiang, and W.K. Fung. 2004. "Sensitivity of score tests for zero-inflation in count data." *Statistics in Medicine* 23:2757–2769.

Lee, A.H., J. Xiao, J.P. Codde, and S.K. Ng. 2002. "Public versus private hospital maternity length of stay: a gamma mixture modelling approach." *Health Services Management Research* 15:46–54.

Lee, A.H., Y. Zhao, K.K.W. Yau, and S.K. Ng. 2009. "A computer graphical user interface for survival mixture modelling of recurrent infections." *Computers in Biology and Medicine* 39:301–307.

Lee, M.-L.T., F.C. Kuo, G.A. Whitmore, and J. Sklar. 2000. "Importance of replication in microarray gene expression studies: statistical methods and evidence from repetitive cDNA hybridizations." *Proceedings of the National Academy of Sciences of the United States of America* 97:9834–9838.

Lee, M.-L.T., and G.A. Whitmore. 2002. "Power and sample size for DNA microarray studies." *Statistics in Medicine* 21:3543–3570.

Lee, S., and G.J. McLachlan. 2013. "On mixtures of skew-normal and skew t-distributions." *Advances in Data Analysis and Classification* 7:241–266.

Lee, S., and G.J. McLachlan. 2014. "Finite mixtures of multivariate skew t-distributions: Some recent and new results." *Statistics and Computing* 24:181–202.

———. 2016. "Finite mixtures of canonical fundamental skew t-distributions: The unification of the restricted and unrestricted skew t-mixture models." *Statistics and Computing* 26:573–589.

Lee, Y., and J.A. Nelder. 1996. "Hierarchical generalized linear models." *Journal of the Royal Statistical Society Series B* 58:619–678.

Lehmann, T., and P. Schlattmann. 2017. "Treatment of nonignorable missing data when modeling unobserved heterogeneity with finite mixture models." *Biometrical Journal* 59:159–171.

Leisch, F. 2004. "FlexMix: A general framework for finite mixture models and latent class regression in R." *Journal of Statistical Software* 11:Issue 8.

Leisch, F., and B. Gruen. 2017. *Task view: Cluster Analysis & Finite Mixture Models.* `https://www.r-pkg.org/ctv/Cluster`.

Lenk, P.J., and W.S. DeSarbo. 2000. "Bayesian inference for finite mixtures of generalized linear models with random effects." *Psychometrika* 65:93–119.

Leroux, B.G. 1992. "Consistent estimation of a mixing distribution." *Annals of Statistics* 20:1350–1360.

Levine, S.Z. 2013. "Evaluating the seven-item Center for Epidemiologic Studies Depression Scale short-form: a longitudinal US community study." *Social Psychiatry and Psychiatric Epidemiology* 48:1519–1526.

Li, H., Y. Luan, F. Hong, and Y. Li. 2002. "Statistical methods for analysis of time course gene expression data." *Frontiers in Bioscience* 7:90–98.

Li, J., X. Tang, W. Zhao, and J. Huang. 2007. "A new framework for identifying differentially expressed genes." *Pattern Recognition* 40:3249–3262.

Li, M., and L. Zhang. 2008. "Multinomial mixture model with feature selection for text clustering." *Knowledge-Based Systems* 21:704–708.

Liang, K.Y., and S.L. Zeger. 1986. "Longitudinal data analysis using generalized linear models." *Biometrika* 73:13–22.

Liang, Z., J.R. MacFall, and D.P. Harrington. 1994. "Parameter estimation and tissue segmentation from multispectral MR images." *IEEE Transactions on Medical Imaging* 13:441–449.

Lillehammer, M., J. Odegard, P. Madsen, B. Gjerde, T. Refstie, and M. Rye. 2013. "Survival, growth and sexual maturation in Atlantic salmon exposed to infectious pancreatic necrosis: a multivariate mixture model approach." *Genetics Selection Evolution* 45:8.

Lin, T.I., J.C. Lee, and H.J. Ho. 2006. "On fast supervised learning for normal mixture models with missing information." *Pattern Recognition* 39:1177–1187.

Lindsay, B.G. 1994. "Efficiency versus robustness – The case for minimum Hellinger distance and related methods." *Annals of Statistics* 22:1081–1114.

Little, R.J.A. 2009. "Selection and pattern-mixture models." In *Longitudinal Data Analysis,* edited by G. Fitzmaurice, M. Davidian, and G. Verbeke, 409–431. Boca Raton, FL: Chapman & Hall/CRC Press.

Little, R.J.A., and D.B. Rubin. 2002. *Statistical Analysis with Missing Data (Second Edition).* New Jersey, NJ: Wiley.

Lo, K., R.R. Brinkman, and R. Gottardo. 2008. "Automated gating of flow cytometry data via robust model-based clutersing." *Cytometry Part A* 73A:321–332.

Lofgren, E.T., S.R. Cole, D.J. Weber, D.J. Anderson, and R.W. Moehring. 2014. "Hospital-acquired clostridium difficile infections estimating all-cause mortality and length of stay." *Epidemiology* 25:570–575.

Love, M.I., W. Huber, and S. Anders. 2014. "Moderated estimation of fold change and dispersion for RNA-seq data with DESeq2." *Genome Biology* 15:550.

Lu, Z., Y.V. Hui, and A.H. Lee. 2003. "Minimum Hellinger distance estimation for finite mixtures of Poisson regression models and its applications." *Biometrics* 59:1016–1026.

Luan, T., J.A. Woolliams, S. Lien, M. Kent, M. Svendsen, and T.H.E. Meuwissen. 2009. "The accuracy of genomic selection in Norwegian red cattle assessed by cross-validation." *Genetics* 183:1119–1126.

Luan, Y., and H. Li. 2003. "Clustering of time-course gene expression data using a mixed-effects model with *B*-splines." *Bioinformatics* 19:474–482.

Lubke, G., and M.C. Neale. 2006. "Distinguishing between latent classes and continuous factors: resolution by maximum likelihood." *Multivariate Behavioral Research* 41:499–532.

Ma, P., C.I. Castillo-Davis, W. Zhong, and J.S. Liu. 2006. "A data driven clustering method for time course gene expression data." *Nucleic Acids Research* 34:1261–1269.

Madabhushi, A., S. Agner, S. Basavanhally, S. Doyle, and G. Lee. 2011. "Predicting patient and disease outcome via quantitative fusion of multi-scale, multi-modal data." *Computerized Medical Imaging and Graphics* 35:506–514.

Maitra, R., and V. Melnykov. 2010. "Simulating data to study performance of finite mixture modeling and clustering algorithms." *Journal of Computational and Graphical Statistics* 19:354–376.

Maller, R.A., and X. Zhou. 1994. "Testing for sufficient follow-up and outliers in survival data." *Journal of the American Statistical Association* 89:1499–1506.

Maller, R.A., and X. Zhou. 1996. *Survival Analysis with Long-term Survivors.* New York: Wiley.

Marengoni, A., D. Rizzuto, H.-X. Wang, B. Winblad, and L. Fratiglioni. 2009. "Patterns of chronic multimorbidity in the elderly population." *Journal of the American Geriatrics Society* 57:225–230.

Martinez, A.M., and J. Vitria. 2000. "Learning mixture models using a genetic version of the EM algorithm." *Pattern Recognition Letters* 21:759–769.

Matsui, S., and H. Noma. 2011. "Estimating effect sizes of differentially expressed genes for power and sample-size assessments in microarray experiments." *Biometrics* 67:1225–1235.

Maxwell, S.E., K. Kelley, and J.R. Rausch. 2008. "Sample size planning for statistical power and accuracy in parameter estimation." *Annual Review of Psychology* 59:537–563.

Mayrose, I., N. Friedman, and T. Pupko. 2005. "A gamma mixture model better accounts for among site rate heterogeneity." *Bioinformatics* 21 (Suppl. 2):ii151–ii158.

McCullagh, P., and J.A. Nelder. 1989. *Generalized Linear Models (2nd Edition).* London: Chapman & Hall.

McCulloch, C.E., and S.R. Searle. 2001. *Generalized, Linear, and Mixed Models.* New York: Wiley.

McGilchrist, C.A. 1993. "REML estimation for survival models with frailty." *Biometrics* 49:221–225.

———. 1994. "Estimation in generalized mixed models." *Journal of the Royal Statistical Society Series B* 56:61–69.

McGilchrist, C.A., and K.K.W. Yau. 1995. "The derivation of BLUP, ML, REML estimation methods for generalised linear mixed models." *Communications in Statistics-Theory and Methods* 24:2963–2980.

McLachlan, G.J. 1987. "On bootstrapping the likelihood ratio test statistic for the number of components in a normal mixture." *Applied Statistics* 36:318–324.

———. 1992. *Discriminant Analysis and Statistical Pattern Recognition.* New York: Wiley.

———. 2012. "Discriminant analysis." *WIREs Computational Statistics* 4:421–431.

McLachlan, G.J., and K.E. Basford. 1988. *Mixture Models: Inference and Applications to Clustering.* New York: Marcel Dekker.

McLachlan, G.J., R.W. Bean, and L. Ben-Tovim Jones. 2006. "A simple implementation of a normal mixture approach to differential gene expression in multiclass microarray." *Bioinformatics* 22:1608–1615.

McLachlan, G.J., R.W. Bean, and S.K. Ng. 2008. "Clustering." In *Bioinformatics, Volume 2: Structure, Function, and Applications (2nd Edition),* edited by J.M. Keith, 423–439. Totowa, New Jersey: Humana Press.

McLachlan, G.J., R.W. Bean, and D. Peel. 2002. "A mixutre model-based approach to the clustering of microarray expression data." *Bioinformatics* 18:413–422.

McLachlan, G.J., K.A. Do, and C. Ambroise. 2004. *Analyzing Microarray Gene Expression Data.* New York: Wiley.

McLachlan, G.J., and T. Krishnan. 2008. *The EM Algorithm and Extensions, 2nd Edition.* Hoboken, New Jersey: Wiley.

McLachlan, G.J., and D.C. McGiffin. 1994. "On the role of finite mixture models in survival analysis." *Statistical Methods in Medical Research* 3:211–226.

McLachlan, G.J., and S.K. Ng. 2009. "EM." In *The Top-Ten Algorithms in Data Mining,* edited by X. Wu and V. Kumar, 93–115. Boca Raton, Florida: Chapman & Hall/CRC.

McLachlan, G.J., S.K. Ng, P. Adams, D.C. McGiffin, and A. Gailbraith. 1997. "An algorithm for fitting mixtures of Gompertz distributions to censored survival data." *Journal of Statistical Software* 2:Issue 7.

McLachlan, G.J., and D. Peel. 2000. *Finite Mixture Models.* New York: Wiley.

McLachlan, G.J., D. Peel, and R.W. Bean. 2003. "Modelling high-dimensional data by mixtures of factor analyzers." *Computational Statistics and Data Analysis* 41:379–388.

McLean, R.A., W.L. Sanders, and W.W. Stroup. 1991. "A unified approach to mixed linear models." *American Statistician* 45:54–64.

McNicholas, P.D. 2016. *Mixture Model-Based Classification.* Boca Raton, Florida: Chapman & Hall/CRC Press.

McParland, D., and I.C. Gormley. 2016. "Model based clustering for mixed data: clustMD." *Advances in Data Analysis and Classification* 10:155–169.

Meng, X.L., and D. van Dyk. 1997. "The EM algorithm – an old folk song sung to a fast new tune." *Journal of the Royal Statistical Society B* 59:511–567.

Meng, X.L., and D.B. Rubin. 1993. "Maximum likelihood estimation via the ECM algorithm: a general framework." *Biometrika* 80:267–278.

Mengersen, K.L., C.P. Robert, and D.M. Titterington. 2011. *Mixtures: Estimation and Applications.* Hoboken, New Jersey: Wiley.

Mihaylova, B., A. Briggs, A. O'Hagan, and S.G. Thompson. 2011. "Review of statistical methods for analysing healthcare resources and costs." *Health Economics* 20:897–916.

Miller, G.E., E. Chen, and E.S. Zhou. 2007. "If it goes up, must it come down? Chronic stress and the hypothalamic-pituitary-adrenocortical axis in humans." *Psychological Bulletin* 133:25–45.

Moertel, C.G., T.R. Fleming, J.S. MacDonald, D.G. Haller, J.A. Laurie, P.J. Goodman, J.S. Ungerleider, et al. 1990. "Levamisole and fluorouracil for adjuvant therapy of resected colon carcinoma." *New England Journal of Medicine* 322:352–358.

Moertel, C.G., T.R. Fleming, J.S. MacDonald, D.G. Haller, J.A. Laurie, C.M. Tangen, J.S. Ungerleider, et al. 1995. "Fluorouracil plus levamisole as effective adjuvant therapy after resection of Stage III colon carcinoma: A final report." *Annals of Internal Medicine* 122:321–326.

Mohammadi, A., M.R. Salehi-Rad, and E.C. Wit. 2013. "Using mixture of Gamma distributions for Bayesian analysis in an M/G/1 queue with optional second service." *Computational Statistics* 28:683–700.

Molenberghs, G., and M.G. Kenward. 2007. *Missing Data in Clinical Studies.* New York: Wiley.

Monheit, A.C., R. Wilson, and R.H. Arnett III (Editors). 1999. *Informing American Health Care Policy.* San Francisco: Jossey-Bass Inc.

Moore, A.W. 1999. "Very fast EM-based mixture model clustering using multiresolution *kd*-tree." In *Advances in Neural Information Processing Systems, Vol. 11,* edited by M.S. Kearns, S.A. Solla, and D.A. Cohn, 543–549. Cambridge, M.A.: MIT Press.

Moore, W., S. Pedlow, P. Krishnamurty, and K. Wolter. 2000. *National Longitudinal Survey of Youth 1997 (NLSY97): Technical Sampling Report.* Chicago, IL: National Opinion Research Center.

Moré, J.J., B.S. Garbow, and K.E. Hillstrom. 1980. *User Guide for MINPACK-1, ANL-80-74.* Argonne National Laboratory, Chicago.

Müller, P., G. Parmigiani, C. Robert, and J. Rousseau. 2004. "Optimal sample size for multiple testing: The case of gene expression microarrays." *Journal of the American Medical Informatics Association* 99:990–1001.

Mulrane, L., E. Rexhepaj, S. Penney, J. Callanan, and W. Gallagher. 2008. "Automatic image analysis in histopathology: A valuable tool in medical diagnostics." *Expert Review of Molecular Diagnostics* 8:707–725.

Munoz, M.A., and J.D. Acuna. 1999. "Sample size requirements of a mixture analysis method with applications in systematic biology." *Journal of Theoretical Biology* 196:263–265.

Muthén, B. 2004. "Latent variable analysis: Growth mixture modeling and related techniques for longitudinal data." In *Handbook of Quantitative Methodology for the Social Sciences,* edited by D. Kaplan, 345–368. CA: Sage Publications.

Muthén, B., and T. Asparouhov. 2009. "Growth mixture modeling: Analysis with non-Gaussian random effects." In *Longitudinal Data Analysis,* edited by G. Fitzmaurice, M. Davidian, G. Verbeke, and G. Molenberghs, 143–165. Boca Raton, Florida: Chapman & Hall/CRC Press.

Muthén, L.K., and B. Muthén. 1998-2015. *Mplus User's Guide (Seventh Edition).* Los Angeles: Muthén & Muthén, Chapters 7 & 8.

———. 2002. *How to use a Monte Carlo study to decide on sample size and determine power.* Los Angeles: Muthén & Muthén. `http://www.statmodel.com/bmuthen/articles/Article_096.pdf`.

Nagin, D.S., and K.C. Land. 1993. "Age, criminal careers, and population heterogeneity: Specific estimation of a nonparametric, mixed Poisson model." *Criminology* 31:327–362.

Neal, R.M., and G.E. Hinton. 1998. "A view of the EM algorithm that justifies incremental, sparse, and other variants." In *Learning in Graphical Models,* edited by M.I. Jordan, 355–368. Dordrecht: Kluwer.

Neema, I., and D. Böhning. 2010. "Improved methods for surveying and monitoring crimes through likelihood based cluster analysis." *International Journal of Criminology & Sociological Theory* 3:477–495.

Nelder, J.A., and R.W.M. Wedderburn. 1972. "Generalized linear models." *Journal of the Royal Statistical Society Series A* 135:370–384.

Ng, S.K. 2013. "Recent developments in expectation-maximization methods for analyzing complex data." *WIREs Computational Statistics* 5:415–431.

———. 2015. "A two-way clustering framework to identify disparities in multimorbidity patterns of mental and physical health conditions among Australians." *Statistics in Medicine* 34:3444–3460.

Ng, S.K., C.M. Cameron, A.P. Hills, R.J. McClure, and P.A. Scuffham. 2014. "Socioeconomic disparities in prepregnancy BMI and impact on maternal and neonatal outcomes and postpartum weight retention: the EFHL longitudinal birth cohort study." *BMC Pregnancy and Childbirth* 14:314.

Ng, S.K., L. Holden, and J. Sun. 2012. "Identifying comorbidity patterns of health conditions via cluster analysis of pairwise concordance statistics." *Statistics in Medicine* 31:3393–3405.

Ng, S.K., T. Krishnan, and G.J. McLachlan. 2012. "The EM algorithm." In *Handbook of Computational Statistics: Concepts and Methods (2nd Edition),* edited by J.E. Gentle, W.K. Hardle, and Y. Mori, 139–172. Heidelberg: Springer-Verlag.

Ng, S.K., and A.K. Lam. 2013. "Automatic segmentation of molecular pathology images using a robust mixture model with Markov random fields." In *Proceedings of DICTA 2013, Conference of Digital Image Computing: Techniques and Applications,* edited by P. de Souza, U. Engelke, and A. Rahman, 420–427. Los Alamitos, California: IEEE Computer Society.

Ng, S.K., and G.J. McLachlan. 2003a. "An EM-based semi-parametric mixture model approach to the regression analysis of competing-risks data." *Statistics in Medicine* 22:1097–1111.

———. 2003b. "On the choice of the number of blocks with the incremental EM algorithm for the fitting of normal mixtures." *Statistics and Computing* 13:45–55.

———. 2004a. "Speeding up the EM algorithm for mixture model-based segmentation of magnetic resonance images." *Pattern Recognition* 37:1573–1589.

———. 2004b. "Using the EM algorithm to train neural networks: Misconceptions and a new algorithm for multiclass classification." *IEEE Transactions on Neural Networks* 15:738–749.

———. 2007. "Extension of mixture-of-experts networks for binary classification of hierarchical data." *Artificial Intelligence in Medicine* 41:57–67.

———. 2013. "Using cluster analysis to improve gene selection in the formation of discriminant rules for the prediction of disease outcomes." In *Proceedings of the 2013 IEEE International Conference on Bioinformatics and Biomedicine (BIBM 2013),* edited by G.-Z. Li, X. Hu, S. Kim, H. Ressorn, M. Hughes, B. Liu, G.J. McLachlan, M. Liebman, and H. Sun, 267–272. Piscataway, New Jersey: IEEE Computer Society.

———. 2014a. "Mixture models for clustering multilevel growth trajectoris." *Computational Statistics and Data Analysis* 71:43–51.

———. 2014b. "Mixture of regression models with latent variables and sparse coefficient parameters." In *Proceedings of the 21st International Conference on Computational Statistics, COMPSTAT,* edited by M. Gilli, G. Gonzalez-Rodriguez, and A. Nieto-Reyes, 223–231. The Hague: The International Statistical Institute/International Association for Statistical Computing.

———. 2016. "Finding group structures in "Big Data" in healthcare research using mixture models." In *Proceedings of the 2016 IEEE International Conference on Bioinformatics and Biomedicine (BIBM 2016), Workshop on Health Informatics and Data Science,* edited by T. Tian, Q. Jiang, Y. Liu, K. Burrage, J. Song, Y. Wang, X. Hu, S. Morishita, Q. Zhu, and G. Wang, 1219–1224. Piscataway, New Jersey: IEEE Computer Society.

———. 2017. "On the identification of correlated differential features for supervised classification of high-dimensional data." In *Studies in Classification, Data Analysis, and Knowledge Organization: Data Secience – Innovative Developments in Data Analysis and Clustering,* edited by F. Palumbo, A. Montanari, and M. Vichi, 43–57. Switzerland: Springer.

Ng, S.K., G.J. McLachlan, D.C. McGiffin, and M.F. O'Brien. 1999. "Constrained mixture models in competing risks problems." *Environmetrics* 10:753–767.

Ng, S.K., G.J. McLachlan, K. Wang, L. Ben-Tovim Jones, and S.W. Ng. 2006. "A mixture model with random-effects components for clustering correlated gene-expression profiles." *Bioinformatics* 22:1745–1752.

Ng, S.K., G.J. McLachlan, K. Wang, Z. Nagymanyoki, S. Liu, and S.W. Ng. 2015. "Inference on differences between classes using cluster-specific contrasts of mixed effects." *Biostatistics* 16:98–112.

Ng, S.K., G.J. McLachlan, K.K.W. Yau, and A.H. Lee. 2004. "Modelling the distribution of ischaemic stroke-specific survival time using an EM-based mixture approach with random effects adjustment." *Statistics in Medicine* 23:2729–2744.

Ng, S.K., A. Olog, A.B. Spinks, C.M. Cameron, J. Searle, and R.J. McClure. 2010. "Risk factors and obstetric complications of large for gestational age births with adjustments for community effects: results from a new cohort study." *BMC Public Health* 10:460.

Ng, S.K., R. Scott, and P.A. Scuffham. 2016. "Contactable non-responders show different characteristics compared to lost to follow-up participants: Insights from an Australian longitudinal birth cohort study." *Maternal and Child Health Journal* 20:1472–1484.

Ng, S.K., R. Tawiah, and G.J. McLachlan. 2019. "Unsupervised pattern recognition of mixed data structures with numerical and categorical features using a mixture regression modelling framework." *Pattern Recognition* 88:261–271.

Ng, S.K., R. Tawiah, M. Sawyer, and P. Scuffham. 2018. "Patterns of multimorbid health conditions: A systematic review of analytical methods and comparison analysis." *International Journal of Epidemiology* 47:1687–1704.

Ng, S.K., K. Wang, and G.J. McLachlan. 2006. "Multilevel modeling for the inference of genetic regulatory networks." In *Proceedings of SPIE 2005, SPIE International Symposium on Microelectronics, MEMS, and Nanotechnology, Vol. 6039,* edited by A. Bender, 60390S-1–60390S-12. Bellingham, Washington: International Society for Optical Engineering.

Nicolle, L.E. 2002. "Urinary tract infection in geriatric and institutionalised patients." *Current Opinion in Urology* 12:51–55.

Nie, L., H. Chu, and S.R. Cole. 2006. "A general approach for sample size and statistical power calculations assessing the effects of interventions using a mixture model in the presence of detection limits." *Contemporary Clinical Trials* 27:483–491.

Oakes, D. 1981. "Survival times: aspects of partial likelihood." *International Statistical Review* 49:235–264.

O'Hagan, A., T.B. Murphy, I.C. Gormley, P.D. McNicholas, and D. Karlis. 2016. "Clustering with the multivariate normal inverse Gaussian distribution." *Computational Statistics and Data Analysis* 93:18–30.

Olsen, C.S., A.E. Clark, A.M. Thomas, and L.J. Cook. 2012. "Comparing least-squares and quantile regression approaches to analyzing median hospital charges." *Academic Emergency Medicine* 19:866–875.

Onwezen, M.C., M.J. Reinders, I.A. van der Lans, S.J. Sijtsema, A. Jasiulewicz, M.D. Guardia, and L. Guerrero. 2012. "A cross-national consumer segmentation based on food benefits: The link with consumption situations and food perceptions." *Food Quality and Preference* 24:276–286.

Pai, M., N. Dendukuri, L. Wang, R. Joshi, S. Kalantri, and H.L. Rieder. 2008. "Improving the estimation of tuberculosis infection prevalence using T-cell-based assay and mixture models." *International Journal of Tuberculosis and Lung Disease* 12:895–902.

Pal, N.R., J.C. Bezdek, and E.C.K. Tsao. 1993. "Generalized clustering networks and Kohonen's self-organizing scheme." *IEEE Transactions on Neural Networks* 4:549–557. Correction, IEEE Transactions on Neural Networks, vol. 6, pp. 521, 1995.

Palardy, G.J., and J.K. Vermunt. 2010. "Multilevel growth mixture models for classifying groups." *Journal of Educational and Behavioral Statistics* 35:532–565.

Pan, W., J. Lin, and C.T. Le. 2002. "Model-based cluster analysis of microarray gene-expression data." *Genome Biology* 3:0009.1–0009.8.

Patterson, H.D., and R. Thompson. 1971. "Recovery of inter-block information when block sizes are unequal." *Biometrika* 58:545–554.

Pavlidis, P., Q. Li, and W.S. Noble. 2003. "The effect of replication on gene expression microarray experiments." *Bioinformatics* 19:1620–1627.

Pawitan, Y., S. Michiels, S. Koscielny, A. Gusnanto, and A. Ploner. 2005. "False discovery rate, sensitivity and sample size for microarray studies." *Bioinformatics* 21:3017–3024.

Peng, Y., K.B.G. Dear, and J.W. Deham. 1998. "A generalized F mixture model for cure rate estimation." *Statistics in Medicine* 17:813–830.

Pepe, M.S. 1991. "Inference for events with dependent risks in multiple endpoint studies." *Journal of the American Statistical Association* 86:770–778.

Pernkopf, F., and D. Bouchaffra. 2005. "Genetic-based EM algorithm for learning Gaussian mixture models." *IEEE Transactions on Pattern Analysis and Machine Intelligence* 27:1344–1348.

Perperoglou, A., A. Keramopoullos, and H.C. van Houwelingen. 2007. "Approaches in modelling long-term survival: an application to breast cancer." *Statistics in Medicine* 26:2666–2685.

Phillips, N., A. Coldman, and M.L. McBride. 2002. "Estimating cancer prevalence using mixture models for cancer survival." *Statistics in Medicine* 21:1257–1270.

Piper, M.E., D.M. Bolt, S.Y. Kim, S.J. Japuntich, S.S. Smith, J. Niederdeppe, D.S. Cannon, and T.B. Baker. 2008. "Refining the tobacco dependence phenotype using the Wisconsin inventory of smoking dependence motives." *Journal of Abnormal Psychology* 117:747–761.

Prados-Torres, A., B. Poblador-Plou, A. Calderón-Larrañaga, L.A. Gimeno-Feliu, F. González-Rubio, A. Poncel-Falcó, A. Sicras-Mainar, and J.T. Alcalá-Nalvaiz. 2012. "Multimorbidity patterns in primary care: interactions among chronic diseases using factor analysis." *PLoS One* 7:e32190.

Prentice, R.L., J.D. Kalbfleisch, A.V. Petersen, N. Flournoy, V.T. Farewell, and N.E. Breslow. 1978. "The analysis of failure times in the presence of competing risks." *Biometrics* 34:541–554.

Pryer, J.A., and S. Rogers. 2009. "Dietary patterns among a national sample of British children aged 1.5 - 4.5 years." *Public Health Nutrition* 12:957–966.

Pyne, S., X. Hu, K. Wang, E. Rossin, T.I. Lin, L.M. Maier, C. Baecher-Allan, et al. 2009. "Automated high-dimensional flow cytometric data analysis." *Proceedings of the National Academy of Sciences of the United States of America* 106:8519–8524.

Pyne, S., S.X. Lee, K. Wang, J. Irish, P. Tamayo, M.-D. Nazaire, T. Duong, et al. 2014. "Joint Modeling and registration of cell populations in cohorts of high-dimensional flow cytometric data." *PLoS One* 9:e100334.

Qi, Y., H. Sun, Q. Sun, and L. Pan. 2011. "Ranking analysis for identifying differentially expressed genes." *Genomics* 97:326–329.

Qiu, W., W. He, X. Wang, and R. Lazarus. 2008. "A marginal mixture model for selecting differentially expressed genes across two types of tissue samples." *International Journal of Biostatistics* 4:Article 20.

Radloff, L.S. 1977. "The CES-D scale: A self report depression scale for research in the general population." *Applied Psychological Measurements* 1:385–401.

Raghupathi, W., and V. Raghupathi. 2014. "Big data analytics in healthcare: promise and potential." *Health Information Science and Systems* 2:3.

Ram, N., and K.J. Grimm. 2009. "Growth mixture modeling: A method for identifying differences in longitudinal change among unobserved groups." *International Journal of Behavioral Development* 33:565–576.

Ramsay, J.O., and B.W. Silverman. 2005. *Functional Data Analysis (2nd edition).* New York: Springer, Chapter 3.

Rao, R.S., A.J. Sigurdson, M.M. Doody, and B.I. Graubard. 2005. "An application of a weighting method to adjust for nonresponse in standardized incidence ratio analysis of cohort studies." *Annals of Epidemiology* 15:129–136.

Raudenbush, S.W., and A.S. Bryk. 2002. *Hierarchical Linear Models: Applications and Data Analysis Methods (2nd edition).* CA: Sage Publications.

Richardson, W.S., and L.M. Doster. 2014. "Comorbidity and multimorbidity need to be placed in the context of a framework of risk, responsiveness, and vulnerability." *International Journal of Epidemiology* 67:244–246.

Ridout, M., J. Hinde, and C.G.B. Demétrio. 2001. "A score test for testing a zero-inflated Poisson regression model against zero-inflated negative binomial alternatives." *Biometrics* 57:219–223.

Rigouste, L., O. Cappé, and F. Yvon. 2007. "Inference and evaluation of the multinomial mixture model for text clustering." *Information Processing & Management* 43:1260–1280.

Riihimaki, J., R. Sund, and A. Vehtari. 2010. "Analysing the length of care episode after hip fracture: a nonparametric and a parametric Bayesian approach." *Health Care Management Science* 13:170–181.

Ringle, C.M., S. Wende, and A. Will. 2010. "Finite mixture partial least squares analysis: Methodology and numerical examples." In *Handbook of Partial Least Squares,* edited by V.E. Vinzi, W.W. Chin, J. Henseler, and H. Wang, 195–218. Heidelberg: Springer.

Ritchie, M.E., B. Phipson, D. Wu, Y. Hu, C.W. Law, W. Shi, and G.K. Smyth. 2015. "limma powers differential expression analyses for RNA-sequencing and microarray studies." *Nucleic Acids Research* 43:e47.

Robert, C.P., and G. Casella. 2004. *Monte Carlo Statistical Methods.* New York: Springer.

Roberts, S.J., D. Husmeier, I. Rezek, and W. Penny. 1998. "Bayesian approaches to Gaussian modeling." *IEEE Transactions on Pattern Analysis and Machine Intelligence* 20:1133–1142.

Robinson, G.K. 1991. "That BLUP is a good thing: the estimation of random effects." *Statistical Science* 6:15–32.

Rosen, O., and M. Tanner. 1999. "Mixtures of proportional hazards regression models." *Statistics in Medicine* 15:1119–1131.

Rotnitzky, A., Q. Lei, M. Sued, and J.M. Robins. 2012. "Improved double-robust estimation in missing data and causal inference models." *Biometrika* 99:439–456.

Rubin, D.B. 1976. "Inference and missing data." *Biometrika* 63:581–592.

———. 1996. "Multiple imputation after 18+ years." *Journal of the American Statistical Association* 91:473–489.

Rubin, D.B., and H.S. Stern. 1998. "Sample size determination using posterior predictive distributions." *Indian Journal of Statistics* 60:161–175.

Rzehak, P., A.H. Wijga, T. Keil, E. Eller, C. Bindslev-Jensen, J. Smit H.A. Weyler, S. Dom, J. Sunyer, and M. et al. Mendez. 2013. "Body mass index trajectory classes and incident asthma in childhood: Results from 8 European birth cohorts - a global allergy and asthma European network initiative." *Journal of Allergy and Clinical Immunology* 131:1528–1536.

Sanchez, L., P. Lorenzo-Luaces, C. Viada, Y. Galan, J. Ballesteros, T. Crombet, and A. Lage. 2014. "Is there a subgroup of long-term evolution among patients with advanced lung cancer?: Hints from the analysis of survival curves from cancer registry data." *BMC Cancer* 14:933.

SAS Institute Inc. 2011. *SAS/STAT 9.3 User's Guide.* North Carolina: SAS Institute Inc.

Schafer, J.L. 1997. *Analysis of Incomplete Multivariate Data.* New York: Chapman & Hall.

Schall, R. 1991. "Estimation in generalized linear models with random effects." *Biometrika* 78:719–727.

Scharfstein, D.O., A. Rotnitzky, and J.M. Robins. 1999. "Adjusting for nonignorbale drop-out using semiparametric nonresponse models." *Journal of the American Statistical Association* 94:1096–1120 (with Rejoinder, 1135–1146).

Schlattmann, P. 2009. *Medical Applications of Finite Mixture Models.* Heidelberg: Springer-Verlag.

Schwarz, G. 1978. "Estimating the dimension of a model." *The Annals of Statistics* 6:461–464.

Scrucca, L., M. Fop, T. Murphy, and A. Raftery. 2016. "mclust 5: Clustering, classification and density estimation using Gaussian finite mixture models." *The R Journal* 8:289–317.

Scuffham, P.A., J.M. Byrnes, C. Pollicino, D. Cross, S. Goldstein, and S.K. Ng. 2019. "The impact of population-based disease management services on health care utilisation and costs: Results of the CAPICHe trial." *Journal of General Internal Medicine* 34:41–48.

Seaman, S.R., and I.R. White. 2013. "Review of inverse probability weighting for dealing with missing data." *Statistical Methods in Medical Research* 22:278–295.

Searle, S.R., G. Casella, and C.E. McCulloch. 1992. *Variance Components*. New York: Wiley.

Seeman, T.E., B. Singer, and P. Charpentier. 1995. "Gender differences in pattern of HPA axis response to challenge: MacArthur studies of successful aging." *Psychoneuroendocrinology* 20:711–725.

Segal, E., R. Yelensky, and D. Koller. 2003. "Genome-wide discovery of transcriptional modules from DNA sequence and gene expression." *Bioinformatics* 19 (Suppl. 1):i273–i282.

Sheikh, K. 2007. "Investigation of selection bias using inverse probability weighting." *European Journal of Epidemiology* 22:349–350.

Simpson, D.G. 1987. "Minimum Hellinger distance estimation for the analysis of count data." *Journal of the American Statistical Association* 82:802–807.

Smith, S.M., H. Soubhi, M. Fortin, C. Hudon, and T. O'Dowd. 2012. "Managing patients with multimorbidity: systematic review of interventions in primary care and community settings." *British Medical Journal* 345:e5205.

Smyth, G. 2004. "Linear models and empirical Bayes methods for assessing differential expression in microarray experiments." *Statistical Applications in Genetics and Molecular Biology* 3:Article 3.

Sovilj, D., E. Eirola, Y. Miche, K.-M. Björk, R. Nian, A. Akusok, and A. Lendasse. 2016. "Extreme learning machine for missing data using multiple imputations." *Neurocomputing* 174:220–231.

Speed, T. 1991. "Comment on Robinson: The estimation of random effects." *Statistical Science* 6:42–44.

Spellman, P.T., G. Sherlock, M.Q. Zhang, V.R. Iyer, K. Anders, M.B. Eisen, P.O. Brown, D. Botstein, and B. Futcher. 1998. "Comprehensive identification of cell cycle-regulated genes of the yeast Saccharomyces cerevisiae by microarray hybridization." *Molecular Biology of the Cell* 9:3273–3297.

Sterne, J.A., I.R. White, J.B. Carlin, M. Spratt, P. Royston, M.G. Kenward, A.M. Wood, and J.R. Carpenter. 2009. "Multiple imputation for missing data in epidemiological and clinical research: potential and pitfalls." *British Medical Journal* 338:b2393.

Stevens, G.A., M.M. Finucane, C.J. Paciorek, S.R. Flaxman, R.A. White, A.J. Donner, and M. Ezzati. 2012. "Trends in mild, moderate, and severe stunting and underweight, and progress towards MDG 1 in 141 developing countries: a systematic analysis of population representative data." *Lancet* 380:824–834.

Storey, J.D. 2007. "The optimal discovery procedure: a new approach to simultaneous significance testing." *Journal of the Royal Statistical Society B* 69:347–368.

Stull, D., K.W. Wyrwich, and F.W. Frueh. 2009. "Methods for personalized medicine: Factor mixture models for investigating differential response to treatment." *Value in Health* 12:A27–A27.

Styczynski, M.P., and G. Stephanopoulos. 2005. "Overview of computational methods for the inference of gene regulatory networks." *Computers & Chemical Engineering* 29:519–534.

Subramaniam, D., G. Periyasamy, S. Ponnurangam, D. Chakrabarti, A. Sugumar, M. Padigaru, S.J. Weir, A. Balakrishnan, S. Sharma, and S. Anant. 2012. "CDK-4 inhibitor P276 sensitizes pancreatic cancer cells to Gemcitabine-induced apoptosis." *Molecular Cancer Therapeutics* 11:1598–1608.

Sy, J.P., and J.M. Taylor. 2000. "Estimation in a Cox proportional hazards cure model." *Biometrics* 56:227–236.

Tai, B., D. Machin, I. White, and V. Gebski. 2001. "Competing risks analysis of patients with osteosarcoma: a comparison of four different approaches." *Statistics in Medicine* 20:661–684.

Tamayo, P., D. Slonim, J. Mesirov, Q. Zhu, S. Kitareewan, E. Domitrovsky, E.S. Lander, and T.R. Golub. 1999. "Interpreting patterns of gene expression pattern with self-organizing maps: methods and application to hematopietic differentiation." *Proceedings of the National Academy of Sciences of the United States of America* 96:2907–2912.

Tang, Y. 2017. "On the multiple imputation variance estimator for control-based and delta-adjusted pattern mixture models." *Biometrics* 73:1379–1387.

Tavazoie, S., J.D. Hughes, M.J. Campbell, R.J. Cho, and G.M. Church. 1999. "Systematic determination of genetic network architecture." *Nature Genetics* 22:281–285.

Tawiah, R., K.K.W. Yau, G.J. McLachlan, Chambers S.K., and S.K. Ng. 2019. "Multilevel model with random effects for clustered survival data with multiple failure outcomes." *Statistics in Medicine* 38:1036–1055. DOI:10.1002/sim.8041.

Tentoni, S., P. Astolfi, A. De Pasquale, and L.A. Zonta. 2004. "Birthweight by gestational age in preterm babies according to a Gaussian mixture model." *British Journal of Obstetrics and Gynaecology* 111:31–37.

Thabane, L., J. Ma, R. Chu, J. Cheng, A. Ismaila, L.P. Rios, R. Robson, M. Thabane, L. Giangregorio, and C.H. Goldsmith. 2010. "A tutorial on pilot studies: the what, why and how." *BMC Medical Research Methodology* 10:1.

Therneau, T.M., and S.A. Hamilton. 1997. "rhDNase as an example of recurrent event analysis." *Statistics in Medicine* 16:2029–2047.

Thijs, H., G. Molenberghs, B. Michiels, G. Verbeke, and D. Curran. 2002. "Strategies to fit pattern-mixture models." *Biostatistics* 3:245–265.

Thompson, C.A., and O.A. Arah. 2014. "Selection bias modeling using observed data augmented with imputed record-level probabilities." *Annals of Epidemiology* 24:747–753.

Thompson, R. 1980. "Maximum likelihood estimation of variance components." *Statistics: A Journal of Theoretical and Applied Statistics* 11:545–561.

Titterington, D.M., A.F.M. Smith, and U.E. Makov. 1985. *Statistical Analysis of Finite Mixture Distributions.* New York: Wiley.

Toh, H., and K. Horimoto. 2002. "Inference of a genetic network by a combined approach of cluster analysis and graphical Gaussian modelling." *Bioinformatics* 18:287–297.

Topchy, A., A.K. Jain, and W. Punch. 2005. "Clustering ensembles: Models of consensus and weak partitions." *IEEE Transactions on Pattern Analysis and Machine Intelligence* 27:1866–1881.

Touchette, E., D. Petit, J.R. Seguin, M. Boivin, R.E. Tremblay, and J.Y. Montplaisir. 2007. "Associations between sleep duration patterns and behavioral/cognitive functioning at school entry." *Sleep* 30:1213–1219.

Tusher, V.G., R. Tibshirani, and G. Chu. 2001. "Significance analysis of microarrays applied to the ionizing radiation response." *Proceedings of the National Academy of Sciences of the United States of America* 98:5116–5121.

Urquia, M.L., R. Moineddin, and J.W. Frank. 2012. "A mixture model to correct misclassification of gestational age." *Annals of Epidemiology* 22:151–159.

Valderas, J.M., B. Starfield, B. Sibbald, C. Salisbury, and M. Roland. 2009. "Defining comorbidity: implications for understanding health and health services." *Annals of Family Medicine* 7:357–363.

van Buuren, S., and K. Groothuis-Oudshoorn. 2011. "mice: Multivariate imputation by chained equations in R." *Journal of Statistical Software* 45:Issue 3.

van den Akker, M., F. Buntinx, and J.A. Knottnerus. 1996. "Comorbidity or multimorbidity: what's in a name? A review of literature." *European Journal of General Practice* 2:65–70.

Van den Broek, J. 1995. "A score test for zero-inflation in a Poisson distribution." *Biometrics* 51:738–743.

Ventura, A.K., E. Loken, and L.L. Birch. 2006. "Risk profiles for metabolic syndrome in a nonclinical sample of adolescent girls." *Pediatrics* 118:2434–2442.

Venturini, S., F. Dominici, and G. Parmigiani. 2008. "Gamma shape mixtures for heavy-tailed distributions." *Annals of Applied Statistics* 2:756–776.

Vermunt, J.K. 2003. "Multilevel latent class models." *Sociological Methodology* 33:213–239.

———. 2007. "Growth models for categorical response variables: Standard, latent-class, and hybrid approaches." In *Longitudinal Models in the Behavioral and Related Sciences,* edited by K. van Montfort, J. Oud, and A. Satorra, 139–158. Mahwah, NJ: Lawrence Erlbaum.

Viroli, C. 2011. "Finite mixtures of matrix normal distributions for classifying three-way data." *Statistics and Computing* 21:511–522.

Vrbik, I., and P.D. McNicholas. 2012. "Analytic calculations for the EM algorithm for multivariate skew-t mixture models." *Statistics & Probability Letters* 82:1169–1174.

Vu, T., C.F. Finch, and L. Day. 2011. "Patterns of comorbidity in community-dwelling older people hospitalised for fall-related injury: a cluster analysis." *BMC Geriatrics* 11:45.

Wallace, C.S., and D.L. Dowe. 1994. "Intrinsic classification by MML – the Snob program." In *Proceedings of the 7th Australian Joint Conference on Artificial Intelligence,* 37–44. Singapore: World Scientific.

Wang, C., and M.J. Daniels. 2011. "A note on MAR, identifying restrictions, model comparison, and sensitivity analysis in pattern mixture models with and without covariates for incomplete data." *Biometrics* 67:810–818.

Wang, K., S.K. Ng, and G.J. McLachlan. 2009. "Multivariate skew t mixture models: applications to fluorescence-activated cell sorting data." In *Proceedings of DICTA 2009, Conference of Digital Image Computing: Techniques and Applications,* edited by H. Shi, Y. Zhang, M.J. Bottema, B.C. Lovell, and A.J. Maeder, 526–531. Los Alamitos, California: IEEE Computer Society.

———. 2012. "Clustering of time-course gene expression profiles using normal mixture models with autoregressive random effects." *BMC Bioinformatics* 13:300.

Wang, K., K.K.W. Yau, and A.H. Lee. 2002a. "A hierarchical Poisson mixture regression model to analyse maternity length of hospital stay." *Statistics in Medicine* 21:3639–3654.

———. 2002b. "A zero-inflated Poisson mixed model to analyze diagnosis related groups with majority of same-day hospital stays." *Computer Methods and Programs in Biomedicine* 68:195–203.

Wang, K., K.K.W. Yau, A.H. Lee, and G.J. McLachlan. 2007a. "Multilevel survival modelling of recurrent urinary tract infections." *Computer Methods and Programs in Biomedicine* 87:225–229.

———. 2007b. "Two-component Poisson mixture regression modelling of count data with bivariate random effects." *Mathematical and Computer Modelling* 46:1468–1476.

Wang, S., J. Zhang, and W. Lu. 2012. "Sample size calculation for the proportional hazards cure model." *Statistics in Medicine* 31:3959–3971.

Wei, C., J. Li, and R.E. Bumgarner. 2004. "Sample size for detecting differentially expressed genes in microarray experiments." *BMC Genomics* 5:87.

Wessman, J., T. Paunio, A. Tuulio-Henriksson, M. Koivisto, T. Partonen, J. Suvisaari, J.A. Turunen, et al. 2009. "Mixture model clustering of phenotype features reveals evidence for association of DTNBP1 to a specific subtype of schizophrenia." *Biological Psychiatry* 66:990–996.

Weuve, J., E.J. Tchetgen, M.M. Glymour, T.L. Beck, N.T. Aggarwal, R.S. Wilson, D.A. Evans, and C.F. Mendes de Leon. 2012. "Accounting for bias due to selective attrition: The example of smoking and congitive decline." *Epidemiology* 23:119–128.

Windle, M., J.A. Grunbaum, M. Elliott, S.R. Tortolero, S. Berry, J. Gilliland, D.E. Kanouse, et al. 2004. "Healthy passages – A multilevel, multimethod longitudinal study of adolescent health." *American Journal of Preventative Medicine* 27:164–172.

Wolf, R.L., D.C. Alsop, I. Levy-Reis, P.T. Meyer, J.A. Maldjian, J. Gonzalez-Atavales, J.A. French, A. Alavi, and J.A. Detre. 2001. "Detection of mesial temporal lobe hypoperfusion in patients with temporal lobe epilepsy by use of arterial spin labeled perfusion MR imaging." *American Journal of Neuroradiology* 22:1334–1341.

Wu, C.F.J. 1983. "On the convergence properties of the EM algorithm." *Annals of Statistics* 11:95–103.

Wu, L.T., G.E. Woody, C. Yang, J.J. Pan, and D.G. Blazer. 2011. "Abuse and dependence on prescription opioids in adults: a mixture categorical and dimensional approach to diagnostic classification." *Psychological Medicine* 41:653–664.

Xiang, L., and A.H. Lee. 2005. "Sensitivity of test for overdispersion in Poisson regression." *Biometrical Journal* 47:167–176.

Xiang, L., A.H. Lee, K.K.W. Yau, and G.J. McLachlan. 2006. "A score test for zero-inflation in correlated count data." *Statistics in Medicine* 25:1660–1671.

———. 2007. "A score test for overdispersion in zero-inflated Poisson mixed regression model." *Statistics in Medicine* 26:1608–1622.

Xiang, L., X. Ma, and K.K.W. Yau. 2011. "Mixture cure model with random effects for clustered interval-censored survival data." *Statistics in Medicine* 30:995–1006.

Xiang, L., and G.S. Teo. 2011. "A note on tests for zero-inflation in correlated count data." *Communications in Statistics – Simulation and Computation* 40:992–1005.

Xiang, L., K.K.W. Yau, Y.V. Hui, and A.H. Lee. 2008. "Minimum Hellinger distance estimation for k-component Poisson mixture with random effects." *Biometrics* 64:508–518.

Xiang, L., K.K.W. Yau, and A.H. Lee. 2012. "The robust estimation method for a finite mixture of Poisson mixed-effect models." *Computational Statistics and Data Analysis* 56:1994–2005.

Xiang, L., K.K.W. Yau, A.H. Lee, and W.K. Fung. 2005. "Influence diagnostics for two-component Poisson mixture regression models: applications in public health." *Statistics in Medicine* 24:3053–3071.

Xue, Q.L., K. Bandeen-Roche, T.J. Mielenz, C.L. Seplaki, S.L. Szanton, R.J. Thorpe, R.R. Kalyani, P.H.M. Chaves, T.T.L. Dam, and K. et al. Omstein. 2012. "Patterns of 12-year change in physical activity levels in community-dwelling older women: Can modest levels of physical activity help older women live longer." *American Journal of Epidemiology* 176:534–543.

Yang, M.-H., D.J. Kriegman, and N. Ahuja. 2002. "Detecting faces in images: A survey." *IEEE Transactions on Pattern Analysis and Machine Intelligence* 24:34–58.

Yau, K.K.W. 2001. "Multilevel models for survival analysis with random effects." *Biometrics* 57:96–102.

Yau, K.K.W., and A.Y.C. Kuk. 2002. "Robust estimation in generalized linear mixed models." *Journal of the Royal Statistical Society Series B* 64:101–117.

Yau, K.K.W., and A.H. Lee. 2001. "Zero-inflated Poisson regression with random effects to evaluate an occupational injury prevention programme." *Statistics in Medicine* 20:2907–2920.

Yau, K.K.W., A.H. Lee, and S.K. Ng. 2003. "Finite mixture regression model with random effects: application to neonatal hospital length of stay." *Computational Statistics and Data Analysis* 41:359–366.

Yau, K.K.W., and C.A. McGilchrist. 1998. "ML and REML estimation in survival analysis with time dependent correlated frailty." *Statistics in Medicine* 17:1201–1213.

———. 1999. "Power family of transformation for Cox's regression with random effects." *Computational Statistics and Data Analysis* 30:57–66.

Yau, K.K.W., and S.K. Ng. 2001. "Long-term survivor mixture model with random effects: application to a multi-centre clinical trial of carcinoma." *Statistics in Medicine* 20:1591–1607.

Yau, K.K.W., K. Wang, and A.H. Lee. 2003. "Zero-inflated negative binomial mixed regression modelling of over-dispersed count data with extra zeros." *Biometrical Journal* 45:437–452.

Yelland, M.J., K.R. Sweeting, J.A. Lyftogt, S.K. Ng, P.A. Scuffham, and K.A. Evans. 2011. "Prolotherapy injections and eccentric loading exercises for painful Achilles tendinosis: a randomised trial." *British Journal of Sports Medicine* 45:421–428.

Yeung, K.Y., C. Fraley, A. Murua, A.E. Raftery, and W.L. Ruzzo. 2001. "Model-based clustering and data transformations for gene expression data." *Bioinformatics* 17:977–987.

Yeung, K.Y., D.R. Haynor, and W.L. Ruzzo. 2001. "Validating clustering for gene expression data." *Bioinformatics* 17:309–318.

Yeung, K.Y., M. Medvedovic, and R.E. Bumgarner. 2003. "Clustering gene-expression data with repeated measurements." *Genome Biology* 4:Article R34.

Young, D., T. Benaglia, D. Chauveau, D. Hunter, T. Hettmansperger, and F. Xuan. 2015. *Package "MIXTOOLS" for R. Tools for analysing finite mixture models.* Lexington, KY: University of Kentucky. R-CRAN.

Yu, J., and S.J. Qin. 2009. "Multiway Gaussian mixutre model based multiphase batch process monitoring." *Industrial and Enginnering Chemistry Research* 48:8585–8594.

Yuan, M., and C. Kendziorski. 2006. "A unified approach for simultaneous gene clustering and differential expression identification." *Biometrics* 62:1089–1098.

Zeger, S.L., and K.Y. Liang. 1986. "Longitudinal data analysis for discrete and continuous outcomes." *Biometrics* 42:121–130.

Zeger, S.L., K.Y. Liang, and P.S. Albert. 1988. "Models for longitudinal data: a generalized estimating equation approach." *Biometrics* 44:1049–1060.

Zhang, J., and Y. Peng. 2009. "Accelerated hazards mixture cure model." *Lifetime Data Analysis* 15:455–467.

Zhang, Z. 2016. "Multiple imputation for time series data with Amelia package." *Annals of Translational Medicine* 4:56.

Zhang, Z., C. Chen, J. Sun, and K.L. Chan. 2003. "EM algorithms for Gaussian mixtures with split-and-merge operation." *Pattern Recognition* 36:1973–1983.

Zhao, Y. 2011. "Posterior probability of discovery and expected rate of discovery for multiple hypothesis testing and high throughput assays." *Journal of the American Statistical Association* 106:984–996.

Zheng, J., and H.C. Frey. 2004. "Quantification of variability and uncertainty using mixture distributions: Evaluation of sample size, mixing weights, and separation between components." *Risk Analysis* 24:553–571.

Zhu, X., and V. Melnykov. 2015. "Probabilistic assessment of model-based clustering." *Advances in Data Analysis and Classification* 9:395–422.

Zill, N., and J.L. Petersen. 1986. *Behavioral Problems Index.* Washington, DC: Child Trends Inc.

Index

For Product Safety Concerns and Information please contact our EU
representative GPSR@taylorandfrancis.com
Taylor & Francis Verlag GmbH, Kaufingerstraße 24, 80331 München, Germany

www.ingramcontent.com/pod-product-compliance
Lightning Source LLC
LaVergne TN
LVHW010541100826
845148LV00001B/257

* 9 7 8 0 3 6 7 7 2 9 3 3 2 *